Recent Advance in Intelligent Vehicle

Recent Advance in Intelligent Vehicle

Editors

Biao Yu
Linglong Lin
Jiajia Chen

MDPI

Basel • Beijing • Wuhan • Barcelona • Belgrade • Novi Sad • Cluj • Manchester

Editors
Biao Yu
Institute of
Intelligent Machines
Hefei Institutes of Physical
Sciences, Chinese Academy
of Sciences
Hefei
China

Linglong Lin
Institute of
Intelligent Machines
Hefei Institutes of Physical
Sciences, Chinese Academy
of Sciences
Hefei
China

Jiajia Chen
School of Automotive and
Transportation Engineering
Hefei University
of Technology
Hefei
China

Editorial Office
MDPI AG
Grosspeteranlage 5
4052 Basel, Switzerland

This is a reprint of articles from the Special Issue published online in the open access journal *World Electric Vehicle Journal* (ISSN 2032-6653) (available at: https://www.mdpi.com/journal/wevj/special_issues/ACLZPBS686).

For citation purposes, cite each article independently as indicated on the article page online and as indicated below:

Lastname, A.A.; Lastname, B.B. Article Title. *Journal Name* **Year**, *Volume Number*, Page Range.

ISBN 978-3-7258-1469-5 (Hbk)
ISBN 978-3-7258-1470-1 (PDF)
doi.org/10.3390/books978-3-7258-1470-1

Cover image courtesy of Biao Yu

Contents

About the Editors vii

Preface ix

Zhaoyan Chen, Xiaolan Wang, Weiwei Zhang, Guodong Yao, Dongdong Li and Li Zeng
Autonomous Parking Space Detection for Electric Vehicles Based on Improved YOLOV5-OBB Algorithm
Reprinted from: *World Electr. Veh. J.* **2023**, *14*, 276, doi:10.3390/wevj14100276 1

Hongjian Wei, Yingping Huang, Qian Zhang and Zhiyang Guo
Online Multiple Object Tracking Using Min-Cost Flow on Temporal Window for Autonomous Driving
Reprinted from: *World Electr. Veh. J.* **2023**, *14*, 243, doi:10.3390/wevj14090243 17

Ke Li, Xuncheng Wu, Weiwei Zhang and Wangpengfei Yu
Bird's-Eye View Semantic Segmentation for Autonomous Driving through the Large Kernel Attention Encoder and Bilinear-Attention Transform Module
Reprinted from: *World Electr. Veh. J.* **2023**, *14*, 239, doi:10.3390/wevj14090239 40

Hao Xu, Xiang Dong, Wenxuan Wu, Biao Yu and Hui Zhu
A Two-Stage Pillar Feature-Encoding Network for Pillar-Based 3D Object Detection
Reprinted from: *World Electr. Veh. J.* **2023**, *14*, 146, doi:10.3390/wevj14060146 54

Wangpengfei Yu, Yubin Qian, Jiejie Xu, Hongtao Sun and Junxiang Wang
Driving Decisions for Autonomous Vehicles in Intersection Environments: Deep Reinforcement Learning Approaches with Risk Assessment
Reprinted from: *World Electr. Veh. J.* **2023**, *14*, 79, doi:10.3390/wevj14040079 68

Ke He, Haitao Ding, Nan Xu and Konghui Guo
Accelerated and Refined Lane-Level Route-Planning Method Based on a New Road Network Model for Autonomous Vehicle Navigation
Reprinted from: *World Electr. Veh. J.* **2023**, *14*, 98, doi:10.3390/wevj14040098 96

Jiajia Chen, Zheng Zhou, Yue Duan and Biao Yu
Research on Reinforcement-Learning-Based Truck Platooning Control Strategies in Highway On-Ramp Regions
Reprinted from: *World Electr. Veh. J.* **2023**, *14*, 273, doi:10.3390/wevj14100273 117

Hua Tao and Baocheng Yang
Coordinated Control of Unmanned Electric Formula Car
Reprinted from: *World Electr. Veh. J.* **2023**, *14*, 58, doi:10.3390/wevj14030058 133

Swapnil Waykole, Nirajan Shiwakoti and Peter Stasinopoulos
Interpolation-Based Framework for Generation of Ground Truth Data for Testing Lane Detection Algorithm for Automated Vehicle
Reprinted from: *World Electr. Veh. J.* **2023**, *14*, 48, doi:10.3390/wevj14020048 150

Sachin B. Chougule, Bharat S. Chaudhari, Sheetal N. Ghorpade and Marco Zennaro
Exploring Computing Paradigms for Electric Vehicles: From Cloud to Edge Intelligence, Challenges and Future Directions
Reprinted from: *World Electr. Veh. J.* **2024**, *15*, 39, doi:10.3390/wevj15020039 165

About the Editors

Biao Yu

Biao Yu (Professor) received his B.S. and Ph.D. degrees from the School of Mechanical and Automotive Engineering, Hefei University of Technology, in 2007 and 2013, respectively. He is currently a Professor with the Institute of Intelligent Machines, Hefei Institutes of Physical Science, Chinese Academy of Sciences, China. His research interests include evolutionary computation, intelligent vehicles, and mobile robot navigation and localization.

Linglong Lin

Linglong Lin (Associate Professor) received his B.S. degree in computer science and technology from the Anhui University of Technology in 2010 and his Ph.D. degree in nuclear science and engineering from the University of Chinese Academy of Sciences in 2016. He has been an Associate Professor with the Institute of Intelligent Machines, Hefei Institutes of Physical Science, Chinese Academy of Sciences, Hefei, China, since 2016. His research interests include self-driving vehicles, such as LiDAR point cloud data processing, object recognition, and tracking and deep learning.

Jiajia Chen

Jiajia Chen (Associate Professor) received his B.S. degree in automation from the Jilin University in 2008 and his Ph.D. degree in precision instruments and machinery from the University of Science and Technology of China in 2015. He has been an Associate Professor with the School of Automotive and Transportation Engineering, Hefei University of Technology, Hefei, China, since 2016. His research interests include aspects of self-driving vehicles, such as decision making, motion planning, and deep reinforcement learning.

Preface

Intelligent vehicles have been considered an essential way to improve urban mobility, as well as reduce emission pollution and traffic accidents. With the development of artificial intelligence, such as deep learning and intelligent vehicle technologies have obtained enormous success. However, due to the unmatured mature of the critical technologies, such as environment perception, motion planning, behavior decisions, and motion control, the intelligent vehicle still cannot be deployed in real and complex scenarios.

The intelligent vehicle is a very complicated technical system. A lot of critical technologies from different disciplines, such as sensor technology, pattern recognition, control engineering, artificial intelligence, and vehicle engineering, can affect its performance. This Special Issue aims to explore the recent progress in these related research fields. The topics include, but are not strictly limited to, the following:

- Imaging and sensor technology, such as LiDAR, camera, millimeter wave radar, and so on;
- Environment perception technology, such as vehicle/pedestrian detection, tracking and prediction, travelable area detection, ground segmentation, and so on;
- Planning and control technology, such as global planning, local planning, behavior decision, motion control, and so on;
- Navigation and localization technology, such as lidar odometry, vision odometry, simultaneous localization and mapping (SLAM), and so on;
- Intelligence test and evaluation.

Biao Yu, Linglong Lin, and Jiajia Chen
Editors

World Electric Vehicle Journal

MDPI

Article

Autonomous Parking Space Detection for Electric Vehicles Based on Improved YOLOV5-OBB Algorithm

Zhaoyan Chen [1], Xiaolan Wang [1,*], Weiwei Zhang [1], Guodong Yao [2], Dongdong Li [2] and Li Zeng [2]

[1] School of Mechanical and Automotive Engineering, Shanghai University of Engineering Science, Shanghai 201620, China
[2] Voyager Technology Inc, Shanghai 201517, China
* Correspondence: m310121423@sues.edu.cn

Abstract: Currently, in the process of autonomous parking, the algorithm detection accuracy and rate of parking spaces are low due to the diversity of parking scenes, changes in lighting conditions, and other unfavorable factors. An improved algorithm based on YOLOv5-OBB is proposed to reduce the computational effort of the model and increase the speed of model detection. Firstly, the backbone module is optimized, the Focus module and SSP (Selective Spatial Perception) module are replaced with the general convolution and SSPF (Selective Search Proposals Fusion) modules, and the GELU activation function is introduced to reduce the number of model parameters and enhance model learning. Secondly, the RFB (Receptive Field Block) module is added to fuse different feature modules and increase the perceptual field to optimize the small target detection. After that, the CA (coordinate attention) mechanism is introduced to enhance the feature representation capability. Finally, the post-processing is optimized using spatial location correlation to improve the accuracy of the vehicle position and bank angle detection. The implementation results show that by using the improved method proposed in this paper, the FPS of the model is improved by 2.87, algorithm size is reduced by 1 M, and the mAP is improved by 8.4% on the homemade dataset compared with the original algorithm. The improved model meets the requirements of perceived accuracy and speed of parking spaces in autonomous parking.

Keywords: autonomous parking; YOLOv5-OBB; parking space detection; coordinate attention mechanism

Citation: Chen, Z.; Wang, X.; Zhang, W.; Yao, G.; Li, D.; Zeng, L. Autonomous Parking Space Detection for Electric Vehicles Based on Improved YOLOV5-OBB Algorithm. *World Electr. Veh. J.* **2023**, *14*, 276. https://doi.org/10.3390/wevj14100276

Academic Editors: Biao Yu, Linglong Lin and Jiajia Chen

Received: 12 August 2023
Revised: 12 September 2023
Accepted: 28 September 2023
Published: 2 October 2023

1. Introduction

Autonomous parking systems for self-driving vehicles are crucial, of which parking space detection [1–3] is a key component. Most of the on-board parking assistance systems on the market today are based on very high-computing-power chips and a wide variety of sensors, etc. In order to develop a lower-cost autonomous parking system, it is necessary to develop it based on a low-computing-power embedded chip. Previous parking space detection methods [4–7] are mainly based on traditional computer vision techniques such as edge detection, corner detection, histograms, and feature matching. For example, Hamada et al. [8] extracted parking space lines using the Hough transform method and inferred parking spaces using geometric constraints, but it is only applicable to parking space scenarios with very good illumination conditions. Bui et al. [9] separated fixed parking space lines by a line segment clustering method. These methods perform poorly when dealing with different car park lighting conditions and variations in the appearance of parking spaces.

In recent years, deep-learning-based methods have been able to extract high-level features from input images and perform location estimation and classification of parking spaces. Methods based on deep learning are mainly divided into target detection methods and semantic segmentation methods, and target detection methods are further divided into one-stage detection and two-stage detection. Li et al. [10] used the deep learning method to predict the location, type, and orientation of parking space corner points and then grouped the corner points using geometrical rules to infer the presence of parking spaces. However, this method can only detect perpendicular and parallel rectangular parking slots. Zhang et al. [11] proposed a two-stage target detection method, DeepPS, which first uses YOLOV2 to detect the corners of parking spaces and then obtains the parking space type and direction matching of parking spaces through local image classification networks and templates. This method can effectively detect different kinds of parking spaces, but it requires two deep neural networks, which makes inference time too slow and the amount of model parameters too large for embedded end deployment.

Zhou et al. [12] proposed an attentional semantic segmentation and instance matching method to improve the accuracy of parking space detection, but it can only be applied to AVP systems, and the attention structure is difficult to deploy to some embedded platforms. Cao et al. [13] proposed a method based on VPS-Net [14] hat can detect different kinds of sign points, but the prediction of parking spaces with different lengths was inaccurate. Li et al. [15] proposed a semantic-segmentation-based method to improve the detection of parking spaces, but the number of arithmetic resources consumed was very high, most vendors are currently trying to deploy autonomous parking systems in low-computing-power embedded platforms with only 1–3 TOPS of arithmetic power, and the arithmetic power is unable to meet the requirement. We are based on a one-stage target detection method, which can not only detect parking spaces with angles but also has low model complexity and a fast detection rate, which are suitable for deploying embedded chips with low computing power.

Other methods [9–17] and datasets [11,15] can only detect and infer a parking space during the autonomous parking process and cannot determine whether there are obstacles (such as ice cream cones, floor locks, etc.) in the parking space. Therefore, the method and dataset we proposed are based on purely visual parking space detection, which can complete the detection of parking spaces and obstacles around the vehicle based on a single image. The method in this article de-distorts the images from the left and right fish-eye cameras and splices them into a bird's-eye view with a size of 128 × 416. The front and rear fish-eye cameras do not need to be de-distorted but directly splice them into an image with a size of 288 × 208 and splice it into an image with a size of 416 × 416. The pictures are passed into the network model for detection, and finally the target detection results are sent to the planning control module. The process is shown in Figure 1.

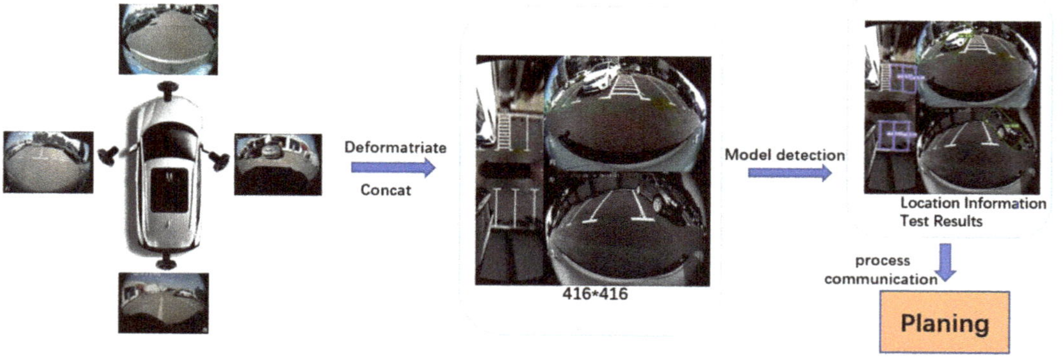

Figure 1. Flow chart for autonomous parking space detection.

The main contributions of our work can be summarized as follows:

1. Improved RFB and CA modules are added to the original yolov5-OBB algorithm to enhance the generalization ability of the model in complex scenarios such as darkness,

while replacing the Focus and SSP structures to reduce the number of parameters in the computation and accelerate the model inference rate.

2. Correlation modeling of the existing a priori knowledge of the simultaneous occurrence of parking spaces and storage corners and setting the penalty factor K to improve the confidence level of the detection of parking spaces and storage corners.
3. A standard evaluation method for target detection was used through comparative experiments and ablation experiments of the original algorithm on a homemade parking space detection dataset as well as on a publicly available dataset, and the results show that our algorithm is competitive in terms of real-time and detection accuracy in complex scenarios such as nighttime.

The rest of this paper is organized as follows. Section 2 introduces the detection method of the rotating target frame based on YOLOv5. Section 3 introduces the improved YOLOv5-OBB algorithm in detail. Section 4 describes the experiments and analysis. We summarize the paper in Section 5.

2. YOLOv5-OBB Detection Algorithm

2.1. YOLOv5s Model

YOLOv5-OBB (You Only Look Once v5-oriented bounding boxes) is based on YOLOv5 with the addition of target box angle prediction to predict the rotated target box. Firstly, the YOLOV5 model has superior performance and has received wide recognition in academia and industry. It has five versions with different model sizes, n, s, m, l, and x, which correspond to different network depths and widths. Here, in order to meet the real-time requirements of the model deployed in the embedded chip platform, the YOLOV5s model was finally selected, and the network structure is shown in Figure 2.

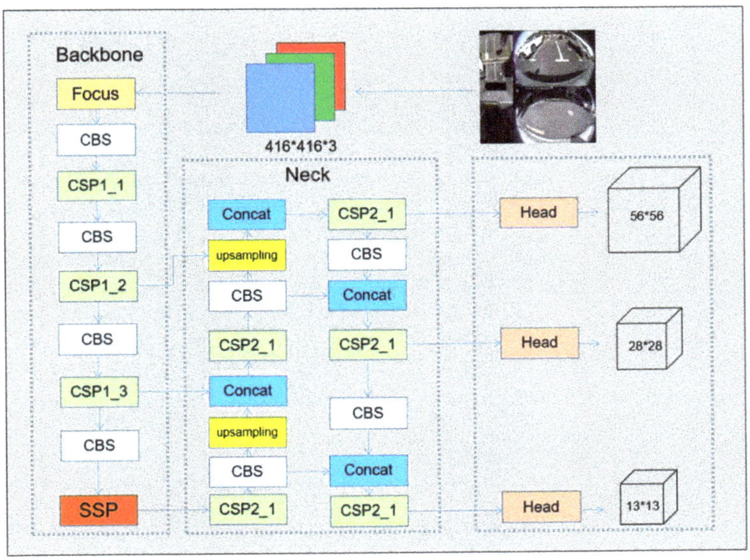

Figure 2. YOLOv5s network overall structure.

Input side: CutMix, Mosic, and other high-level data enhancement methods are used to stitch four pictures into one picture with random adaptive filling. This not only enriches the dataset but also corresponds to the reduction in batch-size and training arithmetic and also optimizes the model's detection effect on small targets, robustness, and generalization of the model. Adaptive anchor frame computation and adaptive image scaling methods are also used.

Backbone: CSPDarknet [18] is used to extract features mainly from the input image. The Focus module is used for feature extraction to reduce the number of computational parameters. The CSP network is used to optimize the problem of huge computation caused by the repetition of gradient information in the CSP network and for better fusion with the features extracted by the previous network.

Neck: The PANet module is used to fuse different feature modules. The FPN delivers high-level semantic features by upsampling, combining high-level semantic information with low-level detail information to achieve cross-scale feature fusion, and the PAN delivers localization features and bottom-level semantic information by downsampling, which delivers and aggregates cross-level information inside the feature pyramid, enabling the network to better capture the target's detail features and contextual information, thus improving the accuracy and robustness of the target detection.

Output: prediction is performed on feature maps of different sizes, in which feature maps of 52×52, 26×26, and 13×13 sizes predict large, medium, and small targets, respectively.

2.2. *Circular Smooth Labels for Angle Classification*

The method of predicting angles using regression can result in predictions outside of our defined range, leading to an angular boundary problem that produces a large loss value. So, YOLOv5-OBB employs the method of circular smoothing labels [19], as shown in Figure 3. The angular regression approach is converted into a classification form, discretizing the continuous problem directly and avoiding the boundary case. This way, since the classification results are finite, they do not go beyond the cases outside the defined range. This also addresses the fact that the classification loss cannot measure the angular distance between the predicted result and the labels; if GT (ground truth) is 0 degrees, the loss value is the same when we predict it as 1 degree and -90 degrees, as shown in Equation (1):

$$\mathrm{CSL}(x) = \begin{cases} g(x), & \theta - r < x < \theta + r \\ 0, & \text{otherwise} \end{cases} \tag{1}$$

where is the window function, which needs to satisfy the properties of periodicity, symmetry, monotonicity, maximum value, etc. It can generally be an impulse function, rectangular function, trigonometric function, and Gaussian function, r is the radius of the window function, and θ denotes the angle of the current enclosing frame. The setting of the window function allows the model to measure the angular distance between the predicted labels and the ground truth labels, i.e., the closer the predicted value is to the true value within a certain range, the smaller the loss value is. Moreover, the problem of angular periodicity is solved by introducing periodicity, i.e., even if the two degrees, 89 and -90, turn out to be near neighbors.

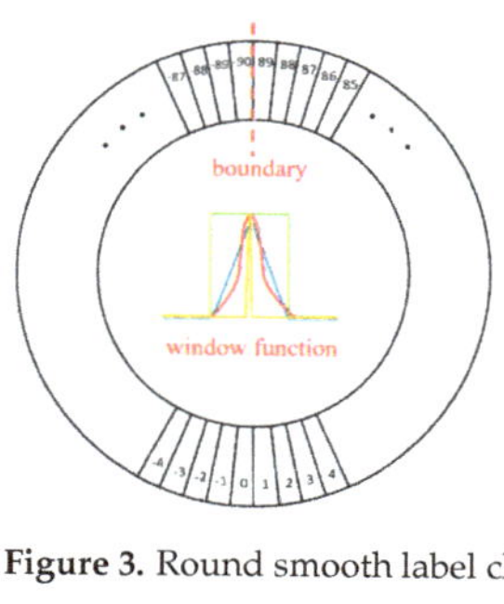

Figure 3. Round smooth label chart.

3. Improvement of YOLOv5-OBB

3.1. Optimizing the Backbone Extraction Module

In YOLOv5, the Focus module is introduced prior to the input layer of the backbone. This module selectively samples every other element from the feature layer of the image, effectively downsampling the image size by two while increasing the number of channels from 3 to 12. Subsequently, these channels are concatenated through a splicing operation. In an effort to optimize computational efficiency and expedite model inference, the Focus operation is tactically replaced with a standard convolution operation featuring a 6 × 6 convolution kernel and a stride of two. This strategic replacement not only addresses potential compilation issues on certain embedded chip platforms associated with the Focus operator but also significantly reduces the computational workload.

Inspired by SPP-net [20], the SPP module is a pooling layer that is used to perform pooling operations on the input feature maps at different scales. Its main purpose is to solve the problem of mismatching in the sensory field size of the CNN when different object sizes appear in the image. The main idea of the SPP module is to create pooling layers of different sizes to capture the feature information at different scales. It is introduced in the YOLOv3-SPP [21] network to achieve feature fusion at different scales, which significantly improves the network detection accuracy. As shown in Figure 4a, the SPP structure achieves feature fusion at different scales by passing the input features through the maximum pooling layers of convolutional kernel sizes 13 × 13, 9 × 9, and 5 × 5 in parallel and then splicing the different output features. The SPPF module is an improved version of the SPP module combined with the FPN (Feature Pyramid Network). The FPN [22] is designed to solve the problem of scale invariance in object detection tasks by fusing different layers of feature maps to deal with objects of different sizes. The difference between the SPPF and the SPP lies in the fact that the SPPF inputs the output features into the three maximum pooling layers of size 5 × 5 in sequence, splices the output results of each layer, and then splices them together. Each layer's output is spliced, as shown in Figure 4b. SPPF is less computationally intensive and faster than SPP. In this paper, the SPP structure is replaced with the more efficient SPPF structure.

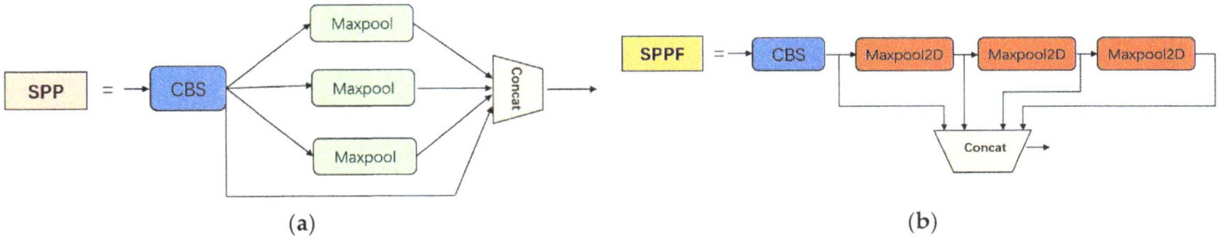

Figure 4. Selective Spatial Perception and Selective Search Proposals Fusion structure diagram. (**a**) SPP structure; (**b**) SPP structure.

Replacing the SiLU (Sigmoid linear unit) activation function in the CBS structure of the backbone network in YOLOv5-OBB with the GELU (Gaussian error linear unit) [23] function improves the generalization ability of the network. The GELU has shown good performance in a variety of tasks and networks, e.g., replacing the activation function ReLU with the GELU in ConvNext [24] improves the performance of the ReLU (rectified linear unit) by 0.7% on the ImageNet dataset, while the number of parameters is also reduced.

The GELU combines the properties of dropout, zoneout, and the ReLU, and its calculation formula is Equation (2). At the input side, the GELU activation function exhibits an approximately linear feature, which can better adapt to most of the features of the input data and can improve the model's learning and expression ability. A comparison of the SiLU and GELU functions is shown in Figure 5.

$$GELU = 0.5x(1 + \tanh(\sqrt{\frac{2}{\pi}}(x + 0.044715x^3))) \quad (2)$$

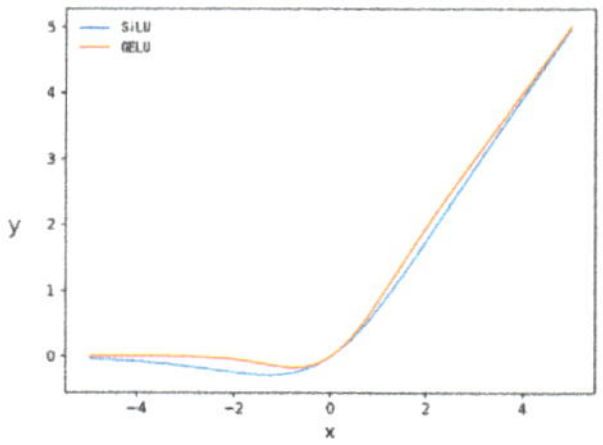

Figure 5. Plot of Gaussian error linear unit activation function and Sigmoid linear unit activation function.

3.2. Introduction of Improved RFB Modules

In the process of autonomous parking, information about the position of the corners is needed. Since the corners are small targets [25], there will be overlapping areas with the rectangular box of the parking space, leading to an overall decrease in the model's detection accuracy of the corners and the parking space. Of the three detection output heads of YOLOv5, the detail information on the 52 × 52 feature map is richer, which is helpful for the detection of small targets such as the corners of the corners. However, due to the small receptive field of the feature map, it lacks richer contextual and semantic information. In order to increase the semantic information of the receptive field and context, the improved RFB (Receptive Field Block) module is introduced in YOLOv5-OBB. Inspired by Inception [26], the RFB [27] module contains convolutional layers with different sizes of convolutional kernels, and these convolutional layers are formed into different multi-branch structures of the improved RFB. In order to increase the receptive field and improve the detection accuracy of small targets, different sizes of cavity convolutions are introduced to give the model a more powerful feature representation.

As shown in Figure 6, the improved RFB module first passes the previously output feature maps through 1 × 1 convolution, changes the number of channels of the feature maps, adjusts the number of output channels, and introduces an activation function to increase the nonlinearity and improve the model's expressive ability. Then, the null convolution with dilation rate = 1, dilation rate = 3, and dilation rate = 5 is mapped in each branch to increase the sensory field of the model. After that, the output of the feature maps of the three branches are concatenated and output through 1 × 1 convolution to achieve the purpose of fusion of different features. Finally, a shortcut operation is added to create jump connections, which are residual connections to prevent gradient vanishing and gradient explosion problems during training.

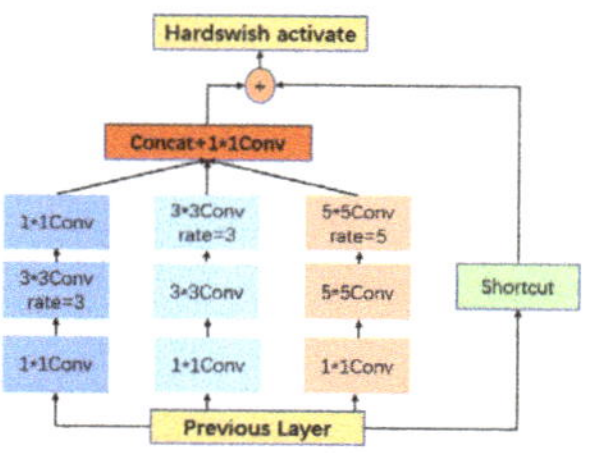

Figure 6. Improved Receptive Field Block structure diagram.

3.3. Increased CA Mechanisms

When in an underground garage with weak lighting conditions or at night, the outline of a parking space cannot be recognized, and the neck network structure of YOLOV5 focuses on deep feature fusion, which leads to a large number of details being lost, thus causing a large number of missed and false detections. In order to improve the recognition rate and reduce the impact of lighting conditions, this paper introduces the CA (coordinate attention) mechanism [28] in the back-end of the backbone network and the up-adoption stage of feature extraction to enhance the feature expression ability of the model. However, most of the current attention mechanisms (e.g., CBAM [29], SE [30]) generally use global maximum pooling or average pooling, which will lose the object spatial information. In contrast, the CA mechanism goes beyond simply incorporating a channel attention mechanism; it also incorporates a spatial attention mechanism. This spatial attention mechanism allows for the incorporation of positional information within the channel attention mechanism.

The CA mechanism consists of two main parts, namely coordinate information embedding and coordinate attention generation. As shown in Figure 7, given input X, two spatial extensions (1 × W) and (1 × H) of the pooling kernel are used to encode each channel along horizontal and vertical coordinates, respectively. The outputs are cascaded and then sent to a shared (1 × 1) convolutional transform. The spliced feature maps are sent to Batchnorm and Nonlinear to encode spatial information in the vertical and horizontal directions. The output then is split into two separate tensors. Using two other (1 × 1) convolutional transforms, they are converted into tensors with the same number of channels to the input X, respectively, to obtain $f \in R^{C\times H\times 1}$ and $f \in R^{C\times 1\times W}$. Then, under the Sigmoid activation function, two attentional weight maps in the spatial direction are obtained, and each attentional weight feature map has a long-term dependency in a particular direction. Finally, the input feature maps are multiplied with the two weights to enhance the representativeness of the feature maps.

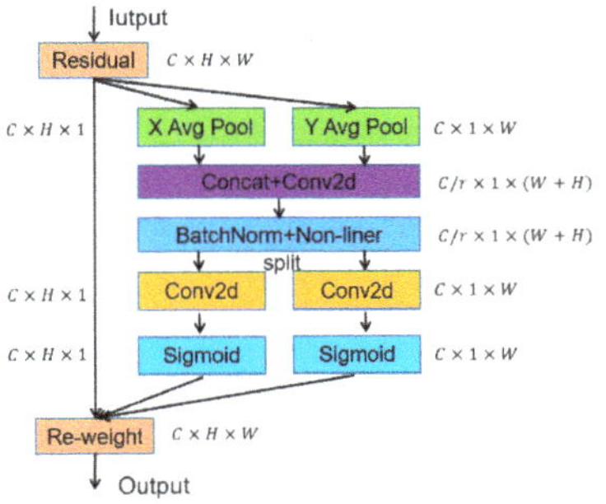

Figure 7. Coordinate attention mechanism calculation process.

In order to reduce the false detection rate and improve the detection accuracy under weak lighting conditions, the algorithm pays more attention to the important features during inference. The CA mechanism is added to the backbone of YOLOV5s, and the improved backbone network is shown in Figure 8. Not only does it not increase the excessive number of parameters and model computation of the network, but it also facilitates the extraction of important feature information. The optimization effect on the dataset is shown in Figure 9, which further improves the prediction score and reduces the false detection rate for data with less feature information.

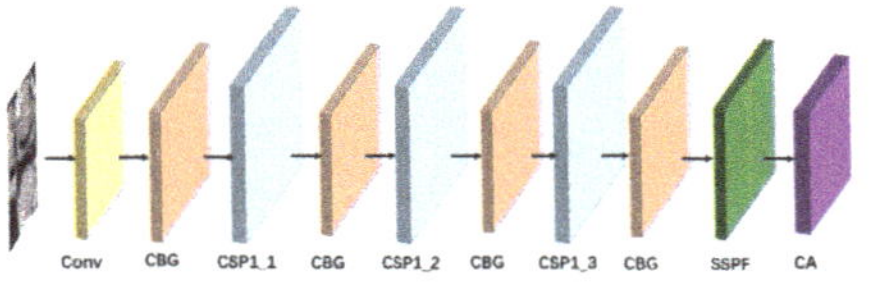

Figure 8. Map of where the coordinate attention mechanism is located in the backbone.

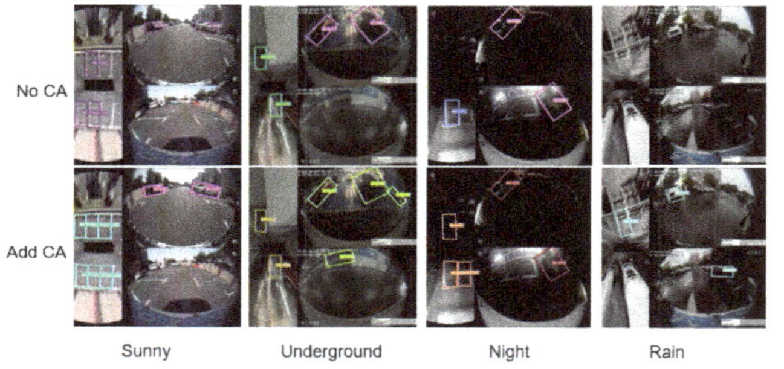

Figure 9. Adding a CA mechanism improves results.

3.4. Location-Rule-Based NMS Improvement

Currently, there is a notable decrease in the detection accuracy of parking spaces when depot corners are included in the training data. This decline in accuracy can be attributed to the spatial overlap between different categories of depot corners and parking spaces. Consequently, the model encounters challenges in accurately defining bounding boxes and category labels for these objects during both training and testing phases. This not only results in reduced bounding box accuracy but also introduces confusion in category labels.

Therefore, it is imperative to optimize the detection of parking spaces and library corners. Presently, post-processing algorithms predominantly emphasize non-maximum suppression methods, which filter out target boxes with confidence scores below a predefined threshold and those with significant positional overlap as determined by the intersection over union (IOU) metric. Leveraging prior knowledge, we have observed that most parking spaces align with library corners and exhibit strong positional correlation. To account for this correlation, we introduced a penalty factor, denoted as K, and incorporated it into existing post-processing algorithms. This strategic inclusion enhances the detection accuracy of both parking spaces and bank corners.

Referring to DIOU [31] loss function in modeling the similar relationship between two target frames in terms of spatial location, the concept of centroid distance is introduced. As shown in Figure 10, the correlation function is constructed by calculating the distance between the centroids of the parking space frame and the library corner frame and the diagonal lengths of the target frames of both. In Equation (3), where d_1 and d_2 represent the diagonal lengths of the two detection frames, respectively, $p(b^1, b^2)$ represents the distance between the center points of the two detection frames. K is the correlation coefficient of the two spatial locations, and the more spatially related the two target frames of different categories are, the larger the calculated value of K. The correlation coefficient of K is the correlation coefficient between the two target frames.

$$\mathrm{K} = e^{-\frac{p(b^1,b^2)}{\max(d_1,d_2)}} \tag{3}$$

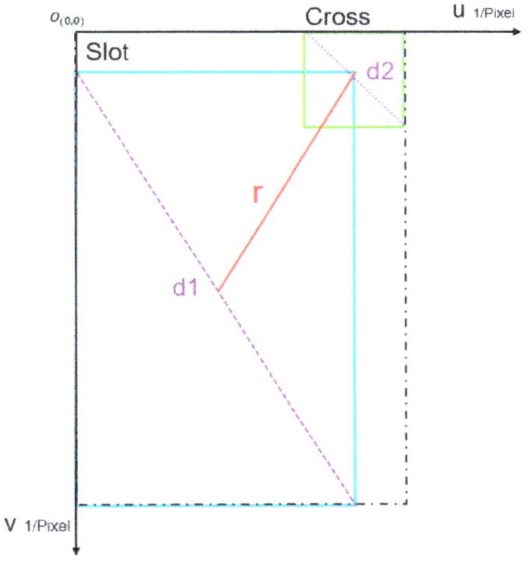

Figure 10. Calculated correlation coefficient K graph.

Then, we need to determine the specific value of the correlation coefficient K. We statistically calculate the distance between the parking space frame and the other category frames on the 20,000 training sets of the homemade dataset, sequentially find the K value between the parking space and the corner of the warehouse, vehicles, pedestrians, and so on, and then sum up and take the average. From Figure 11, we can see that the K-value correlation between the parking space and the corner of the warehouse (cross) is the highest and is much larger than that of the other categories, so we set the K-value to be greater than 0.25 when we process the non-extremely large value suppression based on the location rule and consider that the two target frames are spatially strongly correlated.

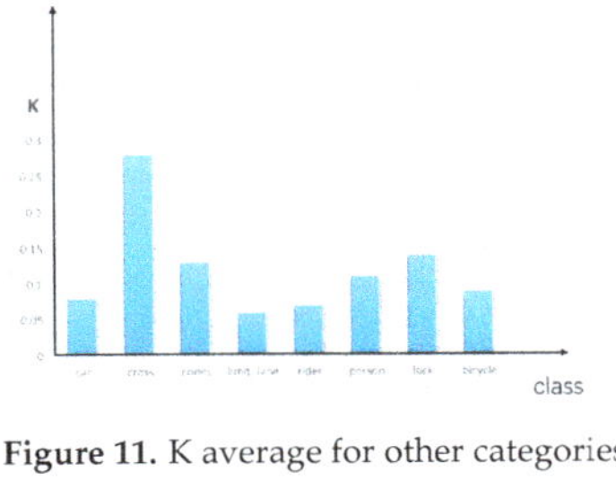

Figure 11. K average for other categories and car parking spaces.

Inspired by the formula of soft-NMS [32], when the correlation score is smaller than the set value of K, no change is made and the original prediction score is retained; when the correlation score is larger than the set value of K, the new prediction score grows linearly, and the optimization function (4) is constructed to be used to optimize the two strongly correlated objective boxes, where $S_{\max}$ represents the detection box with the highest confidence, S_i represents the non-optimization score of the current target box, and S_i^{new} represents the optimized confidence score of the current target frame.

$$s_i^{new} = \begin{cases} s_{\max} \times \mathrm{K} + s_i & \mathrm{K} > 0.25 \\ s_i & other \end{cases} \tag{4}$$

The above optimization method is added to the non-maximal value suppression, called K-NMS, and the algorithm process starts with iteratively traversing to optimize all the target frames, where is the maximum confidence of the candidate target frames in the picture, the confidence of the other target frames, and is the optimized confidence. As shown in Figure 12, if two target frames of different categories are strongly correlated, the confidence of the target frame with lower confidence is optimized according to the correlation coefficient of the two target frames. Firstly, K-value calculation is performed by Equation (3) for all the different categories of target boxes in Figure 12 and is then based on the K-value with the help of Equation (4) optimizing the confidence of the target boxes. The result is shown in Figure 12, which improves the confidence of the car parking spaces.

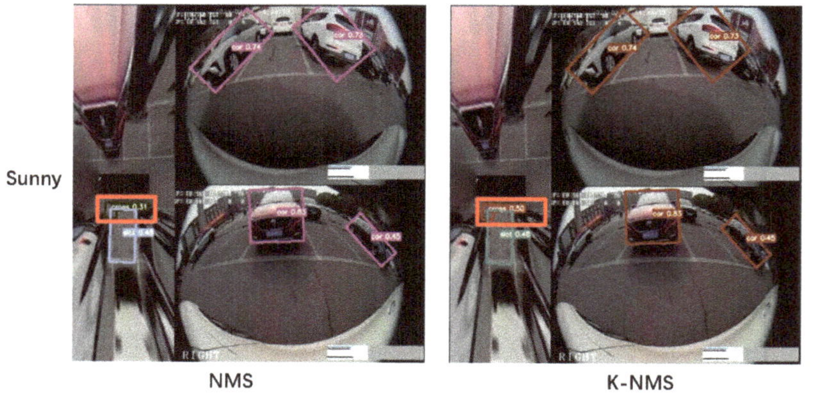

Figure 12. The effect of NMS optimization based on location rules.

4. Experimental Results and Analyses

4.1. Datasets

The experimental dataset used in this paper is a homemade dataset, where each image consists of four images captured by vehicle-mounted fish-eye cameras stitched together with a size of 416 × 416, as shown in Figure 13. A total of 20,000 images were collected in different car park locations and under various weather and lighting conditions, Table 1 is the details of the data set division. Real-time obstacle avoidance is required in autonomous parking; thus, nine categories of target spaces, vehicles, library corners, pedestrians, etc., need to be detected. The number of labels in the dataset is plotted as shown in Figure 14a.

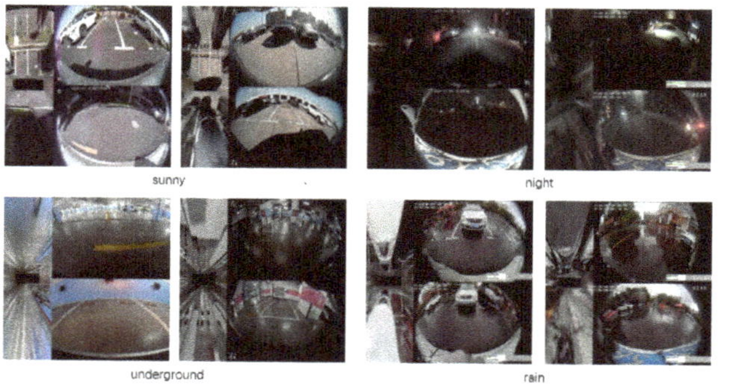

Figure 13. Diagram of the different scenarios of the dataset.

Table 1. Dataset division details.

Sence	Train	Val	Total
Sunny	5500	500	6000
Night	3500	500	4000
Rain	2500	500	3000
Underground	6500	500	7000

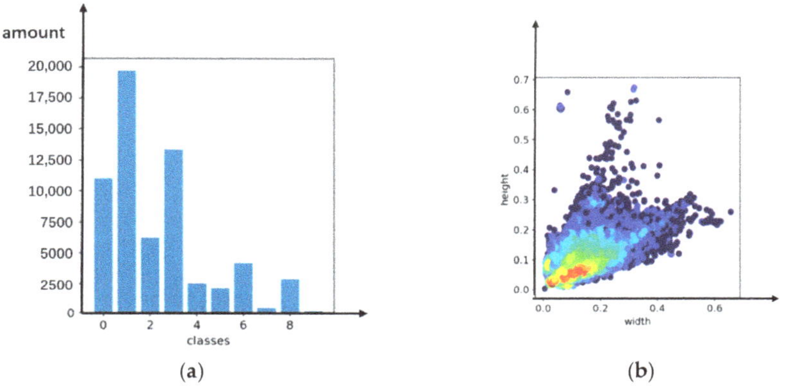

Figure 14. T-map of the dataset label distribution and aspect distribution: (**a**) label distribution; (**b**) aspect distribution.

In Figure 14b above, the target length and width distribution of the dataset indicates that there is a high distribution of small targets and that there is diversity in the size of the targets.

4.2. Experimental Environment

The experimental environment of this paper is Python 3.8, CUDA 11.1, PyTorch 1.10.1, and the graphics card NVIDIA V100 GPU, which performed the training and testing. In this paper, data enhancement techniques such as Mosaic, HSV, and random level flipping were used in the experiments to improve the generalization of the model. The number of training iterations was set to 300 epochs, the batch size was set to 16, the optimizer used SGD (stochastic gradient descent) with a momentum of 0.937, the initial learning rate was set to 0.01, and the EMA (exponential moving average) was used to determine the hybrid exponential sliding average, combined with SGD, making the model more robust. The trained model is converted to ONNX format, and then the model is compiled and converted in the CoreChip platform. Finally, the model is deployed in the CoreChip V9M platform, as shown in Figure 15.

Figure 15. V9M embedded platform development board.

4.3. Evaluation Criteria

In our experiments, we used a specific IoU threshold, which was set to IoU = 0.5 in this experiment. We used the following metrics to evaluate the performance of the model: precision (P), recall (R), average precision (AP), and mean average precision (mAP). These metrics can be calculated using Equations (5)–(8):

$$P = \frac{TP}{TP + FP} \tag{5}$$

$$P = \frac{TP}{TP + FN} \tag{6}$$

$$AP = \int_0^1 P(R)dR \tag{7}$$

$$mAP = \frac{1}{n}\sum_{i=1}^{n} AP_i \tag{8}$$

where TP denotes the number of prediction frames with an IoU greater than the threshold with respect to the target frame, FP denotes the number of prediction frames with an IoU less than the threshold with respect to the target frame, FN denotes the number of target frames that are not predicted, and n is the number of categories in the dataset. By calculating these metrics, we are able to evaluate the performance and accuracy of the model in the target detection task. mAP is an important composite metric that takes into account the average accuracy of the different categories and provides an assessment of the overall model performance.

On embedded devices, the real-time nature of model detection needs to be evaluated, and the size of the model parameter count also needs to be considered. Moreover, the evaluation criterion of FPS (frame per second) is introduced; the larger the FPS, the more frames per second are detected, the faster the detection rate is, and the better the real-time performance of the model is. The number of model parameters is the sum of the parameters in the model, which is directly related to the amount of space required by the model in the disk, affecting the amount of memory occupied by the model inference and also affecting the initialization time of the program.

4.4. Analysis of the Experimental Results

In this paper, different experimental groups are designed to experimentally analyze different improvements using controlled variables, and each group of experiments is tested on different model contents using the same training parameters, so as to analyze the effects of the backbone improvement, the addition of the RFB module and the CA mechanism, and the K-NMS improvement on the model performance. The results of the model testing are shown in Table 2, where "√" represents the strategies used in the improved model, and "×" represents the strategies not used in the improved model.

Table 2. Experimental results of different improved methods.

Improved Name	A	B	C	D	mAP	FPS	Size/MB
No improvement	×	×	×	×	62.32%	49.26	34.1
Improvement 1	√	×	×	×	63.27%	52.66	32.7
Improvement 2	√	√	×	×	66.69%	52.47	32.9
Improvement 3	√	√	√	×	69.65%	52.13	33.1
Improvement 4	√	√	√	√	70.72%	52.13	33.1

Analysis of the results in Table 2 reveals: A is the replacement of Focus and SPP in the YOLOV5-OBB network with more efficient ordinary convolution of size 6 × 6 and SPPF, respectively, and with the replacement of the SiLU activation function with the GELU, the mAP is improved by only 0.95% after improvement 1, but the model inference speed is significantly improved, and the number of model parameters is reduced; B is the introduction of the RFB module, which increases the speed of model inference and reduces the number of model parameters; C is the introduction of the RFB module, which increases the speed of model inference and reduces the number of model parameters. Introduction of the RFB module increases the receptive field, and mAP is improved by 3.42% after improvement 3; C is the addition of the CA module to the YOLOV-OBB network, improving the feature expression ability of the model, at the same time attenuating the transmission of the noise in the network, and mAP is improved by 2.96% after improvement 3; D uses the improved K-NMS algorithm to emphasize the spatial connection between the car parking space and the corner of the depot. Without losing speed and increasing the size of the model, mAP improved by 1.03% after improvement 4.

Compared with the original YOLOv5-OBB, the loss function of the improved training in this paper has a significant decrease, as shown in Figure 16. From the training comparison graph above, it can be clearly seen that with the gradual increase in the number of iterations, the curve of the loss function gradually converges, and the loss value becomes smaller and smaller. When the number of training rounds reaches 270, the loss value basically tends to stabilize. Compared with the original algorithm, the regression accuracy is higher, indicating the effectiveness of the improved algorithm. The enhancement of the detection effect before and after the improvement is shown in Figure 17.

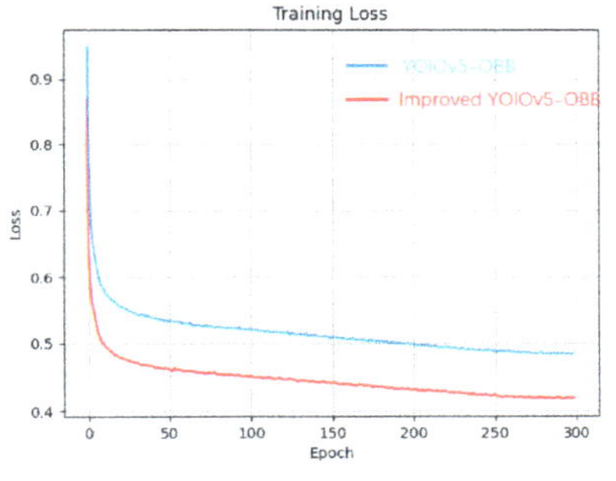

Figure 16. Training loss comparison graph.

To further verify the superiority of the improved YOLOv5-OBB model in terms of accuracy and efficiency, we selected several other parking space detection models for comparative experiments, and the experimental results in our homemade dataset are shown in Table 3. Compared with the VPSNet and DeepPS models in our homemade dataset, the improved YOLOv5-OBB model has 5.73% and 2.03% higher mAP values with fewer parameters and faster detection.

Table 3. Performance comparison of different parking space detection models on our self-made datasets.

Model Name	mAP	FPS	Size/MB
VPSNet [33]	64.99%	41.2	134.1
DeepPS	68.69%	38.6	232.9
YOLOV5-OBB	62.32%	49.26	34.1
Ours	70.72%	52.13	33.1

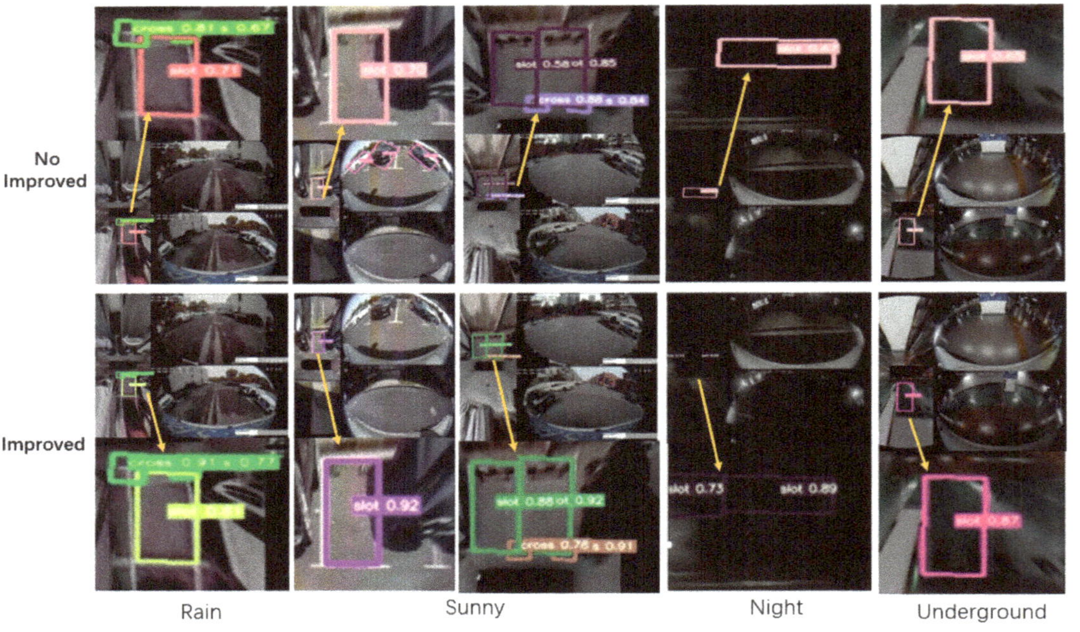

Figure 17. Detection effect before and after improvement in different scenes.

At the same time, we also compared it with some current parking space detection methods based on deep learning in the PSV dataset. As shown in Table 4, among the 1593 real labels in the PSV test set, the precision and recall of our model are both competitive.

Table 4. Parking slot detection performance of different methods in the PSV test set.

Model Name	GT	TP	FP	Precision Rate	Recall Rate
DeepPS	1593	1396	63	95.68%	87.63%
VPSNet [33]	1593	1507	54	96.54%	94.60%
Ours	1593	1510	51	97.21%	95.61%

In summary, the improved YOLOV5-OBB outperforms the previous model in detection environments with small targets and weak lighting environments and has strong robustness, detection, and recognition capabilities. The heat map of the detection results of the improved YOLOV5-OBB in different car park environments is shown in Figure 18.

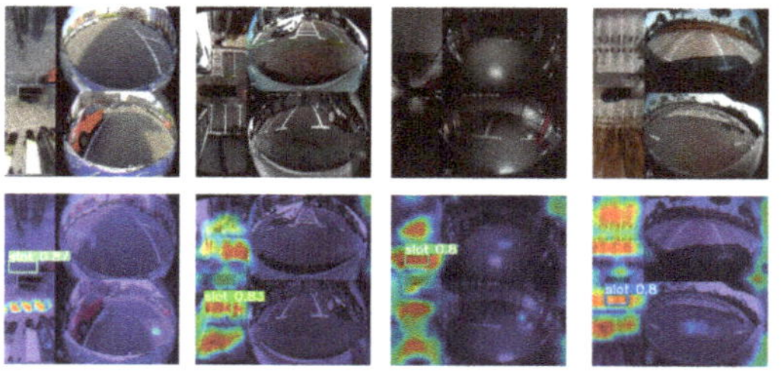

Figure 18. Heat map visualization for car parking detection.

5. Conclusions

In order to solve the problems of low space detection accuracy and slow inference speed in the process of autonomous parking, this paper proposes an improved YOLOv5-OBB algorithm. Firstly, in order to speed up the model inference speed, in the backbone network, the Focus and SSP modules are replaced with more efficient ordinary convolution and SPPF modules, and the SiLU activation function is replaced with the GELU. Secondly, an improved RFB module is introduced to increase the receptive field. After that, the CA mechanism is introduced to improve the effect of off-position detection in environments with weak lighting conditions. Finally, a position-rule-based NMS is proposed to penalize the correlation between the parking space and the corner of the reservoir, which further improves the accuracy of parking space detection. Compared with the original YOLOV5-OBB model, the mAP is improved by 8.4%. When the size of the model is reduced by 1 M, the FPS increases by 2.87, which meets the deployment requirements of automotive embedded platforms. In order to deploy the model in embedded platforms with more limited arithmetic power, subsequent research will carry out optimization of the network structure, using methods such as model pruning or knowledge distillation to reduce the number of parameters of the model and further improve the inference speed.

Author Contributions: Conceptualization, L.Z. and D.L.; methodology, X.W.; software, Z.C.; validation, Z.C. and W.Z.; formal analysis, D.L.; investigation, L.Z.; resources, Z.C.; data curation, Z.C.; writing—original draft preparation, Z.C.; writing—review and editing, G.Y.; visualization, G.Y.; supervision, X.W.; project administration, W.Z.; funding acquisition, G.Y. All authors have read and agreed to the published version of the manuscript.

Funding: This research received no external funding.

Data Availability Statement: Not applicable.

Conflicts of Interest: Guodong Yao, Dongdong Li, and Li Zeng are employees of Voyager Technology Inc. This paper reflects the views of the scientists and not the company.

References

1. Li, H. *Research on Vehicle Detection Based on Improved YOLO and Implementation of Vehicle Position Detection System*; Jilin University: Changchun, China, 2022.
2. Wong, G.S.; Goh, K.O.M.; Tee, C.; Sabri, A.Q.M. Review of Vision-Based Deep Learning Parking Slot Detection on Surround View Images. *Sensors* **2023**, *23*, 6869. [CrossRef] [PubMed]
3. Ma, Y.; Liu, Y.; Shao, S.; Zhao, J.; Tang, J. Review of Research on Vision-Based Parking Space Detection Method. *Int. J. Web Serv. Res.* **2022**, *19*, 1–25. [CrossRef]
4. Suhr, J.K.; Jung, H.G. Fully-automatic recognition of various parking slot markings in Around View Monitor (AVM) image sequences. In Proceedings of the 2012 15th International IEEE Conference on Intelligent Transportation Systems, Anchorage, AK, USA, 16–19 September 2012. [CrossRef]
5. Wang, C.; Zhang, H.; Yang, M.; Wang, X.; Ye, L.; Guo, C. Automatic Parking Based on a Bird's Eye View Vision System. *Adv. Mech. Eng.* **2014**, *6*, 847406. [CrossRef]
6. Li, L.; Li, C.; Zhang, Q.; Guo, T.; Miao, Z. Automatic parking slot detection based on around view monitor (AVM) systems. In Proceedings of the 2017 9th International Conference on Wireless Communications and Signal Processing (WCSP), Nanjing, China, 11–13 October 2017; pp. 1–6.
7. Suhr, J.K.; Jung, H.G. A Universal Vacant Parking Slot Recognition System Using Sensors Mounted on Off-the-Shelf Vehicles. *Sensors* **2018**, *18*, 1213. [CrossRef] [PubMed]
8. Hamada, K.; Hu, Z.; Fan, M.; Chen, H. Surround view based parking lot detection and tracking. In Proceedings of the 2015 IEEE Intelligent Vehicles Symposium (IV), Seoul, Republic of Korea, 28 June–1 July 2015; pp. 1106–1111.
9. Bui, Q.H.; Suhr, J.K. CNN-Based Two-Stage Parking Slot Detection Using Region-Specific Multi-Scale Feature Extraction. *IEEE Access* **2023**, *11*, 58491–58505. [CrossRef]
10. Li, Q.; Lin, C.; Zhao, Y. Geometric features-based parking slot detection. *Sensors* **2018**, *18*, 2821. [CrossRef] [PubMed]
11. Zhang, L.; Huang, J.; Li, X.; Xiong, L. Vision-based parking-slot detection: A DCNN-based approach and a large-scale benchmark dataset. *IEEE Trans. Image Process.* **2018**, *27*, 5350–5364. [CrossRef] [PubMed]
12. Zhou, S.; Yin, D.; Lu, Y. PASSIM: Parking Slot Recognition Using Attentional Semantic Segmentation and Instance Matching. In Proceedings of the IEEE 5th International Conference on Big Data and Artificial Intelligence (BDAI), Fuzhou, China, 8–10 July 2022; pp. 169–175.

13. Cao, L.; Yue, P.; Zhang, Z.; Liu, J.; Huang, M. Automatic parking system based on panoramic image and human-computer interaction. *Automot. Technol.* **2023**, *6*, 24–29.
14. Li, W.; Cao, L.; Yan, L.; Li, C.; Feng, X.; Zhao, P. Vacant parking slot detection in the around view image based on deep learning. *Sensors* **2020**, *20*, 2138. [CrossRef] [PubMed]
15. Li, L.; Zhang, L.; Li, X.; Liu, X.; Shen, Y.; Xiong, L. Vision-based parking-slot detection: A benchmark and a learning-based approach. In Proceedings of the IEEE International Conference on Multimedia and Expo (ICME), Hong Kong, China, 10–14 July 2017; pp. 649–654.
16. Lai, C.; Yang, Q.; Guo, Y.; Bai, F.; Sun, H. Semantic Segmentation of Panoramic Images for Real-Time Parking Slot Detection. *Remote Sens.* **2022**, *14*, 3874. [CrossRef]
17. Do, H.; Choi, J.Y. Context-based parking slot detection with a realistic dataset. *IEEE Access* **2020**, *8*, 171551–171559. [CrossRef]
18. Bochkovskiy, A.; Wang, C.Y.; Liao, H.Y.M. Yolov4: Optimal speed and accuracy of object detection. *arXiv* **2020**, arXiv:2004.10934.
19. Yang, X.; Yan, J. Arbitrary-oriented object detection with circular smooth label. In Proceedings of the Computer Vision–ECCV 2020: 16th European Conference, Glasgow, UK, 23–28 August 2020; Proceedings, Part VIII 16. Springer International Publishing: Berlin/Heidelberg, Germany, 2020; pp. 677–694.
20. He, K.; Zhang, X.; Ren, S.; Sun, J. Spatial Pyramid Pooling in Deep Convolutional Networks for Visual Recognition. *IEEE Trans. Pattern Anal. Mach. Intell.* **2015**, *37*, 1904–1916. [CrossRef] [PubMed]
21. Pebrianto, W.; Mudjirahardjo, P.; Pramono, S.H.; Setyawan, R.A. YOLOv3 with Spatial Pyramid Pooling for Object Detection with Unmanned Aerial Vehicles. *arXiv* **2023**, arXiv:2305.12344.
22. Seferbekov, S.; Iglovikov, V.; Buslaev, A.; Shvets, A. Feature Pyramid Network for Multi-class Land Segmentation. In Proceedings of the IEEE/CVF Conference on Computer Vision and Pattern Recognition Workshops (CVPRW), Salt Lake City, UT, USA, 18–22 June 2018; pp. 272–275.
23. Hendrycks, D.; Gimpel, K. Gaussian error linear units (gelus). *arXiv* **2016**, arXiv:1606.08415.
24. Liu, Z.; Mao, H.; Wu, C.Y.; Feichtenhofer, C.; Darrell, T.; Xie, S. A convnet for the 2020s. In Proceedings of the IEEE/CVF Conference on Computer Vision and Pattern Recognition, New Orleans, LA, USA, 18–24 June 2022; pp. 11976–11986.
25. Xie, C.; Wu, J.; Xu, H. Improved YOLOv5 algorithm for small target detection of UAV images. *Comput. Eng. Appl.* **2023**, *59*, 198–206.
26. Szegedy, C.; Ioffe, S.; Vanhoucke, V.; Alemi, A. Inception-v4, inception-resnet and the impact of residual connections on learning. In Proceedings of the AAAI Conference on Artificial Intelligence, San Francisco, CA, USA, 4–9 February 2017.
27. Liu, S.; Huang, D. Receptive field block net for accurate and fast object detection. In Proceedings of the European Conference on Computer Vision (ECCV), Munich, Germany, 8–14 September 2018; pp. 385–400.
28. Hou, Q.; Zhou, D.; Feng, J. Coordinate attention for efficient mobile network design. In Proceedings of the IEEE/CVF Conference on Computer Vision and Pattern Recognition, Nashville, TN, USA, 20–25 June 2021; pp. 13713–13722.
29. Woo, S.; Park, J.; Lee, J.Y.; Kweon, I.S. Cbam: Convolutional block attention module. In Proceedings of the European Conference on Computer Vision (ECCV), Munich, Germany, 8–14 September 2018; pp. 3–19.
30. Hu, J.; Shen, L.; Sun, G. Squeeze-and-excitation networks. In Proceedings of the IEEE Conference on Computer Vision and Pattern Recognition, Salt Lake City, UT, USA, 18–23 June 2018; pp. 7132–7141.
31. Zheng, Z.; Wang, P.; Liu, W.; Li, J.; Ye, R.; Ren, D. Distance-IoU loss: Faster and better learning for bounding box regression. In Proceedings of the AAAI Conference on Artificial Intelligence, New York, NY, USA, 7–12 February 2020; pp. 12993–13000.
32. Bodla, N.; Singh, B.; Chellappa, R.; Davis, L.S. Soft-NMS--improving object detection with one line of code. In Proceedings of the IEEE International Conference on Computer Vision, Venice, Italy, 22–29 October 2017; pp. 5561–5569.
33. Wu, Y.; Yang, T.; Zhao, J.; Guan, L.; Jiang, W. VH-HFCN Based Parking Slot and Lane Markings Segmentation on Panoramic Surround View. In Proceedings of the 2018 IEEE Intelligent Vehicles Symposium (IV), Changshu, China, 26–30 June 2018; pp. 1767–1772.

World Electric Vehicle Journal

MDPI

Article

Online Multiple Object Tracking Using Min-Cost Flow on Temporal Window for Autonomous Driving

Hongjian Wei [1,2,*], Yingping Huang [2], Qian Zhang [3] and Zhiyang Guo [4]

1 School of Physics and Electronic Engineering, Fuyang Normal University, Fuyang 236037, China
2 School of Optical-Electrical and Computer Engineering, University of Shanghai for Science and Technology, Shanghai 200093, China; yphuangusst06@gmail.com
3 School of Information Technology, Jiangsu Open University, Nanjing 210036, China; zhangqian@jsou.edu.cn
4 School of Traffic Engineering, Jiangsu Shipping College, Nantong 226010, China; gzy@jssc.edu.cn
* Correspondence: 201606017@fynu.edu.cn

Abstract: Multiple object tracking (MOT), as a core technology for environment perception in autonomous driving, has attracted attention from researchers. Combing the advantages of batch global optimization, we present a novel online MOT framework for autonomous driving, consisting of feature extraction and data association on a temporal window. In the feature extraction stage, we design a three-channel appearance feature extraction network based on metric learning by using ResNet50 as the backbone network and the triplet loss function and employ a Kalman Filter with a constant acceleration motion model to optimize and predict the object bounding box information, so as to obtain reliable and discriminative object representation features. For data association, to reduce the ID switches, the min-cost flow of global association is introduced within the temporal window composed of consecutive multi-frame images. The trajectories within the temporal window are divided into two categories, active trajectories and inactive trajectories, and the appearance, motion affinities between each category of trajectories, and detections are calculated, respectively. Based on this, a sparse affinity network is constructed, and the data association is achieved using the min-cost flow problem of the network. Qualitative experimental results on KITTI MOT public benchmark dataset and real-world campus scenario sequences validate the effectiveness and robustness of our method. Compared with the homogeneous, vision-based MOT methods, quantitative experimental results demonstrate that our method has competitive advantages in terms of higher order tracking accuracy, association accuracy, and ID switches.

Keywords: multiple object tracking; min-cost flow; feature extraction; data association on temporal window; autonomous driving

Citation: Wei, H.; Huang, Y.; Zhang, Q.; Guo, Z. Online Multiple Object Tracking Using Min-Cost Flow on Temporal Window for Autonomous Driving. *World Electr. Veh. J.* **2023**, *14*, 243. https://doi.org/10.3390/wevj14090243

Academic Editors: Biao Yu, Linglong Lin and Jiajia Chen

Received: 3 August 2023
Revised: 29 August 2023
Accepted: 31 August 2023
Published: 2 September 2023

1. Introduction

Multiple object tracking (MOT) can be defined as localizing various objects in a scene after obtaining a sequence of data (i.e., a series of RGB images) over a period of time from sensors, combining these with data association techniques to accomplish the correct matching of the same object between data frames, and forming the tracking trajectory of each object. MOT has enormous potential in both academia and industry and has gained increasing attention in the fields of computer vision and artificial intelligence [1–3]. Autonomous driving is the emerging future of the automotive industry, helping to alleviate traffic congestion, reduce traffic accidents, improve driving safety, meet the various needs of different people groups, etc. [4–6]. As a core technology for autonomous driving environment perception, MOT is crucial to tasks such as obstacle avoidance, ego-vehicle route planning, intention recognition, and vehicle control of autonomous driving vehicles.

To realize MOT, two main challenges need to be considered: (1) How to measure the similarity between intra-frame objects, and (2) How to recover identity based on the similarity between objects across frames. Problem (1) involves object feature extraction, while Problem (2) pertains to data association. The object appearance features are important clues for measuring the similarity between objects in MOT. Historically, traditional hand-crafted features like RGB color histograms [7] and Local Binary Pattern Histograms (LBPH) [8] have been commonly used for data association. However, these manually designed features have limitations in fully capturing the semantic information of objects and lack robustness in handling significant appearance variations. As a result, their performance tends to be subpar on MOT benchmarks. Nevertheless, with the emergence and advancement of deep neural networks, specifically Convolutional Neural Networks (CNNs), we have witnessed a remarkable shift in feature extraction. The CNN-based feature extractors have gained widespread popularity in MOT research [9–12]. These feature extractors possess a hierarchical structure, where different convolutional layers characterize the objects from various perspectives. This enables the extracted features to encode more details and semantic information of the objects, thereby distinguishing them from other appearance-similar objects. In addition, due to the high dynamic and complexity of traffic scenes, the main challenges, including object scale changes, light intensity, occlusion, shadows, etc., which are currently faced in MOT for autonomous driving, also increase the difficulty in extracting features with significant discriminability of the object. Therefore, we employ ResNet50 [13] as the backbone network and design an object appearance feature extraction network based on metric learning, which is trained using triplet loss. The network minimizes the distance between the same objects while maximizing the distance between different objects, resulting in discriminative target appearance features. Furthermore, we introduce a Kalman Filter (KF) with constant acceleration (CA) to achieve accurate prediction of object positions by fusing current observations with past state estimates.

Data association is the core of MOT, which aims to correctly match the detections at the current frame with their historical trajectories, and the calculation of feature affinity between objects and trajectories plays an important role in the process. Most existing methods only calculate the feature affinity between the trajectories of the previous frame and the detections of the current frame, ignoring the trajectories that were not matched at the previous frame. This approach is feasible for the continuously detected objects, but for the reappearing objects caused by disappearance or occlusion, they may be assigned to incorrect trajectories, leading to unnecessary ID switches and reducing the tracking performance. To address this issue, we present how to categorize the trajectories within a temporal window composed of consecutive multi-frame images into active trajectories and inactive trajectories and compute the feature affinity between each category of trajectories and detections separately. Active trajectories and inactive trajectories are defined as follows: during the tracking, if no new detections are added to a trajectory in the current frame, its state is set to inactive, otherwise it is set to active. Additionally, we conjoin the concept of the min-cost flow for global association and propose a temporal window data association method. Specifically, when there is a strong (large) affinity between detections and trajectories, our method solves the min-cost flow problem of the network constructed using the affinity between detections and trajectories, enabling optimal matching between detections and trajectories within the temporal window.

To summarize, we present a novel online MOT framework for autonomous driving utilizing min-cost flow on temporal window. The main contributions are summarized as follows:

- By leveraging the feature extraction capabilities of CNN and incorporating metric learning, we design a three-channel neural network with ResNet50 as the backbone network and a triplet loss as learning function. The network aims to extract object appearance features that possess high discriminability. Simultaneously, we employ the KF with CA motion model to optimize and predict the bounding box information of objects. As a result, we obtain robust object representation features.

- The trajectories within the temporal window are divided into active trajectories and inactive trajectories. The affinities between each category of trajectories and detections are computed based on appearance and motion features. By constructing a sparse affinity network and solving the min-cost flow problem, the data association is performed, leading to a reduction in ID switches.
- The extensive experiments have been conducted on the KITTI MOT dataset and our real-world campus scenario. The ablation study confirms the effectiveness of the key modules. The comparison results between our method and the existing homogeneous, vision-based methods using state-of-the-art evaluation measures show that our method exhibits competitive tracking performance.

2. Related Work

Over the past few decades, researchers have proposed a wide range of solutions for MOT. Existing MOT methods can be categorized into two frameworks based on tracking techniques: tracking-by-detection and joint detection and tracking. The main difference lies in whether there is a tracking module integrated with the object detection network.

2.1. Tracking by Detection

The tracking-by-detection methods have become one of the mainstream approaches for MOT [1,7–12,14–16]. It is a paradigm for multi-stage object tracking. In this tracking paradigm, the video sequence frames are first subjected to object detection using a detector. Then, a tracker is employed to extract features, and a data association method is used to establish correspondences between objects and trajectories. By repeating these steps frame by frame, the final tracking results are obtained. The tracking-by-detection paradigm usually consists of three main modules: object detection, feature extraction, and data association.

Owing to the powerful feature extraction capabilities of CNNs, object detection for autonomous driving has made significant breakthroughs [17–23]. Ren et al. [17] proposed Faster R-CNN, which used a Region Proposal Network (RPN) instead of Selective Search to generate Region of Interest (RoI) proposals faster. RPN shared the convolutional layers with Fast R-CNN [18], reducing computational complexity and significantly accelerating object detection. G-RCNN [19] was an innovative object detection model that introduced a unique granule concept in CNN. In unsupervised mode, G-RCNN utilized the granule technique combined with spatiotemporal information to extract more accurate RoIs and efficiently capture the details and contextual information of objects, thereby enhancing the performance of object detection. YOLOv4 [20], with CSPDarknet-53 as the backbone network, introduced the "bag of freebies" and "bag of specials" techniques to improve data augmentation and regularization during training, achieving faster inference. YOLOv4 had been considered to strike the best balance between speed and accuracy. YOLOX [21] utilized an anchor-free, decoupled head techniques that allowed the network to process classification and regression using separate networks, reducing the number of parameters and increasing the inference speed. Edge YOLO [22] was a lightweight object detection framework based on YOLOv4, which was designed for 5G edge computing scenarios. It incorporates channel pruning to significantly reduce the network size and improves the feature fusion method to efficiently reduce GPU resource consumption.

Feature extraction is an essential stage in tracking-by-detection MOT methods, and accurate feature extraction is key to high-quality tracking. The object features in MOT mainly focus on appearance features [7–12], motion models [1,14], aggregation features [15,16], etc. The appearance features extraction can be divided into hand-crafted features and CNN-based feature extraction. Hand-crafted features include RGB histograms [7], LBPH [8], etc. However, these features cannot capture the semantic information of the objects and have limited discriminative ability. Since CNNs were introduced to computer vision, it has been used by many researchers for extracting object appearance features. Mykheievskyi et al. [9] proposed a simple CNN to learn the local feature descriptors of objects, and Gonzalez et al. [10] adopted a Multiple Granularity Network. Quasi-Dense Similarity Learning [11] generates hundreds of region proposals for contrastive learning of appearance features by densely sampling image pairs and employs most of the information regions on the image to obtain high-quality appearance features. Ref. [12] utilized the self-attention mechanisms to focus on key information about the objects to obtain reliable appearance feature representations. Some researchers used motion models to represent the object. For instance, LGM [14] solved the long-term tracking problem in MOT by effectively utilizing object local and global motion information that was modeled by the box and tracklet embedding modules, without object appearance features. If only appearance features or motion models are used in MOT methods, then the spatiotemporal correlation of objects will be neglected, leading to tracking failures. Therefore, aggregation features have been proposed. STURE [15] achieved learning of spatiotemporal representations between current candidate detections and historical sequences in a mutual embedding space. By designing diverse loss functions, it was capable of extracting more discriminative representations for detections and sequences, thereby enhancing the current detection features and eliminating the differences among them. Yang et al. [16] projected the position and motion features of each object into an adaptive search window, which matched based solely on the similarity of appearance features. This ensured a better balance between appearance and motion features.

In tracking-by-detection methods, data association between trajectories and detections is crucial. It first calculates the affinity between the trajectories and the detections using the extracted features, and then employs different strategies for matching based on the affinity. MOTSFusion [24] was a closed-loop method that adopts a strategy of track first, reconstruct, and then reconstruct and track again. It synthesized information from trajectory extraction and object reconstruction to achieve accurate tracking and recovery of objects by combining 2D trajectory with 3D object modeling and can handle occlusion and missing objects, improving the efficiency and robustness of tracking. TripletTrack [25] exploited a CNN trained with the triplet loss and a Long Short-Term Memory to extract appearance features and motion features from the tracked and detected objects, respectively, determined the similarity of their features and constructed an affinity matrix using a small affinity network, and then adopted the Hungarian algorithm for association. FAMNet [26] employed Siamese networks for single object tracking of each object in adjacent frames and implicitly obtained the appearance and position information of the objects. It achieved continuous multi-frame end-to-end joint data association training by performing local associations. ByteTrack [2] separated the objects into two categories, high-confidence and low-confidence, based on their detection confidence. For the first matching, KF was adopted to generate the trajectory positions at the next frame. Based on either motion affinity or appearance affinity, an association matrix was constructed, and the Hungarian algorithm was then used to match the high-confidence detections with the trajectories. For the second matching, the Intersection over Union distance was used to compute the affinity between the low-confidence detections, the remaining tracked objects and trajectories from the previous step. The Hungarian algorithm was employed again to complete the matching.

2.2. *Joint Detection and Tracking*

The joint detection and tracking (JDT) methods typically apply the state-of-the-art detector frameworks as the backbone network. The detection branch and feature extraction branch share the underlying features to accomplish the tasks of object detection and feature extraction, enabling data association and achieving simultaneous object detection and tracking. As the object detection and feature extraction tasks are carried out within the same backbone network, the JDT methods are more efficient. YOLOTracker [3] adopted the Hungarian algorithm for data association. It employed the CSPDarknet-53 network as the backbone network and utilized a path aggregation network to fuse low-resolution and high-resolution features. Additionally, the texture features and semantic information were integrated to reduce inconsistencies in object feature extraction and obtain a more comprehensive object representation. Guo et al. [27] employed ResNet101 as the backbone network and introduced temporal-aware object attention and distractor attention, which enhance object focus and suppress interference. This approach enables better focus on the target and suppression of distractions, leading to collaborative joint optimization between position prediction and embedding association tasks. CenterTrack [28] followed the CenterNet framework and took the current frame, the previous frame, and a heatmap generated based on the centers of tracked objects as inputs. In addition to returning the center point, length, and width of the detected bounding box, the regression branch also returned an extra offset, which was used for tracking prediction. Tokmakov et al. [29] proposed an end-to-end trainable method based on [28], named PermaTrack. It used convolutional gated recurrent units to encode the entire frame for feature mapping estimation. PermaTrack incorporated a spatiotemporal, recurrent memory module to utilize past history and infer the position and identity of objects at the current frame. JDT methods require simultaneous execution of detection and feature extraction tasks, but optimizing these tasks separately can lead to conflicts. To solve it, MOTFR [30] introduced a locally shared information decoupling (LSID) module to separate the detection and feature extraction tasks, effectively addressing their optimization conflicts while ensuring necessary information sharing. MOTFR also incorporated a feature purification module that, combined with the LSID module, utilized extracted object features to guide the optimization of the detection task, further improving tracking performance and enhancing the accuracy of object detection. SegDQ [31] was a Transformer-based method that utilized multi-task learning and dynamic feature queries. It employed a semantic segmentation branch to learn and predict foreground masks, aiding in the extraction of foreground features and bounding box regression for MOT tasks. SegDQ introduced a dynamic query method that generated biased object queries using deep features extracted from the backbone network, enabling more robust prediction of newly detected objects. Cai et al. [32] proposed a Transformer-based method called MeMOT. MeMOT leveraged hypothesis generation to generate object proposals at the current frame, providing an initial point for the tracking task. Additionally, it adopted memory encoding to extract core information from the memory of each tracked object, which was used for feature representation and relational modeling. MeMOT addresses both object detection and data association tasks using memory decoding. By utilizing memory and attention mechanisms, MeMOT can establish long-term connections between objects, resulting in more stable and accurate tracking performance.

3. Method

As shown in Figure 1, our proposed online MOT framework follows tracking-by-detection structures. The framework is divided into two parts: feature extraction and data association on the temporal window. In the first stage, our previous work [33] is utilized to accurately segment the objects in the detected bounding boxes by YOLOX [21]. Then, we design an object appearance feature extraction network based on metric learning to obtain discriminative appearance features and apply a motion model to estimate the position information of each trajectory in the trajectories set at the current frame. In the second stage, within a temporal window composed of consecutive multi-frame images and the affinity between inactive trajectories, active trajectories, and detections is computed based on their appearance features and motion affinity. An affinity network is constructed, and the min-cost flow problem of this network is solved to complete the matching between the current frame's detections and trajectories.

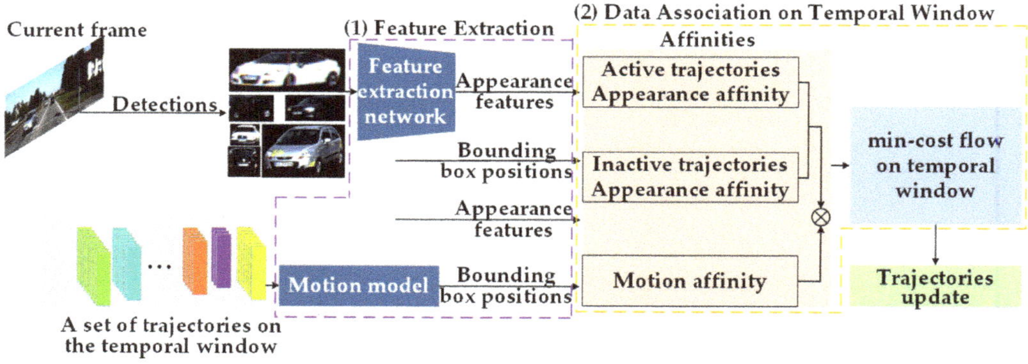

Figure 1. The overall framework of our method. (1) Feature extraction (in the purple dashed box). (2) Data association on temporal window (in the golden dashed box).

3.1. Feature Extraction

The affinity between detected objects and trajectories is the core of data association in MOT. In this paper, both appearance features and motion features of the objects are adopted during the data association. Specifically, the object appearance features are extracted using a metric learning-based CNN, while the motion features are obtained by predicting the position of the object bounding boxes applying KF.

3.1.1. Feature Extraction

ResNet [13] is a deep neural network under the concept of residual learning. By adding skip connections between convolutional layers, ResNet allows information to propagate across multiple hidden layers, effectively alleviating the problems of gradient vanishing and network degradation that commonly occur in traditional deep neural networks. This enables ResNet to have tens or even hundreds of layers. ResNet50 is one of the variations of ResNet, which has been widely adopted as a backbone or base network in various engineering fields, such as object detection, image recognition, autonomous driving, etc., demonstrating well the feature extraction ability.

In MOT, regardless of the method used to extract object appearance features, an important constraint is that the extracted features must be discriminative, meaning that the affinity between the same object should be high while the affinity between different objects should be low. Therefore, considering the number of parameters and the performance of the model, we choose ResNet50 as the backbone network. Incorporating the metric learning, we design an object appearance feature extraction network, which is shown in the red dashed box in Figure 2.

From Figure 2, it can be observed that the designed object appearance feature extraction network based on metric learning consists of ResNet50 and two fully connected layers, and outputs a 1×128-dimensional vector as the object appearance feature. The specific structure is as follows: the last layer used for classification in ResNet50 is removed, and on top of it, a fully connected layer (FC layer) with 1024 nodes and another fully connected layer (appearance feature layer) with 128 nodes is added. Batch normalization and the ReLU activation function are applied between the FC layer and ResNet50, while no activation function is applied between the FC layer and the appearance feature layer, resulting in a 1×128-dimensional object appearance feature vector as the output. During network training, triplets of samples consisting of three images sized 224×224 (anchor sample image, positive sample image, and negative sample image) are input into three identical object appearance feature extraction network models with shared weights. Each network model extracts its respective appearance features, and then the metric loss function is computed and backpropagated to optimize and adjust the network weight parameters.

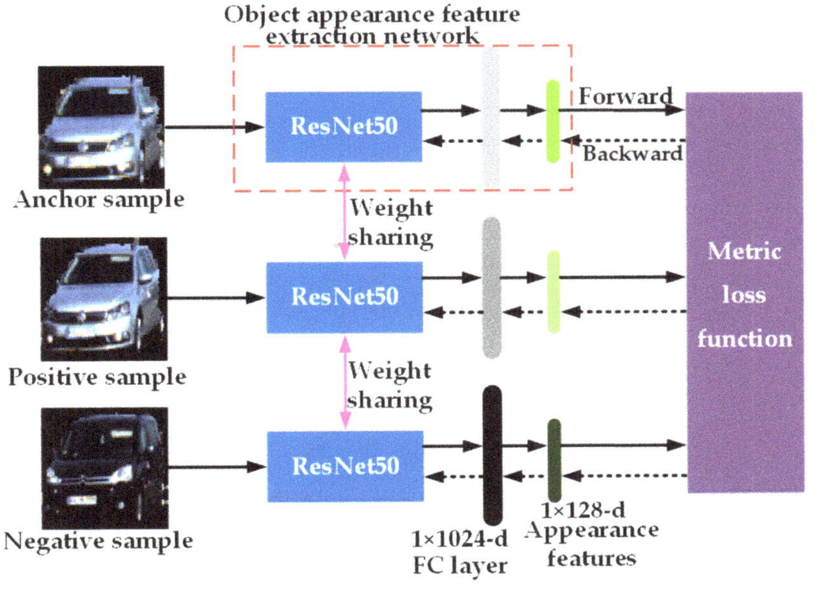

Figure 2. The flowchart of the object appearance feature extraction network based on metric learning. The network in the red dashed box is our designed network.

The objective of training the object appearance feature network is to increase the affinity of features belonging to the same object across different frames, while minimizing the affinity between features of different objects, thus the extracted features are discriminative. Therefore, we introduce metric learning and adopt triplet training to map the object appearance features into a metric space, where the affinity between features of the same object is greater than the affinity between features of different objects.

Let the total number of image triplet samples used for training be N_{sample}. The i-th (i = 1, 2,. . ., N_{sample}) triplet sample is denoted as $G_i = (g_i^{anchor}, g_i^{positive}, g_i^{negative})$, where g_i^{anchor} indicates anchor sample image, $g_i^{positive}$ denotes positive sample image, and $g_i^{negative}$ represents the negative sample image. g_i^{anchor} forms a of the anchor-positive pair with $g_i^{positive}$, while g_i^{anchor} forms an anchor-negative pair with $g_i^{negative}$. The mappings of g_i^{anchor}, $g_i^{positive}$, and $g_i^{negative}$ from the original image space to the learned object appearance feature extraction network feature space are denoted as $\sigma(g_i^{anchor})$, $\sigma(g_i^{positive})$, and $\sigma(g_i^{negative})$, respectively. The network is trained using the triplet loss function, $L_{triploss}$, which is commonly used in metric learning techniques:

$$L_{triploss} = \frac{1}{N_{sample}} \sum_{i=1}^{N_{sample}} max(0, ||\sigma(g_i^{anchor}) - \sigma(g_i^{positive})||_2^2 - ||\sigma(g_i^{anchor}) - \sigma(g_i^{negative})||_2^2 + thr) \quad (1)$$

where, thr refers to the margin value that allows the network to distinguish between positive sample pairs and negative sample pairs. In other words, the triplet loss learning requires the distance between all negative sample pairs to be greater than the distance between positive sample pairs by a positive minimum margin value thr. During the training process, the loss continuously decreases, making the anchor samples closer to the positive samples while keeping a larger distance from the negative samples. A smaller thr means that the anchor samples do not need to be pulled too close to the positive sample set in the feature space, and the anchor samples do not need to be pulled too far from the negative sample set, making it easier to meet the convergence condition. However, since the distance between positive and negative samples is not widened, there is a risk of not being able to effectively distinguish ambiguous data. Conversely, a larger *thr* enables better differentiation of similar images with more certainty. However, it brings challenges during the learning process as it requires pulling the anchor points closer to the positive samples and simultaneously increasing the distance from the negative samples, resulting in a larger loss, severe parameter update oscillation, and training difficulties. Therefore, setting a reasonable *thr* is crucial for training networks based on triplet loss.

When using the triplet loss function of Equation (1) to train the designed object appearance feature extraction network based on metric learning, the learned feature space not only ensures that the distance of the anchor-positive is smaller than the distance between $\sigma(g_i^{anchor})$ and $\sigma(g_i^{negative})$, but also incorporates a predefined boundary *thr*, which can pull samples of the same object closer and push samples belonging to different objects farther in the learned feature space, as illustrated in Figure 3.

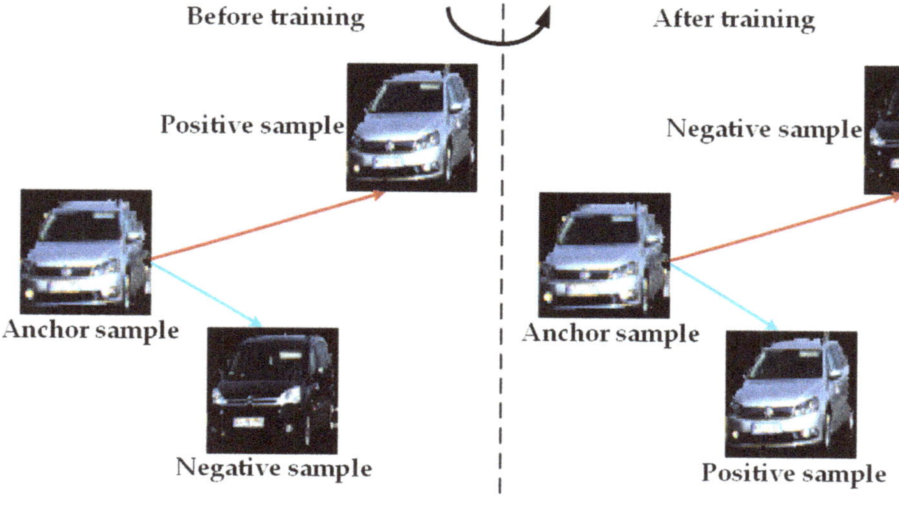

Figure 3. After training, the distance of the anchor-positive decreases and the distance of the anchor-negative increases.

In addition, when training the network, the gradients of $L_{triploss}$ with respect to g_i^{anchor}, $g_i^{positive}$, and $g_i^{negative}$ at the feature extraction layer are given by:

$$\frac{\partial L_{triploss}}{\partial\sigma(g_i^{anchor})} = -\frac{\partial L_{triploss}}{\partial\sigma(g_i^{positive})} - \frac{\partial L_{triploss}}{\partial\sigma(g_i^{negative})} \quad (2)$$

$$\frac{\partial L_{triploss}}{\partial\sigma(g_i^{positive})} = \frac{1}{N_{sample}} \sum_{i=1}^{N_{sample}} 2 \times (\sigma(g_i^{positive}) - \sigma(g_i^{anchor})) \times \mathbf{1}(L_{triploss} > 0) \quad (3)$$

$$\frac{\partial L_{triploss}}{\partial\sigma(g_i^{negative})} = \frac{1}{N_{sample}} \sum_{i=1}^{N_{sample}} 2 \times (\sigma(g_i^{anchor}) - \sigma(g_i^{negative})) \times \mathbf{1}(L_{triploss} > 0) \quad (4)$$

In Equations (3) and (4), $\mathbf{1}(\cdot)$ is an indicator function: if $L_{triploss} > 0$, the output is 1; otherwise, the output is 0.

3.1.2. Motion Model

In previous MOT methods, the majority use a constant velocity (CV) motion model to predict the future motion states of objects and employ filtering algorithms to smooth the predicted states [34,35]. However, the constant velocity motion model neglects acceleration, which may result in double or more motion errors when the object detector misses detections in consecutive frames. In addition, there are also MOT methods that adopt deep learning (i.e., long short-term memory network) to dynamically model the objects and predict their positions at the current frame. However, compared to previous motion models, this approach requires more time cost. In real-world road scenarios, the objects usually exhibit variable-speed motion, and the motion variation between consecutive frames is relatively small. Therefore, the object motion state can be approximated as uniform variable-speed motion. The Kalman Filter, with its simplicity and low complexity, is widely used for state estimation optimization. Based on the past signal information, it utilizes the principle of statistical computation to optimize the minimum mean square error and predict future state variables. Therefore, to strike a better balance between accuracy and speed, we apply a Kalman Filter with a constant acceleration (CA) motion model to predict and optimize the object position (bounding box) information. Specifically, we model the object motion as a CA motion model and use the measurements obtained from the YOLOX object detector at the current frame. Based on the observation and state transition equations, we iteratively predict and update the object state to obtain the predicted object position at the current frame.

Let $x = (u^c, v^c, \gamma^c, h^c, \dot{u}^c, \dot{v}^c, \dot{\gamma}^c, \dot{h}^c, \ddot{u}^c, \ddot{v}^c, \ddot{\gamma}^c, \ddot{h}^c)^T$ be the object state vector, where (u^c, v^c) represents the pixel coordinates of the center point of the object bounding box, γ^c is the aspect ratio of the bounding box, h^c denotes the height of the bounding box, $\dot{u}^c$, $\dot{v}^c$, $\dot{\gamma}^c$, and $\dot{h}^c$ represents the velocities of the corresponding parameters, $\ddot{u}^c$, $\ddot{v}^c$, $\ddot{\gamma}^c$, and $\ddot{h}^c$ represents the accelerations of the corresponding parameters. Assuming the object state vector at the previous frame $t - 1$ is x_{t-1}, and the YOLOX object detector detects the object state observation value at the frame t as $z_t = (u, v, \gamma, h)_t^T$. The state transition equation and observation equation for the object are as follows:

$$x_t = \mathbf{A}x_{t-1} + \omega_{t-1} \tag{5}$$

$$z_t = \mathbf{H}x_t + \mu_t \tag{6}$$

where Equation (5) represents the state transition equation, and the state transition matrix $\mathbf{A}$ indicates the object motion variation. ω is the process noise, which is a comprehensive abstract description of the uncertainty and random disturbances in the establishment of the motion model, following a normal distribution with zero mean and covariance $\mathbf{Q}$. $\mathbf{Q}$ is the process noise covariance matrix, caused by uncertain noise. The smaller the $\mathbf{Q}$, the easier the system converges and the higher confidence in the predicted values of the motion model, but excessively small values may result in divergence. Equation (6) denotes the observation equation, and the observation matrix $\mathbf{H}$ describes the relationship between the object motion state and the observation. μ is the measurement noise, similar to the process noise ω in the state transition equation, obeying a normal distribution with zero mean and covariance $\mathbf{R}$. $\mathbf{R}$ is the measurement noise covariance matrix. In each frame of MOT, it is necessary to predict each object position and update the observation equation and state transition equation corresponding to each object.

The prediction equations are:

$$x_t^- = \mathbf{A}x_{t-1}^+ \tag{7}$$

where, x_{t-1}^+ denotes the posteriori object state estimate of the object at frame $t - 1$, and x_t^- is the prior (predicted) object state estimate at frame t using the CA motion model. Therefore, the predicted state estimate evolves from the optimal estimate (posterior) of the previous state. In this paper, the object motion state is modeled as a CA motion model, and the state transition matrix $\mathbf{A}$ is as follows:

$$\mathbf{A}=\begin{bmatrix}1&0&0&0&\tau&0&0&0&\tau^2/2&0&0&0\\0&1&0&0&0&\tau&0&0&0&\tau^2/2&0&0\\0&0&1&0&0&0&\tau&0&0&0&\tau^2/2&0\\0&0&0&1&0&0&0&\tau&0&0&0&\tau^2/2\\0&0&0&0&1&0&0&0&\tau&0&0&0\\0&0&0&0&0&1&0&0&0&\tau&0&0\\0&0&0&0&0&0&1&0&0&0&\tau&0\\0&0&0&0&0&0&0&1&0&0&0&\tau\\0&0&0&0&0&0&0&0&1&0&0&0\\0&0&0&0&0&0&0&0&0&1&0&0\\0&0&0&0&0&0&0&0&0&0&1&0\\0&0&0&0&0&0&0&0&0&0&0&1\end{bmatrix} \tag{8}$$

where τ denotes the time interval between the capture of two consecutive frames by the camera.

$$\mathbf{P}_t^-=\mathbf{A}\mathbf{P}_{t-1}A^T+\mathbf{Q}_t \tag{9}$$

where $\mathbf{P}_{t-1}$ indicates the 12 × 12-dimensional covariance matrix corresponding to the posteriori object state estimate at frame $t-1$. $\mathbf{P}_t^-$ represents the 12 × 12-dimensional covariance matrix corresponding to the prior object state prediction at frame t, which will be used in the update of the Kalman gain in Equation (10). $\mathbf{Q}$ is the 12 × 12-dimensional process noise covariance matrix.

The updated equations are:

$$\mathbf{K}_t=\mathbf{P}_t^-\mathbf{H}^T(\mathbf{H}\mathbf{P}_t^-\mathbf{H}^T+\mathbf{R}_t)^{-1} \tag{10}$$

$$x_t^+=x_t^-+\mathbf{K}_t(z_t-\mathbf{H}x_t^-) \tag{11}$$

$$\mathbf{P}_t=(\mathbf{I}-\mathbf{K}_t\mathbf{H})\mathbf{P}_t^- \tag{12}$$

where $\mathbf{I}$ is the 12 × 12-dimensional identity matrix, $\mathbf{K}$ is the 12 × 12-dimensional Kalman gain, and $\mathbf{R}$ represents the 4 × 4-dimensional measurement noise covariance matrix. From observing Equations (10)–(12), it can be seen that the KF estimates the current observation by multiplying the prior state estimate with the observation matrix $\mathbf{H}$. The observation residual $(z_t-\mathbf{H}x_t^-)$ is multiplied by the Kalman gain $\mathbf{K}_t$ as a correction to the priori state estimate x_t^- to obtain the object state optimal estimate x_t^+. Finally, the KF employs Equation (12) to update the posterior estimate covariance $\mathbf{P}_t$, which will be used at the next frame. Equation (12) describes the process of changing the covariance matrix of the state vector, and it is this continuously updated mechanism that allows the Kalman filter to overcome the influence of random noise. In this paper, the observation matrix $\mathbf{H}$ is:

$$\mathbf{H}=\begin{bmatrix}1&0&0&0&0&0&0&0&0&0&0&0\\0&1&0&0&0&0&0&0&0&0&0&0\\0&0&1&0&0&0&0&0&0&0&0&0\\0&0&0&1&0&0&0&0&0&0&0&0\end{bmatrix} \tag{13}$$

At this point, a prediction-update iteration process has been completed. In the tracking of each frame, we update the KF for each tracked object, so that we can iteratively predict the each tracked object position (bounding box) information at each frame.

3.2. Data Association by Min-Cost Flow on Temporal Window

We propose an online MOT data association strategy for autonomous driving, which treats trajectories and current detections matching as a min-cost flow problem on a temporal window. First, a sparse affinity network with a finite number of edges is generated using the affinity between trajectories and current detections. Then, the min-cost flow of the network is solved to achieve optimal matching. Finally, the online MOT is realized as the temporal window slides, where the window is composed of consecutive multi-frame images.

3.2.1. Affinity Metrics

Let the set of existing object trajectories in the temporal window be $\mathbf{T} = \mathbf{T}^{act} \cup \mathbf{T}^{inact}$, where $\mathbf{T}^{act}$ represents the active trajectories and $\mathbf{T}^{inact}$ represents the inactive trajectories. At frame t, the detected object set is denoted as $\mathbf{OD}_t$ ($\mathbf{od}_t^{i_1} \in \mathbf{OD}_t$) (i_1=1, 2, 3,. . ., I_1), where I_1 is the total number of detected objects, and $\boldsymbol{od}_t^{i_1}$ for the i_1-th detected object is represented as $\mathbf{od}_t^{i_1} = (\boldsymbol{\rho}_t^{i_1}, \boldsymbol{\varphi}_t^{i_1})$, where $\boldsymbol{\rho}_t^{i_1}$ represents the object appearance features and $\boldsymbol{\varphi}_t^{i_1}$ is the bounding box information.

(1) Appearance Affinity Metric

Most tracking-by-detection MOT methods only calculate the appearance affinity between the detections at the current frame and the active trajectories from the previous frame, while ignoring the affinity with the inactive trajectories. This approach is feasible for continuously detected objects, as their appearance changes are relatively small across adjacent frames. However, for objects that reappear due to occlusion or disappearance, this approach can lead to tracking failures and ID switches, as they may not match with their original trajectories. To address the issue, within the temporal window, we calculate the appearance affinity between the detections at current frame and both the active and inactive trajectories separately.

For the active trajectories $\mathbf{T}^{act}$, the appearance feature vector of the j_1-th trajectory matching detection $\mathbf{od}_{t-1}^{j_1}$ at frame $t - 1$ is $\boldsymbol{\rho}_{t-1}^{j_1}$. The computation of the appearance affinity between $\mathbf{od}_{t-1}^{j_1}$ and the detection $\mathbf{od}_t^{i_1}$ at frame t is as follows:

$$D\_APP_{i_1, j_1}(\mathbf{od}_t^{i_1}, \mathbf{od}_{t-1}^{j_1}) = 1 - \frac{\boldsymbol{\rho}_t^{i_1}\boldsymbol{\rho}_{t-1}^{j_1}}{||\boldsymbol{\rho}_t^{i_1}||_2||\boldsymbol{\rho}_{t-1}^{j_1}||_2} \quad (14)$$

It can be seen that the smaller the D_APP, the more similar the appearance, and vice versa, indicating dissimilarity. It can also be stated that the affinity measurement quantifies dissimilarity, which is determined based on the subsequent requirement of solving the min-cost flow problem.

For the inactive trajectories $\mathbf{T}^{inact}$, calculate the affinity between all N_{j_2} appearance features of the j_2-th inactive trajectory and the detection $\mathbf{od}_t^{i_1}$, and take the average as the appearance affinity (dissimilarity) between the inactive trajectory and the detection $\mathbf{od}_t^{i_1}$:

$$D\_APP_{i_1, j_2} = \frac{1}{N_{j2}}\sum_{a=1}^{N_{j2}}(1 - \frac{\boldsymbol{\rho}_t^{i_1}\boldsymbol{\rho}_{j2}^{a}}{||\boldsymbol{\rho}_t^{i_1}||_2||\boldsymbol{\rho}_{j_2}^{a}||_2}) \quad (15)$$

where, $\boldsymbol{\rho}_{j2}^{a}$ indicates the appearance feature vector of the a-th detection in the j_2-th inactive trajectory. By performing such calculations, it ensures a more reliable estimation of the affinity between $\mathbf{od}_t^{i_1}$ and the inactive trajectory.

(2) Motion affinity metric

We apply the motion model described in Section 3.1.2 to estimate the bounding box positions of all trajectories $\mathbf{T}$ within the temporal window are estimated at frame t, then the motion affinity (dissimilarity) between the trajectory and the detection is measured using the Intersection over Union (IoU) of their respective bounding boxes:

$$D\_BB_{i_1, j_3} = 1 - \mathrm{IoU}(\mathbf{od}_t^{i_1}, \mathbf{op}_t^{j_3}) = 1 - \frac{|\boldsymbol{\varphi}_t^{i_1} \cap \boldsymbol{\varphi}_t^{j_3}|}{|\boldsymbol{\varphi}_t^{i_1} \cup \boldsymbol{\varphi}_t^{j_3}|} \quad (16)$$

where, $\mathbf{op}_t^{j_3}$ represents the prediction of the j_3-th trajectory's detection at frame t, $\boldsymbol{\varphi}_t^{j_3}$ represents the bounding box information of $\mathbf{op}_t^{j_3}$.

3.2.2. Data Association by Min-Cost Flow on Temporal Window

Typically, the min-cost flow is calculated to achieve global optimal matching. In order to apply it in online MOT, we adopt a sliding temporal window approach where global optimal matching is performed within a fixed-length temporal window. Assume that the length of the temporal window is T_L, we obtain the affinities between the detected objects at frames $t - T\_L + 1, t - T\_L + 2, \ldots, t - 1$, and t using Equations (14)–(16). After that, these affinities are utilized to construct an affinity network, which is employed to match the detected objects at frame t. The affinity network is a directed graph composed of the following types of nodes:

Source node: Also known as the trajectory start point, it represents a node connected to all object detection nodes with positive costs and is used to initialize a trajectory with a positive cost.

Sink node: Also known as the trajectory end point, it represents a node connected to all object detection nodes with positive costs and is used to terminate a trajectory.

Object detection nodes: In order to satisfy the constraint of non-overlapping trajectories, each detected object in each frame is split into two nodes (pre-node, post-node) with single flow capacity.

Affinity edges: They refer to the edges that connect the object detection nodes, which have the properties of single flow capacity and negative costs.

At frame t, the affinity of the detected object $\mathbf{od}_t^{i_1}$ with the j_3-th trajectory within the temporal window is calculated as follows:

$$D\_A_{i_1, j_3} = D\_APP_{i_1, j_3} \times D\_BB_{i_1, j_3} \quad (17)$$

where, if the j_3-th trajectory is an active trajectory, Equation (14) is used to calculate $D\_APP_{i_1, j_3}$ and if it is an inactive trajectory, Equation (15) is employed. Figure 4 illustrates the affinity between trajectories and detections used for constructing the affinity network within a temporal window consisting of 5 consecutive frames, based on Equation (17). In Figure 4, each node (the black circle) represents a detection, and the numerical values on the affinity edges (purple edges) indicate the affinity between two nodes.

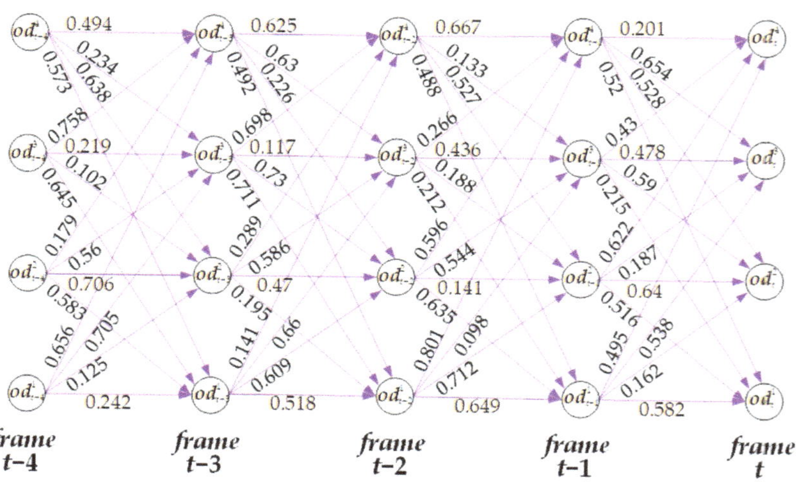

Figure 4. The illustration of the affinity between trajectories and detections within the temporal window of 5 frames. The nodes are the detections, and the number between two nodes is the affinity.

Let THR_E be the affinity edge threshold. In the affinity network, if $D\_APP_{i_1, j_3} \leq THR_E$, a negative-cost affinity edge, $Edge_{i_1, j_3}$, is added between them with a cost of $cost_{i_1, j_3}$:

$$Edge_{i_1, j_3} = cost_{i_1, j_3} = -\frac{1}{D\_A_{i_1, j_3}} \tag{18}$$

Additionally, we set a maximum number of edges T_C that each object detection node can connect to, ensuring that the out-degree of each object detection node. This limitation reduces the overall number of edges and makes the affinity network sparser. Building upon Figure 4, we construct an affinity network for matching trajectories and detections as depicted in Figure 5. In this example, we set THR_E and T_C to 0.3 and 3, respectively.

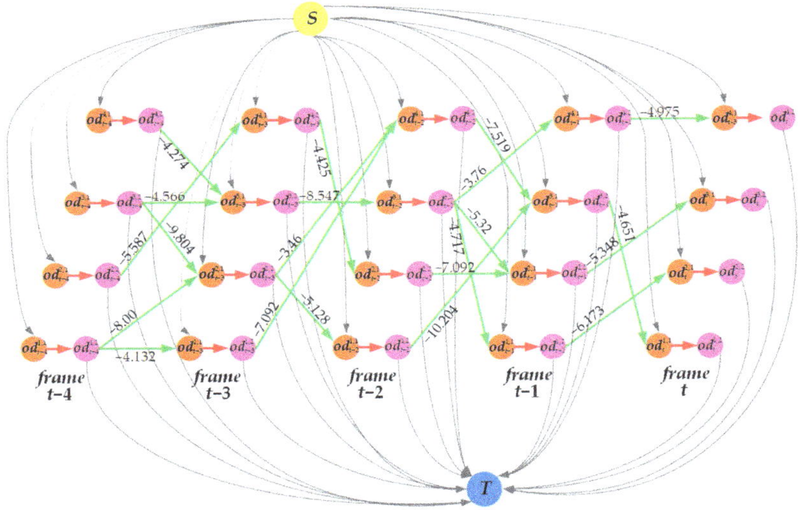

Figure 5. The sparse affinity network flow graph constructed within the temporal window of 5 frames. S is the source node, T is the sink node, the green edges represent affinity edges with negative costs (the numerical values on the green edges), and orange nodes and pink nodes are the object detection nodes.

In Figure 5, each object detection node is composed of a pre-node (orange node) and a post-node (pink node), connected by a red unidirectional edge, ensuring that any detected objects are passed in a single flow direction (i.e., non-overlapping trajectories). The unidirectional links (green lines) connecting the detected objects between consecutive frames, with negative cost, are referred to as affinity edges. The numerical values on the green edges represent the negative cost values of affinity edges calculated according to Equation (18). The yellow node S is the source node used for initializing trajectories, while the blue node T is the sink node that terminates the trajectories.

Summing up the above, we consider the problem of finding the optimal matching between the objects detected at frame t and the trajectories within the temporal window as the task of solving the min-cost flow problem on the constructed sparse affinity network. The min-cost flow is a paradigm widely used to solve data association problems in MOT due to its fast inference and ability to provide globally optimal solutions [36]. Over the years, researchers have developed several effective solutions to solve general min-cost flow problems, including methods like cost scaling, successive shortest path, k-shortest path, etc. However, modifying these methods or directly applying them to MOT may result in computational inefficiency and hinder the application to large-scale problems. To address this, Wang et al. [36] developed a highly efficient and accurate minimum-cost flow solver called minimum-update Successive Shortest Path (muSSP). muSSP updates the shortest path tree only when necessary, i.e., when identifying the shortest *S*-*T* path. Experimental results on various MOT public datasets with different object detection results and graph designs have demonstrated that muSSP provides accurate optimal solutions with high computational efficiency. Experimental results on various public MOT datasets with different object detection results and graph designs have demonstrated that muSSP provides accurate optimal solutions with high computational efficiency. Therefore, we adopt muSSP to solve the min-cost flow in the constructed sparse affinity network. After solving it, if an object detected at frame *t* is matched with a trajectory within the temporal window, the object is assigned to the corresponding trajectory and the trajectory is marked as active. If a trajectory within the temporal window does not have any matches with the detected objects, it is marked as inactive. If a detected object is not connected to any trajectory within the temporal window, a new trajectory is initiated with that object as the starting point.

Figure 6 shows an example of trajectory generation at frame *t*. The temporal window consists of a sequence of 5 consecutive frames (masked with gray). The min-cost flow problem in the sparse affinity network is solved within this window. The existing trajectories within the temporal window are displayed as blue, purple, and red lines. The objects detected at frame t are represented by blank circles.

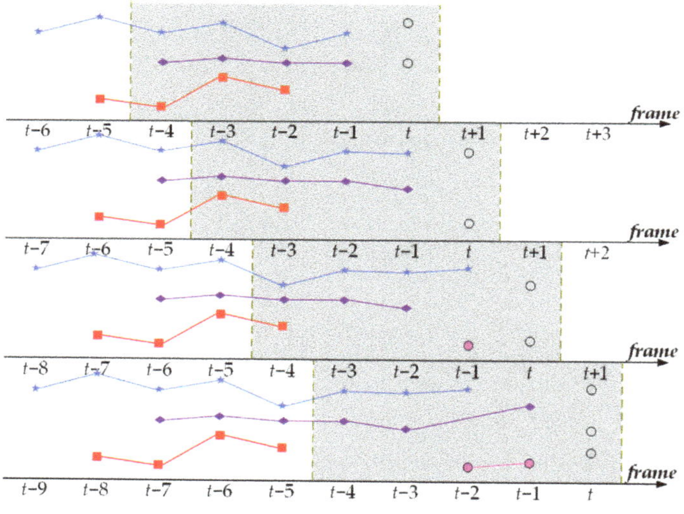

Figure 6. A diagram illustration for the trajectory generation within a moving temporal window (as masked in gray).

From top to bottom, in Figure 6, we demonstrate how to associate new detections with existing trajectories within the temporal window and how the states of the trajectories are set after the association. In the first row, the two new detections (blank circles) are matched with the purple and blue trajectories, marking them as active trajectories, while the red trajectory is marked as an inactive trajectory. In the second row, one detection is matched to the blue trajectory, while the other detection does not match any purple or red trajectories. The unmatched detection is considered as the starting point of a new trajectory, indicated in pink. At this point, the blue and pink trajectories are active, while the purple and red trajectories are inactive. The two detections in the third row, one is matched to the pink trajectory, and another detection is matched to the inactive purple trajectory. Consequently, the purple and pink trajectories become active, the blue trajectory becomes inactive, and the red trajectory is terminated.

4. Experiments

We conduct experiments on the KITTI MOT dataset [37] and the real-world campus scenario sequences to verify the effectiveness of our method. In this section, we first provide a brief overview of the two MOT datasets. Secondly, we introduce the MOT evaluation metrics. Then, we demonstrate the effectiveness of object appearance features, motion models, and triplet loss function through the ablation study. Additionally, we compare and analyze our method with the existing homogeneous, state-of-the-art, visual-based MOT methods on the KITTI MOT dataset. Finally, we present the intuitive visual tracking results of our method on both two MOT datasets.

4.1. Datasets

4.1.1. KITTI MOT Dataset

The KITTI public benchmark is the first and internationally recognized benchmark dataset for evaluating computer vision methods in autonomous driving [37]. The resolution of the rectified images in the dataset is 1242 × 375 pixels, and each image contains a maximum of 15 cars and 30 pedestrians. The KITTI MOT dataset focuses on tracking classes Car and Pedestrian. It consists of 50 sequences, with 21 sequences for training and 29 sequences for testing. For each sequence, it provides LiDAR point clouds, RGB images, and calibration files. The number of frames used for training and testing is 8008 and 11,095, respectively. For the testing dataset, KITTI does not provide any labels to the users but keeps the labels on the server for MOT evaluation. As for the training dataset, it contains 30,601 objects and 636 tracks. The traffic scenes in the KITTI MOT dataset belong to relatively complex tracking scenarios, involving many challenging factors such as low light conditions, significant lighting variations, frequent occlusions between objects, and the ego vehicle making turns.

4.1.2. Real-World Campus Scenario Sequences

The dataset comprises 10 campus scene sequences, where each sequence consists of images captured during different time periods as the experimental vehicle drives within the campus. The real-world campus scenario images were recorded at a rate of 10 frames per second, capturing varying lighting conditions and different locations within the campus. The resolution of the images is 640 × 480 pixels. The experimental vehicle, as shown in Figure 7, is based on the EG6043K Koala sightseeing vehicle and is equipped with several sensors including a millimeter-wave radar (Delphi ESR), a laser radar (IBEO LUX4), a GPS/inertial navigation system (NAV982), an industrial PC (ARK3500), a display monitor, and a grayscale camera (Point Grey Bumblebee XB3 with a focal length of 3.8 mm and a baseline of 120 mm).

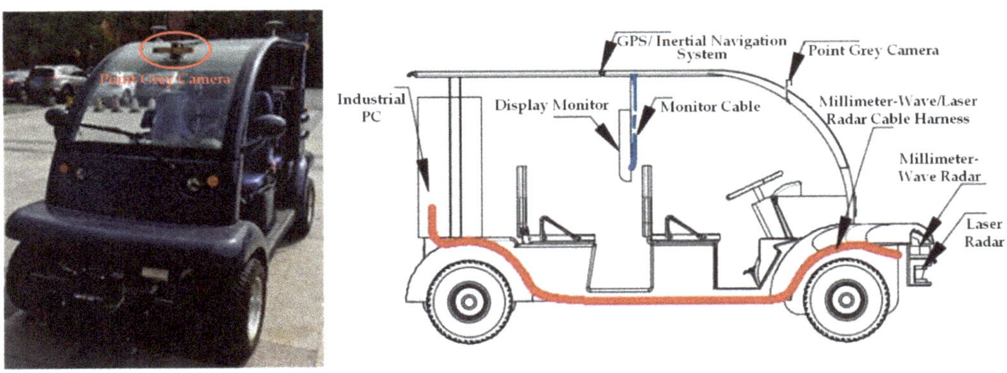

Figure 7. Illustration of experimental vehicle and sensor installation. (**Left**): Experimental vehicle. (**Right**): Diagram of sensor installation.

4.2. MOT Evaluation Metrics

To objectively evaluate the performance of MOT methods, we utilize the following commonly used evaluation metrics in the field of MOT [38–40]:

Identity Switches (***IDSw***↓): The total number of object identity swaps that occurred throughout the entire tracking process.

Multiple Object Tracking Accuracy (***MOTA***↑): This metric is one of the most important evaluation metrics for MOT. It is calculated based on false positives (*FP*), false negatives (*FN*), and identity switches (*IDSw*) as follows:

$$\mathbf{MOTA} = 1 - \sum_{t}(FP_t + FN_t + IDSw_t) / \sum_{t} GT_t \quad (19)$$

where GT_t represents the number of ground truth annotations at frame t. FP_t represents the number of objects in the tracking results at frame t that do not exist in the ground truth annotations. In other words, these are tracking results that cannot be correctly matched to the ground truth annotations. FN_t represents the number of ground truth annotations at frame t that are not detected by the tracking results. These are the objects that exist in the ground truth annotations but are not tracked.

Mostly Tracked (***MT***↑): represents the ratio of the number of tracks in which the tracking results have a matching rate higher than 80% to the total number of ground truth tracks.

Mostly Lost (***ML***↓): represents the ratio of the number of tracks in which the tracking results have a matching rate lower than 20% to the total number of ground truth tracks.

Higher Order Tracking Accuracy (***HOTA***↑): A comprehensive metric used for evaluating the performance of MOT methods. It considers both the accuracy of object detection and tracking and weightedly considers the magnitude of tracking errors, providing a more comprehensive reflection of the detection performance (Detection Accuracy, ***DetA***↑) and tracking performance (Association Accuracy, ***AssA***↑) of MOT methods. HOTA measures the alignment of matched detection trajectories, averages the entire matched detection, and penalizes unmatched detections. It is recommended to refer to [40] for detailed definitions and analysis of ***HOTA***, ***DetA***, and ***AssA***. Since 25 February 2021, the KITTI MOT dataset ranks the MOT methods based on their ***HOTA*** scores in descending order.

For evaluation metrics with quotation marks (↑), a higher score indicates better performance. On the other hand, for evaluation metrics with a hashtag (↓), the opposite is true, where a higher score indicates a worse performance.

4.3. Object Appearance Feature Extraction Network Implementation

We implement the proposed object appearance feature extraction network using the TensorFlow deep learning framework, based on the Windows 10 operating system and Python language. Other parameters of the experimental platform used for training the network weights include CPU Intel® Xeon® Silver 4110 @2.1, 16 GB RAM, and Nvidia GeForce GTX1080TI graphics processor (11 GB VRAM). The training dataset consists of samples from the KITTI MOT training sequences and real-world campus scenario sequences. To train the network, we extract the car category objects from the two datasets and perform scale transformation on the object images to obtain a size of 224 × 224 pixels. Additionally, we augment the training images using rotations of 90° and 180°. For constructing triplet images, the anchor samples are primarily taken from unrotated images, negative samples are preferably selected from rotated negative images, and positive samples are randomly chosen from all positive images.

During the training, we do not load any pre-trained weights. All convolutional layers and fully connected layers in the network are initialized with weights following a Gaussian distribution with mean 0 and standard deviation of 0.01. The bias terms are initialized to 0. We employ the Adaptive Moment Estimation optimizer for weight updates. Based on empirical values, the initial value of the learning rate is set to 0.001, the weight decay rate is set to 0.0005, and the momentum is set to 0.900. In the triplet loss function Equation (1), the threshold *thr* is set to 0.6. Figure 8 shows our final training loss curve. By observing Figure 8, we can see that the loss value starts to increase at around 52,000th epoch, indicating the occurrence of overfitting. Therefore, when using this network to extract target features, the network weights saved at 50,000th epoch are employed.

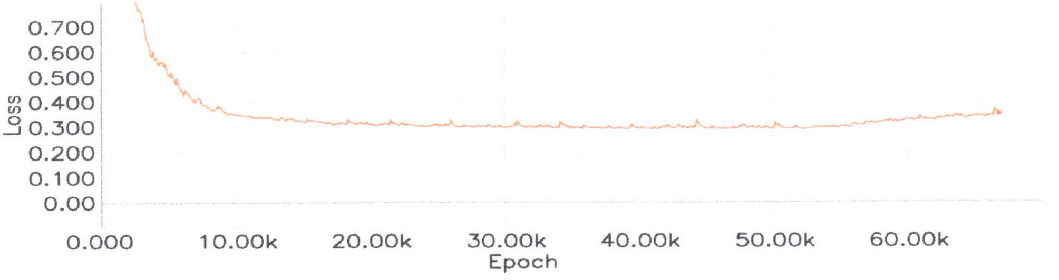

Figure 8. The loss function curve of the feature extraction network.

4.4. Ablation Study

To better analyze the impact of each component of our method on the overall tracking performance, we conducted an ablation study by individually removing each component from the overall framework. Comparative experiments were then conducted with methods that do not use the respective modules to analyze their effects on MOT. The ablation experiments were conducted on the KITTI MOT training sequences. In the ablation study, the 21 KITTI MOT training sequences were split into training/validation sets. Meanwhile, in order to compare the difference between the triplet loss and the conventional binary loss, we designed and utilized a Siamese CNN without altering the network structure of the object appearance model. We calculated the correlation probability between a pair of detection responses and defined the binary cross-entropy loss to train the object appearance model. The results of the ablation study are shown as Table 1, with the best values highlighted in bold. In Table 1, "*seg*" represents the extraction of precise object contours prior to extracting object appearance features, while "*w/o seg*" indicates the absence of object contour extraction, with appearance features directly extracted within the object bounding boxes. "*app_T*" represents the object appearance features using the triplet loss function, "*app_E*" represents the object appearance features using the binary cross-entropy loss function, "*motion*" signifies the object motion information, "*ACT*" denotes the separation of trajectories within the temporal window into active and inactive trajectories, "*w/o ACT*" signifies no differentiation of trajectories within the temporal window. The first row is the overall framework of our method, and other rows are the different versions of our method.

Table 1. Results of ablation study.

Different Versions	*HOTA*(%)↑	*DetA*(%)↑	*AssA*(%)↑	*MOTA*(%)↑	*MT*(%)↑	*ML*(%)↓	IDSw↓
seg + app_T + motion + ACT (**ours**)	**78.20**	**75.36**	**82.24**	**85.70**	**92.72**	**2.84**	**11**
w/o seg + app_T + motion + ACT	76.32	72.18	80.89	81.76	88.39	5.22	18
seg + app_E + motion + ACT	70.74	69.29	73.51	71.98	74.16	9.63	49
seg + app_T + ACT	74.81	70.87	79.26	79.25	86.47	4.75	26
seg + motion + ACT	72.69	68.53	77.30	77.63	79.23	3.26	32
seg + app_T + motion + w/o ACT	75.95	71.74	80.42	82.42	90.82	4.28	21

ours means the standard version of our method with all components employed. The best values are highlighted in bold.

Comparing the first row with the second row, we can see that *HOTA* decreases by 1.88%. This indicates that the approach of first segmenting the objects accurately and then extracting appearance features can significantly improve tracking performance. This is because the segmented bounding boxes only contain the object ontology, without background noise. Thus, the extracted appearance features are less contaminated by noise and can fully represent the object.

Compared to the binary loss (validation loss, corresponding to the third row), using the triplet loss in metric learning techniques can significantly improve the tracking accuracy. *HOTA* increases from 70.74% to 78.20%, which validates the feasibility of training target feature extraction networks with the triplet loss in MOT.

The fourth and fifth rows validate the impact of different object features on MOT performance. The third row uses only appearance features, while the fourth row uses only the object motion model. Comparing the first row with the third and fourth rows, we can see that although using only appearance features has a certain impact on tracking performance, with a decrease in various evaluation metrics, using only the motion model has the largest impact on tracking performance. The *HOTA* experiences the largest drop (approximately 5.51%), and other evaluation metrics also show significant decreases. This indicates that various object features have different degrees of impact on MOT performance, and appearance features play an important role in MOT. It also validates the effectiveness of the proposed object appearance extraction network, which can effectively preserve the visual features of the objects.

The sixth row differs from the first row primarily in that trajectory sets within the temporal window are not differentiated as active or inactive during data association. Comparing the two rows, we can observe that when trajectory states within the temporal window are not differentiated, *HOTA* decreases by approximately 2.25%, while *IDSw* increases by 10. This suggests that distinguishing trajectory states within the temporal window and matching current frame detections with all trajectories in the temporal window can reduce ID switches caused by occlusion or disappearance and improve MOT performance.

4.5. Comparison with the State-of-the-Art Methods

We upload the testing results of our method on the KITTI MOT testing sequences to the official testing platform of KITTI for easier comparison with other state-of-the-art methods. We compare our method with the existing homogeneous, state-of-the-art, visual-based MOT methods [7,8,10,11,14,24–26] in recent years, and the results are shown in Table 2. In Table 2, the best values are highlighted in black bold, while the second-best values are highlighted in blue bold.

Table 2. Performance comparison results of different tracking methods on the KITTI MOT testing sequences.

	HOTA (%)↑	*DetA* (%)↑	*AssA* (%)↑	*MOTA* (%)↑	*MT* (%)↑	*ML* (%)↓	*IDSw*↓	Runtime (ms)↓
FAMNET [26]	52.56	61.00	45.51	75.92	52.46	9.69	521	1500 + D
JCSTD [7]	65.94	65.37	67.03	80.24	57.08	7.85	173	70 + D
MASS [8]	68.25	72.92	64.46	84.64	74.00	2.92	353	10 + D
Quasi-Dense [11]	68.45	72.44	65.49	84.93	69.54	3.85	313	70 + D
MOTSFusion [24]	68.74	72.19	66.16	84.24	72.77	2.92	415	440 + D
LGM [14]	73.14	**74.61**	72.31	**87.60**	**85.08**	**2.46**	448	80 + D
SMAT [10]	71.88	72.13	72.13	83.64	62.77	6.00	198	100 + D
TripletTrack [25]	73.58	73.18	74.66	84.32	69.85	3.85	322	100 + D
Ours	**73.93**	73.35	**75.81**	86.49	78.52	3.17	**126**	63 + D

+ D means the detection time.

As can be seen from Table 2, as the most comprehensive evaluation metric for MOT, our method achieves the highest *HOTA*, which demonstrates the effectiveness of the proposed MOT method. Further analysis reveals that our method also achieves the best performance in terms of the *AssA*, with a score of 75.81%, which is the only method that exceeds 75% among all methods. It means that the proposed data association strategy exhibits strong association performance and can maintain the object IDs as much as possible. This is mainly attributed to the following factors: (1) The proposed object appearance extraction network can extract appearance features that adequately represent the objects, and the object motion model based on CA smooths and optimizes the object bounding box information. (2) The data association by min-cost flow on the temporal window considers the classification of inactive and active trajectory states. Unlike other methods that only consider the matching between the current frame detections and the previous frame trajectories, our method matches all trajectories within the temporal window with the current frame detections, allowing reappearing objects to be matched with their historical trajectories to maintain consistent IDs. Additionally, applying the min-cost flow within the sliding temporal window achieves global optimal matching within the temporal window.

Moreover, our method obtains the fewest *IDsw* (126), which further demonstrates the effectiveness of the proposed MOT method. The data association by min-cost flow on the temporal window reduces ID switches caused by occlusion or disappearance, resulting in stronger robustness for the MOT method. Our method does not achieve the highest DetA score (73.35), which is lower than LGM [14] (74.61) but higher than other comparative methods. This is mainly due to the fact that different methods employ different object detectors, and the performance of various object detectors varies. Despite the relatively lower performance of the object detector utilized in our method, we also acquire better results in terms of HOTA, AssA, and IDsw compared to LGM. It further validates the superiority of our proposed data association strategy by global association on the temporal window. In other words, even in situations where the object detector performance is lower, our method still can effectively address the matching problem between trajectories and detections, leading to improved performance.

From Table 2, among all the MOT methods, our method has a per-frame runtime of 63 ms, which is faster than FAMNET [26], LGM [14], JCSTD [7], SMAT [10], Quasi-Dense [11], TripletTrack [25], and MOTSFusion [24], but slower than MASS [8]. Since the KITTI MOT dataset is captured at a rate of 10 frames per second, selecting an object detection method with a per-frame runtime of no more than 37 ms would ensure real-time performance for the MOT method.

4.6. Visually Intuitive Evaluation

To provide a more intuitive illustration of the performance of our proposed MOT method, in addition to the aforementioned quantitative results, we showcase some qualitative results of our method on the KITTI MOT testing sequences and real-world campus scenario sequences in Figure 9.

Figure 9a depicts a challenging intersection scene from the KITTI MOT testing sequences, which often involves crossing, overlapping, and occluded objects. It can be observed that at frame 82, Obj. 7 is severely occluded by Obj. 0. However, at frame 86, when Obj. 7 reappears, our proposed method successfully associates it with its historical trajectory. Similarly, in Figure 9b, Obj. 5 is completely occluded, but our method still assigns the correct ID when the object reappears. In Figure 9d, in a real-world campus scenario with poor lighting conditions, Obj. 13, 14 are completely occluded at frame 29. However, our method correctly matches the trajectories with the objects when they reappear. In Figure 9e, Obj. 20 is at a distant distance from the ego-vehicle, but even after being heavily occluded by pedestrians and reappearing, our method maintains its ID. These four examples demonstrate the robustness of our proposed MOT method in various traffic scenarios. Figure 9c shows an example where our method fails. The scenario depicts the ego-vehicle waiting for traffic lights at a traffic intersection. Obj. 23 is under a tree shade and is considerably far from the ego-vehicle. It is completely occluded by a tanker truck at frame 599 and reappears at frame 609. However, our method assigns it the ID 26, resulting in an ID switch. The main reason for this failure is that, during the experiment, the chosen temporal window consists of seven consecutive frames, while Obj. 23 reappears after ten frames. By then, its trajectory is no longer part of the trajectory set within the temporal window, making it impossible to match it with historical trajectory. Since our method uses min-cost flow to match trajectories with detections within the temporal window, the size of the temporal window cannot be arbitrarily increased. A larger temporal window leads to a more complex network, longer computation time to solve the network, and increased runtime of the MOT method. Therefore, the size of the temporal window limits the tracking performance of our MOT method.

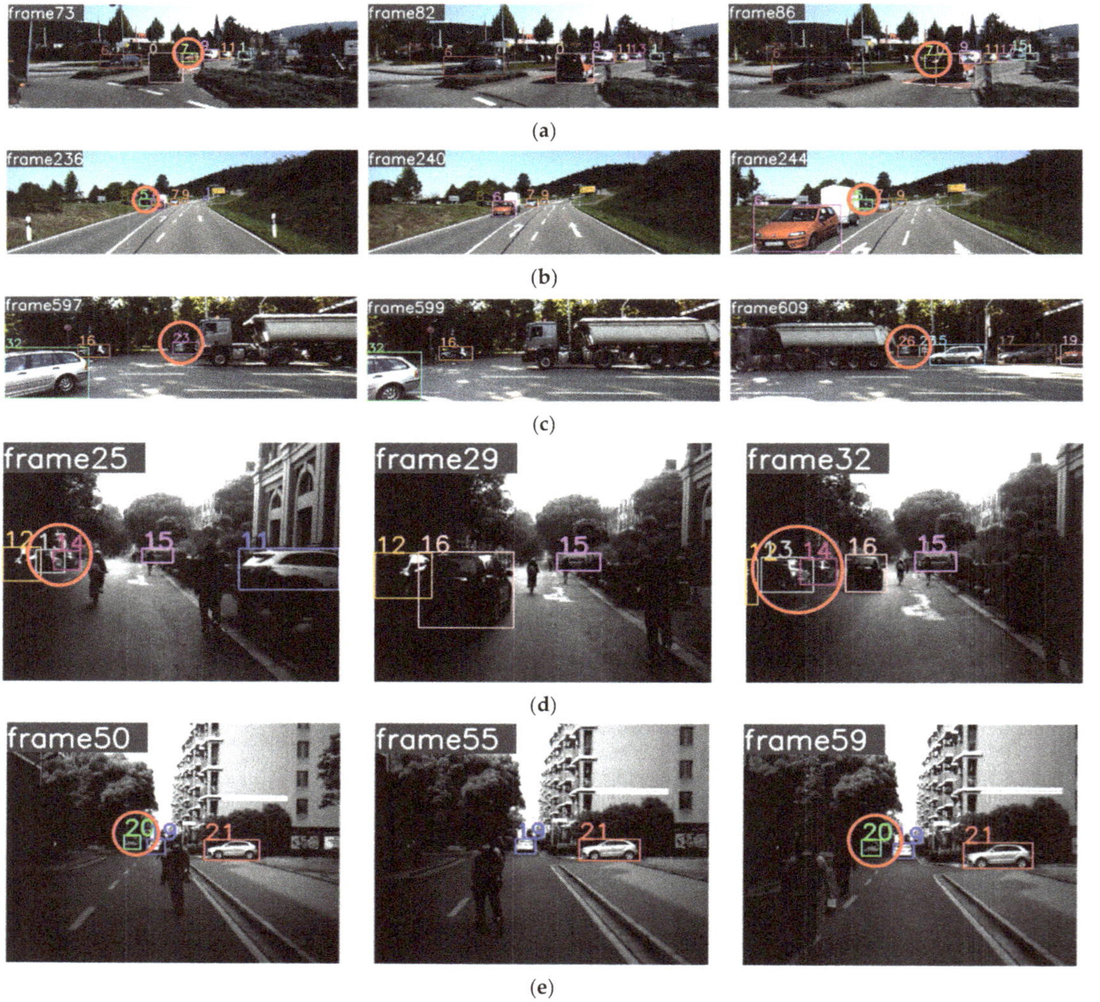

Figure 9. Tracking results. In each sequence, the tracked cars' IDs are located at the top of their bounding box, and the objects that will undergo severe or complete occlusion in subsequent frames are marked with red circles. (**a**) KITTI MOT testing-0008 tracking results. (**b**) KITTI MOT testing-0009 tracking results. (**c**) KITTI MOT testing-0015 tracking results. (**d**) Real-world campus scenarios-0001 tracking results. (**e**) Real-world campus scenarios-0003 tracking results.

In addition, observation of Figure 9 shows that our proposed MOT method accurately associates continuously appearing objects with their corresponding trajectories. It indicates that our method exhibits good association capabilities when dealing with the continuous motion of objects, which helps maintain the uniqueness and continuity of the tracking. This is particularly important in handling objects that temporarily disappear or experience partial occlusion in the sequences. Therefore, our MOT method demonstrates high reliability and accuracy when dealing with continuously appearing objects.

5. Conclusions

In this paper, following the tracking-by-detection pipeline, we present an online multiple object tracking using min-cost flow on temporal window for autonomous driving, which mainly composes of two parts: feature extraction and data association on the temporal window. We apply ResNet50 as the backbone network and design a three-channel network based on metric learning to extract discriminative object appearance features and employ a KF with a CA motion model to optimize the object bounding box information, resulting in reliable and discriminative object representations. In the temporal window composed of consecutive frames, we compute the affinities between the current frame detection and active/inactive trajectories. Based on this, we construct a sparse affinity network and solve the min-cost flow problem on the network to obtain the MOT results. Qualitative and quantitative experiments on the KITTI MOT testing sequences and our real-world campus scenario sequences show that the proposed method outperforms existing visual-based MOT methods of the same type. In the future, we will further optimize the data association strategy to reduce the impact of temporal window length on tracking performance.

Author Contributions: Conceptualization, H.W. and Y.H.; methodology, H.W. and Y.H.; software, H.W.; validation, H.W. and Y.H.; formal analysis, H.W.; investigation, H.W.; writing—original draft preparation, H.W. and Y.H.; writing—review and editing, H.W., Y.H., Q.Z. and Z.G.; project administration, Y.H.; funding acquisition, Y.H. and Q.Z. All authors have read and agreed to the published version of the manuscript.

Funding: This research was funded by the Shanghai Nature Science Foundation of Shanghai Science and Technology Commission, China, grant number 20ZR1437900, and the National Natural Science Foundation of China, grant number 62206114.

Institutional Review Board Statement: Not applicable.

Informed Consent Statement: Not applicable.

Data Availability Statement: All data generated or analyzed during this study are included in this manuscript.

Conflicts of Interest: The authors declare no conflict of interest.

References

1. Guo, G.; Zhao, S. 3D multi-object tracking with adaptive cubature kalman filter for autonomous driving. *IEEE Trans. Intell. Veh.* **2023**, *8*, 512–519. [CrossRef]
2. Zhang, Y.; Sun, P.; Jiang, Y.; Yu, D.; Weng, F.; Yuan, Z.; Luo, P.; Liu, W.; Wang, X. Bytetrack: Multi-object tracking by associating every detection box. In Proceedings of the European Conference on Computer Vision (ECCV), Tel Aviv, Israel, 23–27 October 2022. [CrossRef]
3. Chan, S.; Jia, Y.; Zhou, X.; Bai, C.; Chen, S.; Zhang, X. Online multiple object tracking using joint detection and embedding network. *Pattern Recognit.* **2022**, *130*, 108793. [CrossRef]
4. Abudayyeh, D.; Almomani, M.; Almomani, O.; Alsoud, H.; Alsalman, F. Perceptions of autonomous vehicles: A case study of Jordan. *World Electr. Veh. J.* **2023**, *14*, 133. [CrossRef]
5. Alqarqaz, M.; Bani Younes, M.; Qaddoura, R. An Object Classification Approach for Autonomous Vehicles Using Machine Learning Techniques. *World Electr. Veh. J.* **2023**, *14*, 41. [CrossRef]
6. Liu, Y.; Li, G.; Hao, L.; Yang, Q.; Zhang, D. Research on a Lightweight Panoramic Perception Algorithm for Electric Autonomous Mini-Buses. *World Electr. Veh. J.* **2023**, *14*, 179. [CrossRef]
7. Tian, W.; Lauer, M.; Chen, L. Online multi-object tracking using joint domain information in traffic scenarios. *IEEE Trans. Intell. Transp. Syst.* **2020**, *21*, 374–384. [CrossRef]
8. Karunasekera, H.; Wang, H.; Zhang, H. Multiple object tracking with attention to appearance, structure, motion and size. *IEEE Access* **2019**, *7*, 104423–104434. [CrossRef]
9. Mykheievskyi, D.; Borysenko, D.; Porokhonskyy, V. Learning local feature descriptors for multiple object tracking. In Proceedings of the Asian Conference on Computer Vision (ACCV), Kyoto, Japan, 30 November–4 December 2020. [CrossRef]

10. Gonzalez, N.F.; Ospina, A.; Calvez, P. SMAT: Smart multiple affinity metrics for multiple object tracking. In Proceedings of the International Conference on Image Analysis and Recognition (ICIAR), Póvoa de Varzim, Portugal, 24–26 June 2020. [CrossRef]
11. Pang, J.; Qiu, L.; Li, X.; Chen, H.; Li, Q.; Darrell, T.; Yu, F. Quasi-dense similarity learning for multiple object tracking. In Proceedings of the IEEE/CVF Conference on Computer Vision and Pattern Recognition (CVPR), Nashville, TN, USA, 19–25 June 2021. [CrossRef]
12. Qin, W.; Du, H.; Zhang, X.; Ren, X. End to end multi-object tracking algorithm applied to vehicle tracking. In Proceedings of the Asia Conference on Algorithms, Computing and Machine Learning (CACML), Hangzhou, China, 19–25 November 2022. [CrossRef]
13. He, K.; Zhang, X.; Ren, S.; Sun, J. Deep residual learning for image recognition. In Proceedings of the IEEE Conference on Computer Vision and Pattern Recognition (CVPR), Las Vegas, NV, USA, 27–30 June 2016. [CrossRef]
14. Wang, G.; Gu, R.; Liu, Z.; Hu, W.; Song, M.; Hwang, J. Track without appearance: Learn box and tracklet embedding with local and global motion patterns for vehicle tracking. In Proceedings of the IEEE/CVF International Conference on Computer Vision (ICCV), Montreal, QC, Canada, 10–17 October 2021. [CrossRef]
15. Wang, H.; Li, Z.; Li, Y.; Nai, K.; Wen, M. Sture: Spatial–temporal mutual representation learning for robust data association in online multi-object tracking. *Comput. Vis. Image Underst.* **2022**, *220*, 103433. [CrossRef]
16. Yang, F.; Wang, Z.; Wu, Y.; Sakti, S.; Nakamura, S. Tackling multiple object tracking with complicated motions–Re–designing the integration of motion and appearance. *Image Vis. Comput.* **2022**, *124*, 104514. [CrossRef]
17. Ren, S.; He, K.; Girshick, R.; Sun, J. Faster r-cnn: Towards real-time object detection with region proposal networks. *IEEE Trans. Pattern Anal. Mach. Intell.* **2017**, *39*, 1137–1149. [CrossRef]
18. Girshick, R. Fast r-cnn. In Proceedings of the IEEE International Conference on Computer Vision (ICCV), Santiago, Chile, 7–13 December 2015. [CrossRef]
19. Pramanik, A.; Pal, S.; Maiti, J.; Mitra, P. Granulated rcnn and multi-class deep sort for multi-object detection and tracking. *IEEE Trans. Emerg. Top. Comput. Intell.* **2022**, *6*, 171–181. [CrossRef]
20. Bochkovskiy, A.; Wang, C.; Liao, H. Yolov4: Optimal speed and accuracy of object detection. *arXiv* **2020**, arXiv:2004.10934. [CrossRef]
21. Ge, Z.; Liu, S.; Wang, F.; Li, Z.; Sun, J. Yolox: Exceeding yolo series in 2021. *arXiv* **2021**, arXiv:2107.08430. [CrossRef]
22. Liang, S.; Wu, H.; Zhen, L.; Hua, Q.; Garg, S.; Kaddoum, G.; Hassan, M.M.; Yu, K. Edge yolo: Real-time intelligent object detection system based on edge-cloud cooperation in autonomous vehicles. *IEEE Trans. Intell. Transp. Syst.* **2022**, *23*, 25345–25360. [CrossRef]
23. Xu, H.; Dong, X.; Wu, W.; Yu, B.; Zhu, H. A two-stage pillar feature-encoding network for pillar-based 3D object detection. *World Electr. Veh. J.* **2023**, *14*, 146. [CrossRef]
24. Luiten, J.; Fischer, T.; Leibe, B. Track to reconstruct and reconstruct to track. *IEEE Robot. Autom. Lett.* **2020**, *5*, 1803–1810. [CrossRef]
25. Marinello, N.; Proesmans, M.; Gool, L.V. Triplettrack: 3D object tracking using triplet embeddings and LSTM. In Proceedings of the IEEE/CVF Conference on Computer Vision and Pattern Recognition Workshops (CVPRW), New Orleans, LA, USA, 19–20 June 2022. [CrossRef]
26. Chu, P.; Ling, H. Famnet: Joint learning of feature, affinity and multi-dimensional assignment for online multiple object tracking. In Proceedings of the IEEE/CVF International Conference on Computer Vision (ICCV), Seoul, Republic of Korea, 27 October–2 November 2019. [CrossRef]
27. Guo, S.; Wang, J.; Wang, X.; Tao, D. Online multiple object tracking with cross-task synergy. In Proceedings of the IEEE/CVF Conference on Computer Vision and Pattern Recognition (CVPR), Nashville, TN, USA, 20–25 June 2021. [CrossRef]
28. Zhou, X.; Koltun, V.; Krähenbühl, P. Tracking objects as points. In Proceedings of the European Conference on Computer Vision (ECCV), Virtual Platform, 23–28 August 2020. [CrossRef]
29. Tokmakov, P.; Li, J.; Burgard, W.; Gaidon, A. Learning to track with object permanence. In Proceedings of the IEEE/CVF International Conference on Computer Vision (ICCV), Montreal, QC, Canada, 10–17 October 2021. [CrossRef]
30. Kong, J.; Mo, E.; Jiang, M.; Liu, T. Motfr: Multiple object tracking based on feature recoding. *IEEE Trans. Circuits. Syst. Video Technol.* **2022**, *32*, 7746–7757. [CrossRef]
31. Liu, Y.; Bai, T.; Tian, Y.; Wang, Y.; Wang, J.; Wang, X.; Wang, F. Segdq: Segmentation assisted multi-object tracking with dynamic query-based transformers. *Neurocomputing* **2022**, *48*, 91–101. [CrossRef]
32. Cai, J.; Xu, M.; Li, W.; Xiong, Z.; Xia, W.; Tu, Z.; Soatto, S. Memot: Multi-object tracking with memory. In Proceedings of the IEEE/CVF Conference on Computer Vision and Pattern Recognition (CVPR), New Orleans, LA, USA, 18–24 June 2022. [CrossRef]
33. Wei, H.; Huang, Y.; Hu, F.; Zhao, B.; Guo, Z.; Zhang, R. Motion Estimation Using Region-Level Segmentation and Extended Kalman Filter for Autonomous Driving. *Remote Sens.* **2021**, *13*, 1828. [CrossRef]
34. Wojke, N.; Bewley, A.; Paulus, D. Simple online and realtime tracking with a deep association metric. In Proceedings of the IEEE International Conference on Image Processing (ICIP), Beijing, China, 17–20 September 2017. [CrossRef]
35. Bewley, A.; Ge, Z.; Ott, L.; Ramos, F.; Upcroft, B. Simple online and realtime tracking. In Proceedings of the IEEE International Conference on Image Processing (ICIP), Phoenix, AZ, USA, 25–28 September 2016. [CrossRef]
36. Wang, C.; Wang, Y.; Wang, Y.; Wu, C.; Yu, G. muSSP: Efficient min-cost flow algorithm for multi-object tracking. In Proceedings of the Annual Conference on Neural Information Processing Systems (NeurIPS), Vancouver, BC, Canada, 8–14 December 2019.
37. Geiger, A.; Lenz, P.; Stiller, C. Vision meets robotics: The KITTI dataset. *Int. J. Robot. Res.* **2013**, *32*, 1231–1237. [CrossRef]

38. Bernardin, K.; Stiefelhagen, R. Evaluating multiple object tracking performance: The CLEAR MOT metrics. *EURASIP J. Image Video Process.* **2008**, *2008*, 246309. [CrossRef]
39. Li, Y.; Huang, C.; Nevatia, R. Learning to associate: HybridBoosted multi-target tracker for crowded scene. In Proceedings of the IEEE Conference on Computer Vision and Pattern Recognition (CVPR), Miami, FL, USA, 20–25 June 2009. [CrossRef]
40. Luiten, J.; Ošep, A.; Dendorfer, P.; Torr, P.; Geiger, A.; Leal-Taixé, L.; Leibe, B. HOTA: A higher order metric for evaluating multi-object tracking. *Int. J. Comput. Vis.* **2021**, *129*, 548–578. [CrossRef] [PubMed]

World Electric Vehicle Journal

MDPI

Article

Bird's-Eye View Semantic Segmentation for Autonomous Driving through the Large Kernel Attention Encoder and Bilinear-Attention Transform Module

Ke Li [1], Xuncheng Wu [1,*], Weiwei Zhang [2] and Wangpengfei Yu [2]

[1] School of Mechanical and Automotive Engineering, Shanghai University of Engineering Science, Shanghai 201620, China; jxgz-like@foxmail.com
[2] Shanghai Smart Vehicle Cooperating Innovation Center Co., Ltd., Shanghai 201805, China
* Correspondence: wuxunchengsues@163.com

Abstract: Building an autonomous driving system requires a detailed and unified semantic representation from multiple cameras. The bird's eye view (BEV) has demonstrated remarkable potential as a comprehensive and unified perspective. However, most current research focuses on innovating the view transform module, ignoring whether the crucial image encoder can construct long-range feature relationships. Hence, we redesign an image encoder with a large kernel attention mechanism to encode image features. Considering the performance gains obtained by the complex view transform module are insignificant, we propose a simple and effective Bilinear-Attention Transform module to lift the dimension completely. Finally, we redesign a BEV encoder with a CNN block of a larger kernel size to reduce the distortion of BEV features away from the ego vehicle. The results on the nuScenes dataset confirm that our model outperforms other models with equivalent training settings on the segmentation task and approaches state-of-the-art performance.

Keywords: camera; bird's eye view; autonomous driving; view transformation; semantic segmentation

Citation: Li, K.; Wu, X.; Zhang, W.; Yu, W. Bird's-Eye View Semantic Segmentation for Autonomous Driving through the Large Kernel Attention Encoder and Bilinear-Attention Transform Module. *World Electr. Veh. J.* **2023**, *14*, 239. https://doi.org/10.3390/wevj14090239

Academic Editors: Biao Yu, Linglong Lin and Jiajia Chen

Received: 23 July 2023
Revised: 14 August 2023
Accepted: 21 August 2023
Published: 1 September 2023

1. Introduction

The development of autonomous driving has become a highly dynamic area of research. To ensure safety, autonomous driving needs access to a robust, detailed, and rich representation of its surroundings, especially in urban driving scenarios. As one of the crucial technologies of autonomous driving, environment perception [1,2] is mainly achieved through the combination of cameras, radar, LIDAR, and other sensors to capture information about the environment around the vehicle, which is also a prerequisite and basis for the realization of autonomous driving. In recent years, offline High-Definition maps combined with environmental awareness have become a viable solution to achieve high-level autonomous driving functions as soon as possible. However, the High-Definition map is limited by the update frequency and update cost and is not the perfect choice for the solution. In this context, Bird's-Eye View (BEV) perception is gradually developed as an alternative solution that provides a more comprehensive and detailed representation of the surroundings in a top-down view of the scene and facilitates downstream tasks such as navigation and control of autonomous vehicles.

Due to the advantages of BEV in downstream tasks, the field of BEV sensing has grown rapidly in the past three years and has produced many excellent research studies. The history of the development of BEV perception can be traced back to [3], which proposes the Inverse Perspective Transformation (IMP) to accomplish the transformation of image views to BEV and is a pioneering work in view transformation. Existing BEV perception methods can be divided into four categories according to the view transform module: IMP-based methods, depth-based or voxel-based methods, MLP-based methods, and Transformer-based methods. However, the existing methods have some limitations and areas for improvement. The IMP-based methods, such as [4–6], are unable to generate high-quality BEV representations due to the large differences and severe deformations between the two views. In addition, these methods rely heavily on the horizon assumption, and the models perform very poorly when the environment does not conform to that assumption or when no prior information is available. The MLP-based methods, such as [7,8], ignore the geometric prior for calibrating the camera and usually transform the multi-view images separately, which cannot fully exploit the information embedded in the overlapping parts of the images. Transformer-based methods, such as [9,10], have undergone rapid development in the last two years and have shown excellent model results. However, these methods have efficiency problems in the training and inference processes, which limit their practical application. Moreover, these methods still rely on deep pretraining, indicating that depth information is still crucial for view transformation in such methods. In contrast, depth-based methods, such as [11–14], have higher computational efficiency and flexibility in multi-view image processing. However, there is still a gap between these methods and the state-of-the-art LiDAR-based models. To bridge this gap, we need to explore performance improvement paths further while maintaining computational efficiency.

Summarizing previous work, depth- or voxel-based BEV perception exhibits a high degree of modularity and reusability with a paradigm with three essential components: an image encoder, a view transform module, and a BEV encoder. These different modules form a comprehensive BEV perception pipeline. The image encoder is the foundation and provides the necessary feature extraction capabilities. The view transform module converts input data from multiple camera views into a uniform BEV representation for consistent and coherent processing. The BEV encoder module encodes the 3D voxel data, capturing the intrinsic spatial and semantic features for subsequent analysis. This modular design allows the decoupling of specific functions, increasing flexibility and facilitating the reuse of individual components in different BEV-perception systems. Most current research focuses on improving or innovating view transform modules, ignoring the image encoder and the BEV encoder. However, the state-of-the-art view transform module yields only a four-point performance improvement [15]. While it is true that the view transformation module is an integral part of the model both intuitively and practically, using simple bilinear sampling to perform the view transformation work is equally effective, though not at the most advanced level. The view transform module has much less impact on the model's performance in the current architecture than the selection of the appropriate input resolution and batch size [15]. Meanwhile, since the region size of BEV features is artificially set, the data features out of range in the original image will be discarded. Therefore, in the initial stage of the model, using a backbone network with stronger modeling capability to learn long-range relationships can further exploit the information in the images and improve the model's performance. In addition, the 2D to 3D transform module causes much information loss, especially distortion of features at long distances.

Following the above, we investigate three components: the image encoder, the view transform module, and the BEV encoder. The traditional CNN image encoders, such as ResNet [16] and ConvNeXt [17], cannot model long-range feature relationships. In contrast, the Transformer-based image encoder ignores more feature relationships between channels, while model training is more difficult and data requirements are greater. In addition, the Large Kernel Attention [18] module combines the advantages of convolution and self-attentive mechanisms, including local structural information, remote-dependent modeling capability, and adaptability. Therefore, we utilize the Large Kernel Attention in combination with the ConvNeXt module, where ConvNeXt can improve the modeling ability of local structural information in the model, and the Large Kernel Attention can further complement the modeling ability of the model for remote features while retaining the modeling ability of the model for channels. For the view transform module, a single bilinear sampling is used to complete the view transform task, which does not pay enough attention to the local information of image features. We combine bilinear sampling and attention mechanisms to design a view transform module to ensure that it can complete the view transformation simply and efficiently while having better transform performance. Meanwhile, some depth-based studies [15,19,20] all follow the same BEV features map settings yet simply use ResNet-18 as the BEV features encoder, which also leads to the fact that the BEV features map itself suffers from distortion of long-range features after the view transform module, which is not well solved, and therefore the model is less effective in the regions farther away from the center of the ego vehicle. To alleviate this contradiction, we redesigned a module for encoding BEV features using a combination of large kernel-size convolutional blocks [21], which can efficiently model the remote relations of BEV features. In this study, we select semantic segmentation as the evaluation task of our proposed model. We also conducted experiments on the nuScenes dataset to evaluate our proposed model's performance. Our proposed model shows good performance with a mIoU of 45.6 in the experiments. The results also illustrate the effectiveness of each component.

In conclusion, our contributions are summarized as follows:

1. We redesign an image encoder combined with the Large Kernel Attention to address the conventional encoder's lack of remote feature modeling capability.
2. To overcome the great difference between image features and BEV features, a view transform module is designed by combining bilinear sampling and attention mechanisms to ensure that BEV features pay more attention to image features that are closely related.
3. To address the problem of distortion of BEV features at long distances, a BEV encoder with a large kernel size for BEV features is redesigned to obtain a larger receptive area.

The paper is organized as follows: Section 2 reviews the related work in BEV perception methods. Section 3 introduces our proposed model's overall architecture and each component's composition. Next, Section 4 describes the experimental setup, experiment results, comparative results, and detailed component analysis. The conclusion of this study is summarized in Section 5.

2. Related Work

In this section, we classify BEV perception into two categories according to the view transformation methods: geometry-based and network-based methods, and we describe the work related to each of these two categories.

2.1. Geometry-Based Method

The IMP-based method utilizes the homographic matrix derived from the internal and external parameters of the camera. Cam2BEV [4] first employs IPM to transform multiple image features and finally obtains the BEV semantic map. To alleviate the distortion problem of the IPM-based method, TrafCam3D [5] proposes a dual-view network structure. For the pedestrian prediction problem, SHOT [6] projects each part of the pedestrian at different ground levels, respectively. The above studies demonstrate that IMP is effective enough under the condition that the flat-ground assumption is satisfied.

However, it is obvious that real-world driving scenarios cannot always satisfy the flat-ground assumption. Therefore, researchers began working on predicting the exact depth needed to accomplish the task of perspective view to BEV transformation. First, the Depth-based methods lift 2D features into 3D space by adding depth. Specifically, [22,23] predict each pixel's depth and directly utilize the existing LiDAR-based task head after transforming the 2D features into a pseudo-point cloud type. CaDDN [12] proposes a similar approach, but instead of directly generating a pseudo-point cloud, the pixels with predicted depth distribution are projected to the BEV, while the process uses depth supervision from the LiDAR. In addition, LSS [11] proposes to predict an explicit depth distribution for each 2D feature and then project the 2D features into BEV features. Based on the LSS, BEVDet [19] proposes a multi-camera model for 3D object detection tasks. BEVDet4D [24] exploits the previous camera frames to enhance the model's performance. BEVDepth [20] demonstrates that the performance of BEV-perception models can benefit from depth supervision and proposes a faster pooling operation.

2.2. Network-Based Method

The network-based methods start with using MLP for the view transformation task, transforming the perspective view to BEV. VED [25] first proposes an end-to-end monocular real-time prediction model with a MLP layer to transform the perspective view to BEV. VPN [7] further applies the MLP-based view conversion module to scenarios with multiple camera inputs. Specifically, VPN first transforms the encoded image features from multiple cameras into BEV features using MLP and then fuses all BEV features. FISHING [26] then introduces LIDAR and radar features based on VPN to complete the post-fusion and achieve multimodal perception. To address the problem of spatial information loss caused by MLP, PON [8] uses feature pyramids to obtain multi-scale image features and then uses MLP for view transformation. HFT [27] makes a further comparison between the advantages and disadvantages of using the camera parameter-based MLP method.

The Transformer-based methods utilize the currently popular Transformer to design the view transform module without the camera parameters. Unlike the MLP-based method that starts with 2D features, this method, in turn, uses an attention mechanism to capture the corresponding 2D features. DETR [28] and STSU [29] accomplish the 2D detection task by capturing the corresponding features with the pre-defined query. DETR3D [30], on the other hand, extends DETR to 3D object detection by geometric feature sampling. For autonomous driving, PETR [10] and PETRv2 [31] further employ camera parameters to generate position encoding and utilize temporal cues, respectively.

3. Method

This section presents the detailed design of our proposed model for BEV semantic segmentation. In Section 3.1, we first describe the overall architectural details of our proposed model. In Section 3.2, we introduce the image encoder for extracting 2D features from multi-camera images and illustrate why and how to redesign the image encoder. Next, we explain in detail how our designed view transform module generates 3D voxel features from 2D features in Section 3.3. In Section 3.4, we give detailed illustrations of the BEV encoder of our proposed model.

3.1. Overall Architecture

In this study, our model pipeline takes N RGB images $\mathbf{F} = \{\mathbf{F}_i \in \mathbb{R}^{3\times H\times W}, i = 1,2,\ldots,N\}$ from multiple cameras and the corresponding internal and external parameters as input. The output is a BEV segmentation map obtained by a specific task head. Specifically, our proposed model consists of three core components: the image encoder, the view transform module named the Bilinear-Attention module, and the BEV encoder. In particular, the Bilinear-Attention module consists of two submodules: the Bilinear Sampling module and the Deformable Attention module, as shown in Figure 1. The multi-camera images are first fed to the image encoder, which outputs 2D features $\mathbf{F}^{2d} = \left\{\mathbf{F}_i^{2d} \in \mathbb{R}^{C_{2d}\times H_{2d}\times W_{2d}}, i = 1,2,\ldots,N\right\}$. Next, the 3D voxel $\mathbf{F}^{vox} \in \mathbb{R}^{C_{vox}\times Z_{vox}\times X_{vox}}$ and corresponding 3D coordinates generated in advance are projected onto the 2D features and constructed by bilinear sampling. The 3D voxel is then further refined by the Deformable Attention module. Finally, the 3D voxel is encoded into BEV features $\mathbf{F}^{bev} \in \mathbb{R}^{C_{bev}\times H_{bev}\times W_{bev}}$ by the BEV encoder. The details of each component in the pipeline are described in the following sections.

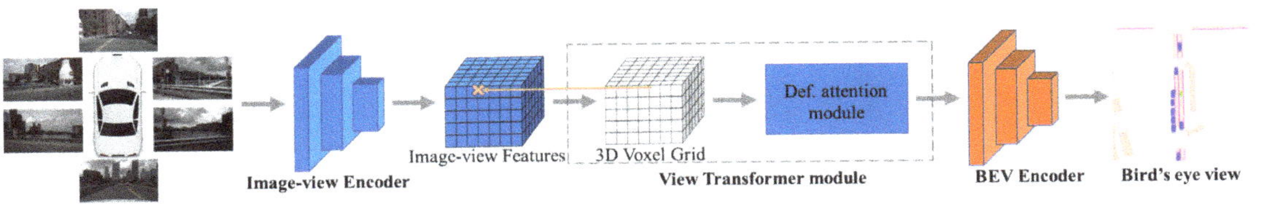

Figure 1. The pipeline of our proposed model.

3.2. Image Encoder

A high-performance image feature encoder is crucial for computer vision tasks. Much research in BEV perception uses ResNet or EfficientNet as the backbone network of the image encoder because both ResNet and EfficientNet are proven and mature networks with excellent performance. However, when training the model, the backbone network is the last stage to update the parameters and is less affected because the model is updated by backpropagation. Theoretically, the performance of the model is better if a backbone network with better pre-training performance and more robustness is selected. In the past, the powerful modeling capability of Visual Transformer has greatly impacted the image field. ConvNeXt [17], on the other hand, proposes a new pure convolution that provides stronger performance by emulating the Visual Transformer model. However, ConvNeXt is limited by the convolutional kernel size, cannot model remote dependencies, and cannot provide a good balance between local and global modeling capabilities. Although the excellent Transformer-based backbone network has excellent remote feature modeling capability, its huge data demand, higher training difficulty, and more computational resources required than convolutional networks make it not applicable. The Large Kernel Attention module [18] overcomes the abovementioned drawbacks and nicely combines the advantages of self-attention and large kernel convolution. The Large Kernel Attention module consists of three components: a spatial local convolution (depth-wise convolution), a spatial long-range convolution (depth-wise dilation convolution), and a channel convolution (1×1 convolution), as shown in Figure 2. Specifically, a $K \times K$ convolution is decomposed into a $\lceil \frac{K}{d} \rceil \times \lceil \frac{K}{d} \rceil$ depth-wise dilation convolution with dilation d, a $(2d-1) \times (2d-1)$ depth-wise convolution, and a 1×1 convolution.

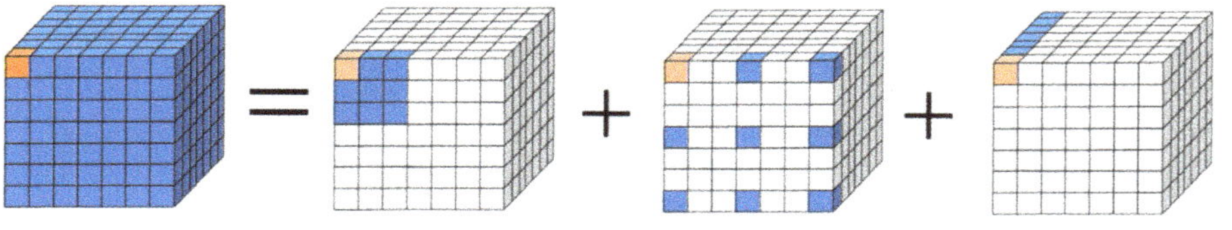

Figure 2. Decomposition diagram of large-kernel convolution from [18].

Through the above decomposition, the Large Kernel Attention module captures long-range relationships with slight computational cost and parameters, estimates the importance of a point, and generates an attention map. The Large Kernel Attention module can be written as

$$\textbf{Attention} = \text{Conv}_{1\times1}(\text{DW-D-Conv}(\text{DW-Conv}(\text{F}))) \tag{1}$$

$$\textbf{Output} = \textbf{Attention} \otimes \textbf{F} + \textbf{F} \tag{2}$$

where $\mathbf{F} \in \mathbb{R}^{C\times H\times W}$ is the input feature, $\textbf{Attention} \in \mathbb{R}^{C\times H\times W}$ denotes attention map, which indicates the importance of each feature, and $\otimes$ means element-wise product. We redesign an image encoder combining the Large Kernel Attention and ConvNeXt blocks to address the above situation. The network architecture of our image encoder is shown in Figure 3.

Given N images of the size (H, W), we construct the image encoder shown in Figure 3a to downsample the input image by a factor of 8 to obtain the output features with a resolution of $(H/8, W/8)$. Specifically, the network architecture of the image encoder consists of five parts: the stem part, three downsampling blocks, four stages of ConvNeXt blocks, two blocks with the Large Kernel Attention, and a bilinear upsampling block. We first employ a convolutional layer with a kernel size of 4 and a stride of 4 and choose layer normalization to do the normalization operation in the stem part, as shown in Figure 3b. In this way, we can obtain the output feature map of the stem part as $(H/4, W/4)$. Next, we adopt three convolution stages of ConvNeXt [17], which contain convolution blocks in the order of (3, 3, 9). The specific composition of the convolution block is shown in Figure 3d.

It contains a depthwise convolution layer with a kernel size of 7 responsible for mixing information in the spatial dimension, layer normalization, the GELU activation function, and two pointwise convolution layers for mixing information in the channel dimension, respectively. To further obtain more global features, we employ the Large Kernel Attention mechanism as described before after the first and second convolutional stages, whose specific structure is shown in Figure 3e. In addition, a separate downsampling layer is added in the middle of each convolutional stage, consisting of a convolutional layer with a kernel size of 2 and a stride of 2, and a layer normalization to achieve spatial downsampling. Finally, to obtain richer features, we concatenate the output of the third convolution block with an upsampling multiplier of two through an upsampling block with the output of the fourth convolution block to obtain the final 2D features $\mathbf{F}^{2d}$. This up-convolution block consists of one up-sampling layer and two convolution layers with a kernel size of 3, as shown in Figure 3c.

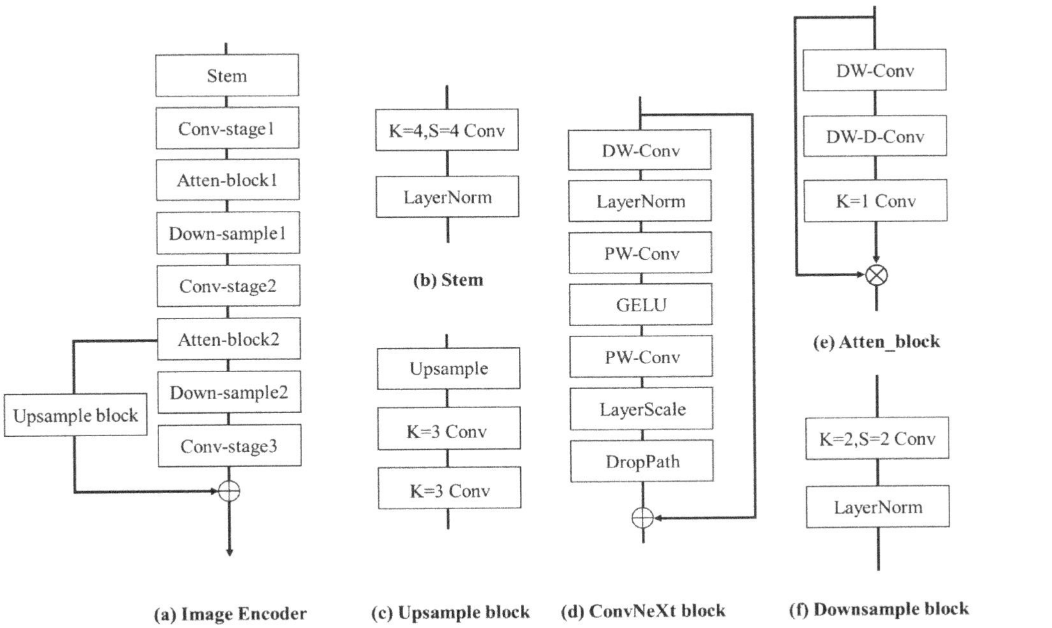

Figure 3. The structure of our redesigned image encoder. (**a**) The overall architecture of the image-view encoder. (**b**) The structure of the Stem part. (**c**) The structure of the Upsample block. (**d**) The structure of the ConvNeXt block. (**e**) The structure of the Large Kernel Attention block. (**f**) The structure of the downsample block.

3.3. View Transform Module

The depth-based approach achieves dimensionality lifting of 2D features by predicting a corresponding set of depth values for each 2D feature, then puts all features into a pre-generated view frustum, and finally forms a BEV feature map by pooling calculations. Although the method achieves satisfactory results, the predicted depth values depend on the ground plane assumption, and the accuracy of depth prediction seriously affects the model's overall performance. In addition, the Simple-BEV [15] also demonstrates the substitutability of the Lift-Splat method [11]. For the lifting operation of 2D features, [15] employs a pre-generated set of 3D voxels to obtain sub-pixel features by projecting each voxel in the 2D feature map using bilinear sampling for each voxel. It has been experimentally demonstrated that this method is more efficient due to the absence of hyperparameters while maintaining its effectiveness. However, there are better choices than this view transform method because simple sampling implies a lack of global information modeling capability, which impacts model performance. To address the above situation, we designed a view transform module named the Bilinear-Attention module, as shown in Figure 4. The structure of the proposed Bilinear-Attention module is composed of a Bilinear Sample module and a Compressed Attention module.

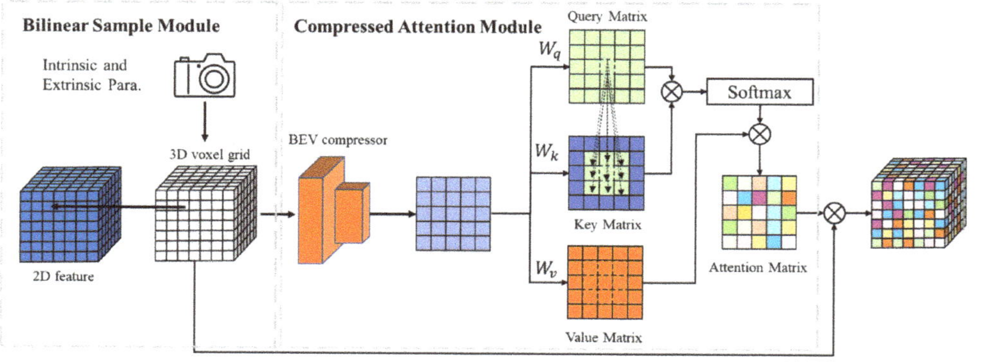

Figure 4. The overall architecture of the Bilinear-Attention module. The view transform module has two components: the Bilinear Sample module to complete the 2D feature lifting dimension operation and the Compressed Attention module to refine the 3D voxel features.

We assume that the input feature map size for the view transform module is (1, 2). We first generate a 3D feature for each voxel grid by bilinear sampling using a pre-defined 3D voxel grid with dimensions (Z, Y, X) and its corresponding 3D coordinate information. Specifically, according to the set hyperparameters (Z, Y, X), a 3D voxel grid is generated along with the 3D coordinate information corresponding to each grid, which is then converted to the corresponding 2D coordinates using the coordinate system conversion formula based on 2D and 3D space, as follows:

$$\lambda \mathbf{p} = \left[\mathbf{K} \mid \mathbf{0}_3 \right] \begin{bmatrix} \mathbf{R} & 0 \\ \mathbf{0}_3 & 1 \end{bmatrix} \begin{bmatrix} \mathbf{0}_3^T & -\mathbf{C} \\ 0 & 1 \end{bmatrix} \mathbf{P} \tag{3}$$

where $\mathbf{p} = \begin{pmatrix} x & y & 1 \end{pmatrix}^T$ denotes the 2D pixel position. $\mathbf{P} = \begin{pmatrix} X & Y & Z & 1 \end{pmatrix}^T$ being a 3D point defined with homogeneous coordinates. The projection matrix that incorporates the intrinsic parameters is denoted as $\mathbf{K}$ throughout this thesis. Mathematically, the position and orientation of the camera are defined by a 3×1 vector $\mathbf{C}$ and a 3×3 rotation matrix $\mathbf{R}$. Next, bilinear sampling is performed according to the corresponding 2D coordinates to generate the corresponding voxels for each 3D grid to obtain the sampled 3D voxel grid $\mathbf{F}'$, whose dimensional size is (N, Y, Z, X). This method, however, has a limited receptive field for mapping the generated 3D features onto the 2D feature maps. To address this problem, we then compress the dimension using a convolutional layer with a kernel size of 3 to obtain the output features X used to generate the Query Q and Key K on the basis of a 3D voxel grid filled with features of size $(N, 1, Z, X)$, and the process can be expressed as follows:

$$Q = W_q X \tag{4}$$

$$K = W_k X \tag{5}$$

$$V = W_v X \tag{6}$$

where the W_q, W_k, and W_v are the learnable parameters. Meanwhile, the dimensions of Q and K are set to be the same $(N, 1, Z, X)$ and the dimensions of value V are the same as

3D features $\mathbf{F}'$. On top of the obtained Query, Key, and Value, we adopt a self-attentive mechanism to refine the 3D features further and increase the global interaction of the features. We first compute the attention map using the dot product of Query and Key. In the next step, we employ the computed attention map and V to generate the final voxel features. The process can be expressed as follows:

$$\mathbf{Attention} = \mathrm{Softmax}(QK^T) \tag{7}$$

$$\mathbf{F}^{vox} = \mathbf{Attention} \otimes V \oplus \mathbf{F}' \tag{8}$$

where $Attention$ denotes the obtained attentional map. After our proposed feature diffusion module, the final output $\mathbf{F}^{vox}$ of the view transform module, a 3D voxel grid with rich features, is obtained.

3.4. BEV Encoder

The BEV feature encoder of many depth-based research studies is ResNet-18 with small receptive fields, while traditional convolution methods are less capable of modeling the long-range relationships of BEV features due to the limitations of the moving window, which cannot focus on the long-range features outside the center of the ego vehicle. In addition, the view transform module causes much information loss, especially distortion of features at long distances. Although the current BEV perception algorithms are generally set within 100 m of the surrounding area, the problem of information loss at long distances is still apparent. This means that many studies using ResNet-18 as a standard BEV feature encoder are weakened in their ability to model the long-range relationships of BEV features by the loss of information at long distances, thus leading to significantly further deterioration of the long-range results compared to the near-range results. Therefore, to enhance the spatial long-range modeling capability, we redesigned a BEV feature encoder with large kernel-size convolutional blocks [21], as shown in Figure 5, to try to focus on the features away from the center of the BEV features.

We assume that the 3D voxel features after the view transform module is $\mathbf{F}^{vox} \in \mathbb{R}^{C_{vox} \times Z_{vox} \times X_{vox}}$, where C_{bev}, Z_{vox}, X_{vox} denote the number of channels, height, and width of the 3D voxel features initially obtained, respectively. As shown in Figure 5a, the network architecture of our BEV feature encoder consists of three components: a stem part, two stages consisting of ResNet-18 blocks, and a stage consisting of RepLK blocks. In particular, we first employ a convolutional layer with a kernel size of 7, followed by a batch normalization layer and a ReLU activation layer in the stem part, as shown in Figure 5b. After the stem part, we can get the output features as $(Z_{vox}/2, X_{vox}/2)$. Furthermore, we connect the two Res-stages to encode the feature maps further and downsample to $(Z_{vox}/4, X_{vox}/4)$, where the Res-stage consists of the ResNet-18 block shown in Figure 5d. Finally, we adopt several RepLK blocks to form a RepLK stage, as shown in Figure 5e, to sample the feature map $(H_{vox}/8, W_{vox}/8)$. The core components of the RepLK block are a large kernel convolution part and a feedforward part, where a depth-wise convolutional layer of the kernel size of 31 further encodes BEV features, and a 1×1 convolutional layer in the feedforward part is responsible for changing the number of channels or feature map size. To obtain richer BEV features, we upsample the output features after each downsampling to their original size and then perform channel dimension stitching. Finally, the final output of the BEV encoder is fed into a specific task head to obtain the prediction results.

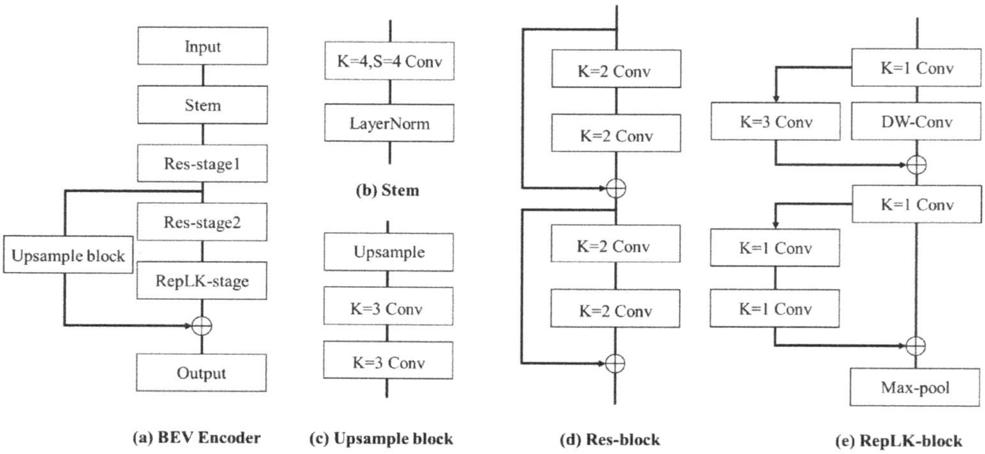

Figure 5. The network architecture of the BEV feature encoder. (**a**) The structure of the BEV encoder. (**b**) The structure of the Stem part. (**c**) The structure of the Upsample block. (**d**) The structure of the ResNet-18 blocks. (**e**) The structure of the RepLK blocks.

4. Experiments

Authors should discuss the results and how they can be interpreted from the perspective of previous studies and the working hypotheses. The findings and their implications should be discussed in the broadest possible context. Future research directions may also be highlighted.

4.1. Setup

Dataset. We evaluated our proposed model on the nuScenes dataset. The nuScenes dataset contains 1000 scenes, of which 850 are used for training and validation purposes, and the remaining 150 are reserved for testing. Each scene lasts 20 s, providing much temporal information for analysis. The nuScenes dataset contains a comprehensive sensor suite, including six cameras, one LiDAR sensor, and five radar sensors, where each camera has known corresponding internal and external parameters. In total, the dataset contains 40,000 keyframes capturing scenes from multiple angles and sensor modes. The camera images in the dataset have a resolution of 1600 × 900 pixels, ensuring a high level of detail and visual fidelity for visual perception tasks.

Evaluation Metrices. We follow the evaluation metrics of traditional segmentation tasks and measure the intersection-over-union (IoU) between the segmentation results and the ground truth. The IoU for each class can be written as follows:

$$\mathrm{IoU}(S_p, S_g) = \left| \frac{(S_p \cap S_g)}{(S_p \cup S_g)} \right| \tag{9}$$

And the average IoU for all classes can be written as:

$$\mathrm{mIoU}(S_p, S_g) = \frac{1}{N}\sum_{n=1}^{N} \mathrm{IoU}(S_p, S_g) \tag{10}$$

where $S_p \in \mathbb{R}^{H_g \times H_g \times N}$ and $S_g \in \mathbb{R}^{H_g \times H_g \times N}$ denote the segmentation prediction results and the ground truth, respectively. H_g and W_g denote the height and width of the ground truth, respectively. N is the number of dataset categories.

Details. For the image encoder, we employ the ConvNeXt block that has been pre-trained on the ImageNet dataset in advance. Our proposed model and reproduced model are trained on two NVIDIA GeForce RTX 3060 12G GPUs. Except for the specially stated hyperparameters, we follow the settings in ConvNeXt [17] and VAN [18]. For training, we use the AdamW optimizer, whose learning rate is set to $1 \times e^{-3}$ and the weight decay is set to $1 \times e^{-2}$. The loss function is computed using binary-cross-entropy loss functions. For the hyperparameters (Z, Y, X) of the view transform module, we follow the same settings $(200, 8, 200)$ as in the baselines of this task.

4.2. Experiment Result

In this section, to evaluate the performance of the proposed model, we comprehensively compare our proposed model with other state-of-the-art methods, including FISHING [26], LSS [11], FIERY [13], CVT [32], GKT [33], TIIM [34], BEVFormer [9], and Simple-BEV [15], as shown in Table 1. For the LSS and the Simple-BEV, we show the retraining results using the same configuration as the results reported in the original papers. The results of CVT and GKT are as reported in the original papers. For the model performance of other methods, we use all the data from this study [15]. For a fair comparison, we use only the single time step model without considering the time model and only consider the model's performance with multi-camera images as input. The results of evaluating the models on nuScenes are shown in the table. The proposed model achieves 45.4 mIoU on the nuScenes dataset, which outperforms most current segmentation methods and is similar to state-of-the-art performance. For the LSS and the Simple-BEV, we retrain with eight batch size settings, and two batch size settings and obtain results of 33.0 mIoU and 43.8 mIoU, respectively. In addition, to further demonstrate the performance of our proposed model, we visualized four key frames of the nuScenes dataset, as shown in Figure 6.

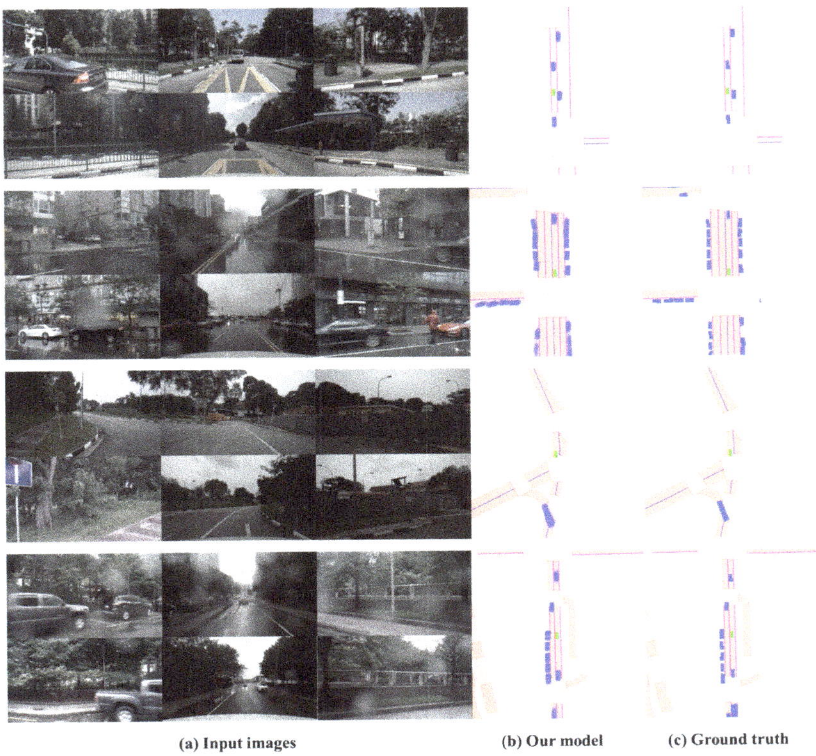

Figure 6. The visualization of results includes (**a**) the multi-camera input images, (**b**) the prediction results of our proposed model, and (**c**) the ground truth.

Table 1. Comparison of results of BEV segmentation on nuScenes. () denotes our reproduced results of our same setting.

Method	Lifting	Batch Size	mIoU
FISHING [26]	MLP	-	30.0
LSS [11]	Depth Estimation	8 (4)	33.0 (32.1)
FIERY [13]	Depth Estimation	12	35.8
CVT [32]	Deformable Attn.	16	36.0
GKT [33]	Geometry Attn.	16	37.2
TIIM [34]	Ray Attn.	8	38.9
BEVFormer [9]	Deformable Attn.	1	44.4
Simple-BEV [15]	Bilinear	2 (40)	42.5 (47.4)
Ours	Bilinear-Attn.	2	45.6

4.3. Detailed Analysis

In this section, we test the nuScenes dataset using different model combinations to validate our proposed components' effectiveness. Here we emphasize in advance that, without special instructions, our experimental settings are all batch size 2 and input resolution 448 $\times$ 800. In order to compare the segmentation performance of the components in detail, we compare the experimental results according to two categories: encoders and view transform modules.

Image encoder and BEV encoder: We employ an experimental comparison using different encoder combinations, and the view transform module defaults to our proposed Bilinear-Attention module. Specifically, we select the classical ResNet-101 and ResNet-18 as an image encoder and a BEV encoder, respectively, together with our proposed two encoders, and combine them into four experimental setups for our experiments. As shown in Table 2, Conv-LKA and Res-RepLK denote our proposed image encoder and BEV encoder, respectively. It can be observed that our proposed image encoder and BEV encoder can improve the performance with an increase of 1.8 and 1.3 in mIoU, respectively. Finally, when both of our proposed encoders are used, the model's overall performance is improved by 2.6 mIoU. The growth of FLOPs is also obvious when our proposed encoders are used. As shown in Table 3, we perform comparison experiments on BEV encoders with different kernel sizes. We can observe that a larger convolutional kernel size is beneficial for the model's performance but leads to performance degradation when the kernel size exceeds a certain limit. Experimental results show that the optimal kernel size is 13 $\times$ 13, while the larger kernel size does not significantly impact the overall number of parameters in our proposed model. Finally, we use the retrained simple-BEV to compare it with our proposed method and visualize the results, as shown in Figure 7.

Table 2. Ablations of the different encoder combinations.

ResNet-101	Conv-LKA	ResNet-18	Res-RepLK	Parameters	FLOPs	mIoU
√	-	√	-	42.1 M	428.3 G	43.0
√	-	-	√	40.6 M	512.7 G	44.3
-	√	√	-	38.1 M	552.1 G	44.8
-	√	-	√	36.6 M	653.5 G	45.6

Table 3. Ablations of the different kernel sizes of the BEV encoder.

Kernel Size	Parameters	mIoU
7 $\times$ 7	36.5 M	43.9
9 $\times$ 9	36.5 M	44.5
13 $\times$ 13	36.6 M	45.6
31 $\times$ 31	36.7 M	45.4

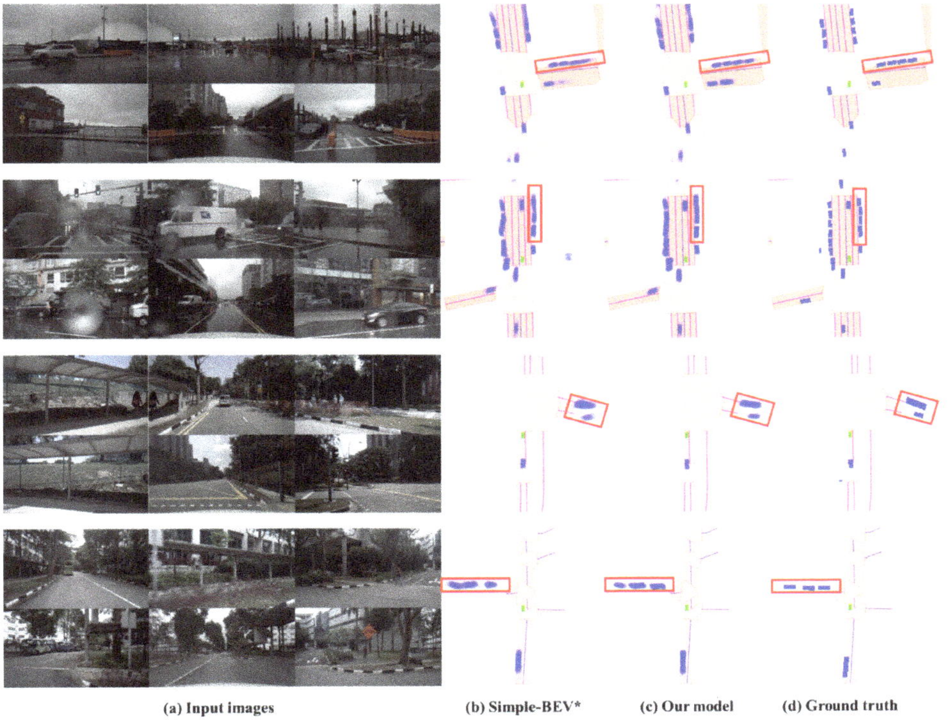

Figure 7. The comparison results of our proposed model and Simple-BEV. "*" denotes re-training with equivalent training settings. (**a**) The multi-camera input images; (**b**) the prediction results of retrained Simple-BEV; (**c**) the prediction results of our proposed model; and (**d**) the ground truth.

View transform module: We introduce another commonly used depth-based view transform method to compare and analyze with our proposed view transform module. Our proposed Bilinear-Attention module is split into a Bilinear Sample module and a Compressed Attention module for the ablation experiments. As shown in Table 4, Bilinear and Attention denote the Bilinear Sample module and the Compressed Attention module, respectively. LKA-RepLK indicates that both of our proposed encoders are used. We can observe that using the Bilinear Sample module alone does not perform as well as the depth prediction, but it is very close. Moreover, it can be observed that our proposed view transform module achieves an 8.2 improvement over the MLP approach in mIoU. Using the Bilinear Sample module alone also yields a 7.5 improvement in mIoU.

Table 4. Ablations of the different view transform module combinations.

Encoder	MLP	Depth	Bilinear	Attention	mIoU
LKA-RepLK	√	-	-	-	37.2
LKA-RepLK	-	√	-	-	44.8
LKA-RepLK	-	-	√	-	44.7
LKA-RepLK	-	-	√	√	45.6

5. Conclusions

This study proposes a camera-based model to accomplish the semantic segmentation task from the BEV perspective. To obtain an image view encoder with more powerful encoding performance and capable of capturing long-distance relationships, we redesign the backbone network with the Large Kernel Attention module. In addition, we propose the Bilinear Sample module to complete the lifting operation instead of directly predicting the depth, and then refine the 3D features with our proposed Compressed Attention module. We redesign the structure of the BEV encoder with RepLK to address the problem of long-range distortion of BEV features. We evaluated our proposed model on the nuScenes dataset. Our experiment results demonstrate that our model outperforms other models with equivalent training settings on the segmentation task while approaching state-of-the-art performance. While our current work focuses on semantic segmentation, we recognize the significance of expanding our evaluation to include object detection tasks. We are committed to enhancing computational efficiency and model size, ensuring that our approach remains practical for real-world applications. This optimization will contribute to the scalability and deployability of our model. Meanwhile, we recognize the challenges associated with detecting small or distant objects in the Bird's-Eye view. Our future work will involve dedicated optimization strategies to address these challenges and improve the model's performance in such scenarios. Finally, point clouds provide invaluable depth and spatial information, we envision integrating point cloud data alongside other sensor modalities to augment our model's understanding of the environment.

Author Contributions: Conceptualization, K.L. and X.W.; methodology, K.L.; software, K.L.; validation, K.L.; formal analysis, K.L.; investigation, X.W.; resources, W.Z.; data curation, X.W.; writing—original draft preparation, K.L.; writing—review and editing, X.W.; visualization, X.W.; supervision, W.Y.; project administration, W.Y.; funding acquisition, W.Z. All authors have read and agreed to the published version of the manuscript.

Funding: This research received no external funding.

Data Availability Statement: Not applicable.

Conflicts of Interest: The authors declare no conflict of interest.

References

1. Lang, A.H.; Vora, S.; Caesar, H.; Zhou, L.; Yang, J.; Beijbom, O. Pointpillars: Fast encoders for object detection from point clouds. In Proceedings of the IEEE/CVF Conference on Computer Vision And Pattern Recognition, Long Beach, CA, USA, 15–20 June 2019; pp. 12697–12705.
2. Li, Q.; Wang, Y.; Wang, Y.; Zhao, H. Hdmapnet: An online hd map construction and evaluation framework. In Proceedings of the 2022 International Conference on Robotics and Automation (ICRA), Philadelphia, PA, USA, 23–27 May 2022; pp. 4628–4634.
3. Mallot, H.A.; Bülthoff, H.H.; Little, J.; Bohrer, S. Inverse perspective mapping simplifies optical flow computation and obstacle detection. *Biol. Cybern.* **1991**, *64*, 177–185. [CrossRef] [PubMed]
4. Reiher, L.; Lampe, B.; Eckstein, L. A sim2real deep learning approach for the transformation of images from multiple vehicle-mounted cameras to a semantically segmented image in bird's eye view. In Proceedings of the 2020 IEEE 23rd International Conference on Intelligent Transportation Systems (ITSC), Rhodes, Greece, 20–23 September 2020; pp. 1–7.
5. Zhu, M.; Zhang, S.; Zhong, Y.; Lu, P.; Peng, H.; Lenneman, J. Monocular 3D vehicle detection using uncalibrated traffic cameras through homography. In Proceedings of the 2021 IEEE/RSJ International Conference on Intelligent Robots and Systems (IROS), Prague, Czech Republic, 27 September–1 October 2021; pp. 3814–3821.
6. Song, L.; Wu, J.; Yang, M.; Zhang, Q.; Li, Y.; Yuan, J. Stacked homography transformations for multi-view pedestrian detection. In Proceedings of the IEEE/CVF International Conference on Computer Vision, Montreal, BC, Canada, 11–17 October 2021; pp. 6049–6057.
7. Pan, B.; Sun, J.; Leung, H.Y.T.; Andonian, A.; Zhou, B. Cross-view semantic segmentation for sensing surroundings. *IEEE Robot. Autom. Lett.* **2020**, *5*, 4867–4873. [CrossRef]
8. Roddick, T.; Cipolla, R. Predicting semantic map representations from images using pyramid occupancy networks. In Proceedings of the IEEE/CVF Conference on Computer Vision and Pattern Recognition, Seattle, WA, USA, 13–19 June 2020; pp. 11138–11147.
9. Li, Z.; Wang, W.; Li, H.; Xie, E.; Sima, C.; Lu, T.; Qiao, Y.; Dai, J. Bevformer: Learning bird's-eye-view representation from multi-camera images via spatiotemporal transformers. In Proceedings of the European Conference on Computer Vision, Tel Aviv, Israel, 23–27 October 2022; pp. 1–18.
10. Liu, Y.; Wang, T.; Zhang, X.; Sun, J. Petr: Position embedding transformation for multi-view 3d object detection. In Proceedings of the European Conference on Computer Vision, Tel Aviv, Israel, 23–27 October 2022; pp. 531–548.
11. Philion, J.; Fidler, S. Lift, splat, shoot: Encoding images from arbitrary camera rigs by implicitly unprojecting to 3D. In Proceedings of the Computer Vision–ECCV 2020: 16th European Conference, Glasgow, UK, 23–28 August 2020; Proceedings, Part XIV 16, 2020. pp. 194–210.

12. Reading, C.; Harakeh, A.; Chae, J.; Waslander, S.L. Categorical depth distribution network for monocular 3D object detection. In Proceedings of the IEEE/CVF Conference on Computer Vision and Pattern Recognition, Virtual. 19–25 June 2021; pp. 8555–8564.
13. Hu, A.; Murez, Z.; Mohan, N.; Dudas, S.; Hawke, J.; Badrinarayanan, V.; Cipolla, R.; Kendall, A. Fiery: Future instance prediction in bird's-eye view from surround monocular cameras. In Proceedings of the IEEE/CVF International Conference on Computer Vision, Montreal, BC, Canada, 11–17 October 2021; pp. 15273–15282.
14. Xie, E.; Yu, Z.; Zhou, D.; Philion, J.; Anandkumar, A.; Fidler, S.; Luo, P.; Alvarez, J.M. M^2BEV: Multi-Camera Joint 3D Detection and Segmentation with Unified Birds-Eye View Representation. *arXiv* **2022**, arXiv:2204.05088.
15. Harley, A.W.; Fang, Z.; Li, J.; Ambrus, R.; Fragkiadaki, K. Simple-BEV: What really matters for multi-sensor bev perception? In Proceedings of the 2023 IEEE International Conference on Robotics and Automation (ICRA), London, UK, 29 May–2 June 2023; pp. 2759–2765.
16. He, K.; Zhang, X.; Ren, S.; Sun, J. Deep residual learning for image recognition. In Proceedings of the Conference on Computer Vision and Pattern Recognition, Las Vegas, NV, USA, 27–30 June 2016; pp. 770–778.
17. Liu, Z.; Mao, H.; Wu, C.-Y.; Feichtenhofer, C.; Darrell, T.; Xie, S. A convnet for the 2020s. In Proceedings of the IEEE/CVF Conference on Computer Vision and Pattern Recognition, New Orleans, LA, USA, 18–24 June 2022; pp. 11976–11986.
18. Guo, M.-H.; Lu, C.-Z.; Liu, Z.-N.; Cheng, M.-M.; Hu, S.-M. Visual attention network. *arXiv* **2022**, arXiv:2202.09741. [CrossRef]
19. Huang, J.; Huang, G.; Zhu, Z.; Ye, Y.; Du, D. Bevdet: High-performance multi-camera 3D object detection in bird-eye-view. *arXiv* **2021**, arXiv:2112.11790.
20. Li, Y.; Ge, Z.; Yu, G.; Yang, J.; Wang, Z.; Shi, Y.; Sun, J.; Li, Z. Bevdepth: Acquisition of reliable depth for multi-view 3D object detection. In Proceedings of the AAAI Conference on Artificial Intelligence, Washington, DC, USA, 7–14 February 2023; pp. 1477–1485.
21. Ding, X.; Zhang, X.; Han, J.; Ding, G. Scaling up your kernels to 31 × 31: Revisiting large kernel design in cnns. In Proceedings of the IEEE/CVF Conference on Computer Vision and Pattern Recognition, New Orleans, LA, USA, 18–24 June 2022; pp. 11963–11975.
22. Wang, Y.; Chao, W.-L.; Garg, D.; Hariharan, B.; Campbell, M.; Weinberger, K.Q. Pseudo-lidar from visual depth estimation: Bridging the gap in 3d object detection for autonomous driving. In Proceedings of the IEEE/CVF Conference on Computer Vision and Pattern Recognition, Long Beach, CA, USA, 15–20 June 2019; pp. 8445–8453.
23. You, Y.; Wang, Y.; Chao, W.-L.; Garg, D.; Pleiss, G.; Hariharan, B.; Campbell, M.; Weinberger, K.Q. Pseudo-lidar++: Accurate depth for 3D object detection in autonomous driving. *arXiv* **2019**, arXiv:1906.06310.
24. Huang, J.; Huang, G. Bevdet4d: Exploit temporal cues in multi-camera 3D object detection. *arXiv* **2022**, arXiv:2203.17054.
25. Lu, C.; van de Molengraft, M.J.G.; Dubbelman, G. Monocular semantic occupancy grid mapping with convolutional variational encoder-decoder networks. *IEEE Robot. Autom. Lett.* **2019**, *4*, 445–452. [CrossRef]
26. Hendy, N.; Sloan, C.; Tian, F.; Duan, P.; Charchut, N.; Xie, Y.; Wang, C.; Philbin, J. Fishing net: Future inference of semantic heatmaps in grids. *arXiv* **2020**, arXiv:2006.09917.
27. Zou, J.; Zhu, Z.; Huang, J.; Yang, T.; Huang, G.; Wang, X. HFT: Lifting Perspective Representations via Hybrid Feature Transformation for BEV Perception. In Proceedings of the 2023 IEEE International Conference on Robotics and Automation (ICRA), London, UK, 29 May–2 June 2023; pp. 7046–7053.
28. Carion, N.; Massa, F.; Synnaeve, G.; Usunier, N.; Kirillov, A.; Zagoruyko, S. End-to-end object detection with transformers. In Proceedings of the European Conference on Computer Vision, Glasgow, UK, 23–28 August 2020; pp. 213–229.
29. Can, Y.B.; Liniger, A.; Paudel, D.P.; Van Gool, L. Structured bird's-eye-view traffic scene understanding from onboard images. In Proceedings of the IEEE/CVF International Conference on Computer Vision, Montreal, BC, Canada, 11–17 October 2021; pp. 15661–15670.
30. Wang, Y.; Guizilini, V.C.; Zhang, T.; Wang, Y.; Zhao, H.; Solomon, J. Detr3d: 3D object detection from multi-view images via 3D-to-2D queries. In Proceedings of the Conference on Robot Learning, Auckland, New Zealand, 14–18 December 2022; pp. 180–191.
31. Liu, Y.; Yan, J.; Jia, F.; Li, S.; Gao, Q.; Wang, T.; Zhang, X.; Sun, J. Petrv2: A unified framework for 3D perception from multi-camera images. *arXiv* **2022**, arXiv:2206.01256.
32. Zhou, B.; Krähenbühl, P. Cross-view transformers for real-time map-view semantic segmentation. In Proceedings of the IEEE/CVF Conference on Computer Vision and Pattern Recognition, New Orleans, LA, USA, 18–24 June 2022; pp. 13760–13769.
33. Chen, S.; Cheng, T.; Wang, X.; Meng, W.; Zhang, Q.; Liu, W. Efficient and robust 2D-to-bev representation learning via geometry-guided kernel transformer. *arXiv* **2022**, arXiv:2206.04584.
34. Saha, A.; Mendez, O.; Russell, C.; Bowden, R. Translating images into maps. In Proceedings of the 2022 International Conference on Robotics and Automation (ICRA), Philadelphia, PA, USA, 23–27 May 2022; pp. 9200–9206.

World Electric Vehicle Journal

MDPI

Article

A Two-Stage Pillar Feature-Encoding Network for Pillar-Based 3D Object Detection

Hao Xu [1], Xiang Dong [1,*], Wenxuan Wu [1], Biao Yu [2] and Hui Zhu [2]

[1] School of Electrical Engineering and Automation, Anhui University, Hefei 230601, China; z21201036@stu.ahu.edu.cn (H.X.); z21301055@stu.ahu.edu.cn (W.W.)
[2] Hefei Institutes of Physical Science, Chinese Academy of Sciences, Hefei 230031, China; byu@hfcas.ac.cn (B.Y.); hzhu@iim.ac.cn (H.Z.)
* Correspondence: xdong@ahu.edu.cn

Abstract: Three-dimensional object detection plays a vital role in the field of environment perception in autonomous driving, and its results are crucial for the subsequent processes. Pillar-based 3D object detection is a method to detect objects in 3D by dividing point cloud data into pillars and extracting features from each pillar. However, the current pillar-based 3D object-detection methods suffer from problems such as "under-segmentation" and false detections in overlapping and occluded scenes. To address these challenges, we propose an improved pillar-based 3D object-detection network with a two-stage pillar feature-encoding (Ts-PFE) module that considers both inter- and intra-relational features among and in the pillars. This novel approach enhances the model's ability to identify the local structure and global distribution of the data, which improves the distinction between objects in occluded and overlapping scenes and ultimately reduces under-segmentation and false detection problems. Furthermore, we use the attention mechanism to improve the backbone and make it focus on important features. The proposed approach is evaluated on the KITTI dataset. The experimental results show that the detection accuracy of the proposed approach are significantly improved on the benchmarks of BEV and 3D. The improvement of AP for car, pedestrian, and cyclist 3D detection are 1.1%, 3.78%, and 2.23% over PointPillars.

Keywords: point cloud; autonomous vehicles; 3D object detection; pillar; LiDAR

Citation: Xu, H.; Dong, X.; Wu, W.; Yu, B.; Zhu, H. A Two-Stage Pillar Feature-Encoding Network for Pillar-Based 3D Object Detection. *World Electr. Veh. J.* **2023**, *14*, 146. https://doi.org/10.3390/wevj14060146

Academic Editor: Grzegorz Sierpiński

Received: 6 May 2023
Revised: 1 June 2023
Accepted: 2 June 2023
Published: 3 June 2023

1. Introduction

With expanding application scenarios such as autonomous driving, intelligent robotics, and virtual reality, 3D object-detection technology has emerged as an important research direction in the realm of computer vision [1]. The main task of point cloud 3D object detection is to detect various objects, such as cars, pedestrians, and cyclists, from 3D point cloud data [2]. Unlike traditional image-based object detection, point cloud 3D object detection needs to deal with disordered point cloud data and consider the 3D information of the objects [3,4]. Therefore, the key challenge of this task is how to extract the information concerned, such as the object's position, size, orientation, and so on, from the point cloud data [5]. To solve this problem, researchers usually employ deep learning techniques to build point cloud 3D object-detection models and optimize the performance of the models by training on a large amount of point cloud data [6,7].

Regarding point cloud representation methods, current 3D object -detection techniques can be divided into three categories: range image-based, point-based, and voxel-based methods [8]. The range image-based method is a technique that involves the projection of point cloud data onto a plane to generate a depth image, which is then utilized for object detection. This method can process large-scale point cloud data efficiently and be easily integrated with 2D image-processing algorithms. Nevertheless, at greater distances or lower resolutions, this method may be prone to inaccuracies [9,10]. The point-based methods, represented by PointNet [11], directly process the raw data and treat the point cloud as a disordered set of points. It processes each point independently, then aggregates the features of the points to obtain the entire point cloud's feature representation. The improved PointNet++ [12] can better manage point cloud local information. However, processing all point clouds leads to a huge computation complexity and high hardware requirements. In contrast, the voxel-based method first converts the point cloud data into a 3D voxel grid form, then processes and extracts the features of each voxel [13]. The pillar-based method is a special voxel-based method which simplifies voxels further by disregarding the z dimension, thereby turning the 3D problem into a 2D problem and reducing the computation complexity to a certain extent. However, it leads to a loss of information, such as the features of spatial location and relative position of each element [14]. PointPillars [15] is a typical 3D object-detection network of a pillar-based method which relies solely on point-wise aggregated features to represent pillar characteristics. This approach fails to accurately represent the features of the pillar itself and the relationship between pillars and the entire point cloud. This absence may give rise to several challenges. Firstly, in the presence of mutually occluded and overlapping objects, it could lead to the under-segmentation phenomenon, where the pillars of distinct objects are incorrectly classified as belonging to the same entity [16]. Secondly, these issues could significantly undermine the accuracy and reliability of object detection and localization, leading to missed or false detection [17]. Additionally, changes of the relative relationships between the pillars and the overall point cloud in different datasets or scenarios can lead to the increased instability of detection results [18]. Therefore, it is crucial to consider the features of the pillar itself and the relationship between the pillars and the overall point cloud.

In this paper, we propose an improved pillar-based network for 3D object detection based on pillars. It comprises three essential components: the pillar feature-encoding module, a backbone consisting of a region proposal network (RPN), squeeze-and-excitation networks (SeNet) for feature extraction, and a head for 3D boxes' regression.For pillar feature encoding, we propose a two-stage pillar feature-encoding (Ts-PFE) module that considers both inter- and intra-relational features among and in the pillars which can help the model better understand the spatial location and relative position of each pillar. On one hand, the model can better identify the local structure and global distribution of the pillars, thus improving the distinction between objects and reducing the under-segmentation phenomenon that arises in the presence of mutually occluded and overlapping objects. On the other hand, more accurate detection and localization can alleviate the issue of missed and false detections. Moreover, the stability and generalization capabilities of the model in different datasets or scenarios are enhanced. Additionally, we have incorporated SeNet into the backbone module for improved performance. This module enhances key features in pseudo-images and suppresses irrelevant features through attention mechanism, thereby bolstering the network's ability to extract important features of objects for detection.

The experimental evaluation on the KITTI dataset demonstrates the effectiveness of the proposed approach. Specifically, our method achieves significant improvements in object-detection performance while reducing under-segmentation problems in occluded and overlapping scenes. Our contributions can be summarized as follows:

- To solve the problem of under-segmentation due to missing features, compared with other pillar-based approaches that only consider the intra-relational features, we propose Ts-PFE, a feature-encoding network which considers both inter- and intra-relational features among and in the pillars. It improves the distinction between objects and reduces the under-segmentation problems in occluded and overlapping scenes.
- We improved the backbone by integrating SeNet, enhancing key features in pseudo-images, and suppressing irrelevant information to enhance the network's ability to extract important features of objects to be detected. By leveraging the power of SeNet, the proposed approach exhibits superior performance in object detection compared to prior works.
- Evaluated on the KITTI dataset, the experiments show that the detection accuracy of the proposed approach are significantly improved; the improvement of AP for car,

pedestrian, and cyclist 3D detection are 1.1%, 3.78%, and 2.23% over the baseline. The results of qualitative evaluation show that the under-segmentation problem is reduced in the occlusion and overlapping scenes.

2. Related Work

2.1. 3D Object Detection from Point Cloud Based on Voxel/Pillar

In recent years, voxel- and pillar-based methods have been widely developed and applied. These methods first transform the point cloud into fixed-shaped voxels or pillars, followed by feature extraction using 2D/3D convolution, and then implement object-detection tasks through classification and regression [19]. This approach provides significant computational advantages over processing each point individually, making it particularly suitable for real-time applications such as autonomous driving [15]. VoxelNet [13] is among the pioneering works in this field, which innovatively uses voxel feature encoding (VFE) and 3D convolution to encode and extract features from the voxels. However, the huge number of voxels lead to a large computation of 3D convolution and a slow inference speed. To address this issue, SECOND [20] proposed 3D sparse convolution to accelerate the training and inference process by eliminating the empty voxels. Despite its improved efficiency, sparse convolution is not deployment friendly. PointPillars [15], instead of using the traditional voxel-based method, simplify the voxel representation by generating a pseudo-image that can be processed by 2D convolution neural networks. This approach has efficient computational performance and became one of the dominant methods in practice. More recently, CenterPoint [21] was proposed, which converts the point cloud into a voxel network, predicts the object center, size, and orientation at the center of each voxel, and uses a 2D convolutional network to extract local features, which achieves a good balance between accuracy and speed, and can also handle small targets with high detection accuracy.

The voxel/pillar-based algorithm, while effective in some regards, has some notable limitations [19]. First, the uniform voxelization method can result in a loss of point cloud information and uneven sampling [22]. Secondly, when dealing with some small objects or dense point cloud, there may be a high rate of false detection and missed detection [23]. In addition, ignoring the relationship between the local and the global feature can lead to the under-segmentation issue [24]. To overcome these limitations, several improvements have been proposed, including improvements to voxelization methods, attention mechanisms, multi-scale features, and data-enhancement methods. For instance, in SCNet [25], each grid is divided into smaller sub-grids to preserve more point cloud information and alleviate the under- and over-segmentation. SA-Det3D [26] proposes two variants of self-attention for contextual modeling in 3D object detection by adding convolutional features and self-attentive features to model in 3D object detection. PV-RCNN [27] uses a voxel feature-encoding network to extract features of voxels, and then fuses the voxel features with point cloud features, which can better handle objects of different sizes and shapes.

2.2. Attention Mechanisms in Object Detection

As a crucial technique in neural networks, the attention mechanism has been heavily employed in computer vision, including tasks such as classification, detection, and segmentation [28]. Through this mechanism, the attention of the network is directed to key details and focusing more on the region or feature of interest, thus improving detection accuracy and robustness [29].

The attention mechanism has also been widely applied in the field of point cloud object detection. The commonly used methods are channel attention, spatial attention, and their combination. Spatial attention allows the network to prioritize information from certain locations in the point cloud, while channel attention means that the network can pay more attention to the features of certain channels [30]. The self-attention method [31], transformer [32], and SeNet [33], etc., are other types of attention mechanisms. Ref. [34] delves into the role of attention mechanisms in 3D point cloud object detection and shows that self-attention modules are not preferable in processing 3D point cloud data, and that

SE and CBAM enable the effectiveness and efficiency of 3D point cloud feature refinement. In [35], three modified attention mechanisms (SOCA, SOPA, and SAPI) are used to improve the effectiveness of feature extraction. Ref. [26] proposed two self-attention variants, FSA and DSA, for contextual modeling in 3D object detection, incorporating convolutional and self-attentive features in the model. Moreover, the attention mechanism can be used for point cloud segmentation, as seen in [36], where the network structure is based on the point-attention mechanism, effectively improving point cloud segmentation performance.

3. Our Approach

In this section, we introduce our approach, the pillar-based 3D object-detection network using a two-stage feature encoding module. The network architecture is illustrated in Figure 1. It comprises three components: the pillar feature-encoding module, a backbone which contains RPN and SeNet for feature extraction, and a head for 3D boxes' regression.We provide a detailed description of each block in the remainder of this section.

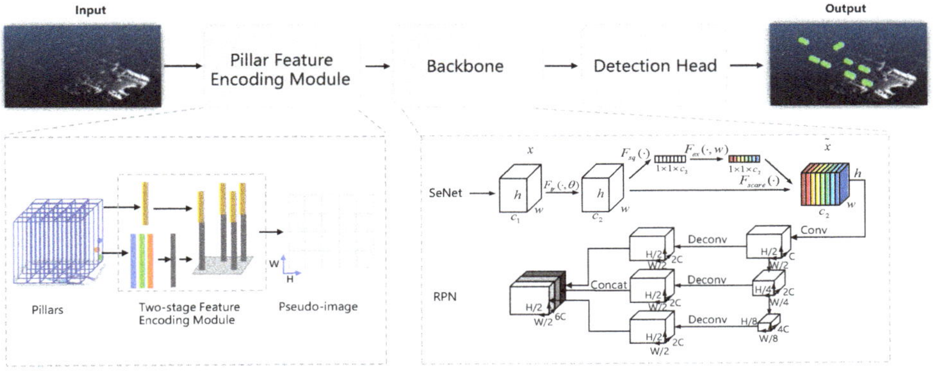

Figure 1. The overall network architecture of the proposed approach. The main components of the network are a pillar feature-encoding module, a backbone, and a detection head. The raw point cloud is converted to pillars. The encoder uses the pillars to learn a set of features that can be scattered back to a 2D pseudo-image for a convolutional neural network. The features from the backbone are used by the detection head to predict 3D bounding boxes for objects.

3.1. Two-Stage Pillar Feature Encoding

The voxel and pillar encoding approach has significantly impacted the development of 3D object detection based on point cloud by redefining the organization and processing of point clouds. By employing aggregated features to characterize voxel and pillar features, it has substantially reduced computational requirements while increasing operational speed [13]. However, this approach has a limitation; it only captures local features, failing to accurately represent the features of the pillar itself and the relationship between pillars and the entire point cloud [15]. In this paper, we propose a novel two-stage pillar feature-encoding (Ts-PFE) module as a solution to this issue. It considers both inter- and intra-relational features among and in the pillars, which can help the model better identify the local structure and global distribution of the pillars, thus improving the distinction between objects and reducing the under-segmentation problems in the presence of mutually occluded and overlapping objects. In addition, more accurate detection and localization can reduce the problems of missed and false detections. Furthermore, the inclusion of a wider range of features enhances the model's expressiveness and generalization capabilities. As shown in Figure 2, the proposed Ts-PFE module consists of three units: (1) the point feature-encoding unit, (2) the pillar feature-encoding unit, (3) the feature-fusion unit.

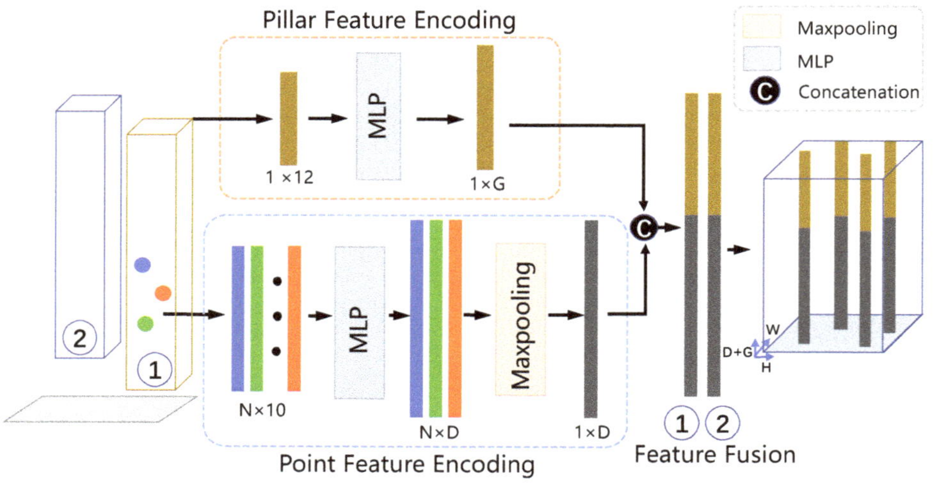

Figure 2. The Two-stage Pillar Feature Encoding (Ts-PFE) module architecture. It consists of three units: the point feature-encoding unit, the pillar feature-encoding unit, and the feature-fusion unit. The point feature encoding is used to obtain the point-wise aggregated features. The pillar feature encoding is used to obtain the pillar-wise features. Finally, the fusion unit concatenates the features together.

We first pillarize the point cloud data by considering only the X, Y direction and ignoring the information in the Z axis direction. Specifically, the point cloud P in 3D space has a range of L, W, and H along the X, Y, and Z axes, and is divided into a set of pillars of size v_L, v_W, and H. Each pillar is represented by $v = \{pi = [xi, yi, zi, ri] \in R^{N\times 4}\}$. x_i, y_i, and z_i denote the coordinates of each point along X, Y, and Z axes, respectively, while r_i represents the laser reflection intensity.

3.1.1. Point Feature Encoding

In this unit, we aggregate the features of the points to represent the features of the pillars. First, the four-dimensional feature vector of the points is expanded to a ten-dimensional vector, $p_i^n = \{\left[x_i, y_i, z_i, r_i, x_i^c, y_i^c, z_i^c, x_i^p, y_i^p, z_i^p\right] \in R^{N\times 10}\}$. Here, $[x_i, y_i, z_i, r_i]$ represent the coordinates and reflectance of the points, while $[x_i^c, y_i^c, z_i^c]$ represent the distance to the arithmetic mean of all points in the pillar, and $\left[x_i^p, y_i^p, z_i^p\right]$ is the offset of each point from the pillar center. Subsequently, the feature is enhanced and mapped to a higher dimension by an MLP layer network, which is defined as

$$p_i^m = m(p_i^n; w_m) \tag{1}$$

where the function $m(\cdot)$ signifies a stack of multiple MLP layers which contain batch normalization (BN) layers and rectified linear unit (ReLU) layers, w_m represents the learnable weights, and $p_i^m \in R^{N\times D}$ is the point-wise feature. Next, max-pooling is used to aggregate the features of all points in pillar j into a single-feature vector used to characterize the information of this pillar. It is given by

$$p_j^m = MAX(p_i^m) \tag{2}$$

where $MAX(\cdot)$ denotes the max-pooling operation across these points' features, and $p_j^m \in R^D$ is the resultant feature vector for the pillar j.

3.1.2. Pillar Feature Encoding

It is not enough to obtain the features of the pillar aggregated from the points. Therefore, we utilize the second-stage feature encoding, which is the pillar feature-encoding unit, to encode the pillar's inherent features and its relationship with the entire point cloud.

First, we use the arithmetic mean $\left[x_j^c,\ y_j^c,\ z_j^c\right]$ of all the points in pillar j to represent the center of gravity of the pillar, and the center value $\left[x_j^p,\ y_j^p,\ z_j^p\right]$ of pillar j as the pillar's center. Next, we find the arithmetic mean $\overline{x}_j^c,\ \overline{y}_j^c,\ \overline{z}_j^c$ of all points as the center of gravity of the overall point cloud, and the coordinates $\overline{x}_j^p,\ \overline{y}_j^p,\ \overline{z}_j^p$ of the center pillar as the pillar center of the overall point cloud, and then we can obtain the offset $\left[x_j^c-\overline{x}_j^c,\ y_j^c-\overline{y}_j^c,\ z_j^c-\overline{z}_j^c\right]$ from each pillar to the center of gravity and the offset $\left[x_j^p-\overline{x}_j^p,\ y_j^p-\overline{y}_j^p,\ z_j^p-\overline{z}_j^p\right]$ from each pillar to the pillar center of the overall pillar. At this time, we have the following features: $v_j^n=\{[x_j^c,\ y_j^c,\ z_j^c,\ x_j^p,\ y_j^p,\ z_j^p,\ x_j^c-\overline{x}_j^c,\ y_j^c-\overline{y}_j^c,\ z_j^c-\overline{z}_j^c,\ x_j^p-\overline{x}_j^p,\ y_j^p-\overline{y}_j^p,\ z_j^p-\overline{z}_j^p]\in R^{12}\}$. Then, the features are augmented and mapped to a higher dimension by an MLP layer network. The equation is

$$v_j^m=m(v_j^n;\ w_m) \tag{3}$$

where the function $m(\cdot)$ is a stack of multiple MLP layers which contain batch normalization (BN) layers and rectified linear unit (ReLU) layers, w_m represents the learnable weights, and $v_j^m\in R^G$ is the pillar-wise feature, which contains the features of the pillar itself, the relationship between the pillars, and the overall point cloud.

3.1.3. Feature Fusion

The aforementioned features can be categorized into two types: the former is extracted from the points and the latter is calculated using the pillar's own features. They are both vectors used to characterize the features of the pillar. In this fusion module, we fuse the two types of features to obtain the complete features of the pillar. Its formula is as follows:

$$V_j^c=fusion(p_j^m;\ v_j^m) \tag{4}$$

where $fusion(\cdot)$ is the feature concatenation function and $V_j^c\in R^{D+G}$ is the fused pillar feature. After obtaining the fused features, we only need to map each pillar to a plane according to the coordinates of the pillar to obtain a pseudo-image U whose dimension is $H\times W\times(D+G)$. For convenience, we consider $(D+G)$ as C.

3.2. Backbone

After obtaining the pseudo-image, the following part is a convolutional network which is used for feature extraction. We improved the backbone with SeNet, which can highlight the key information while ignoring the irrelevant information; its structure is as illustrated in Figure 1. The backbone consists of two parts: one of them is the SeNet and the other is an RPN connected in series behind.

3.2.1. SeNet

Squeeze-and-excitation networks (SeNet) are a channel-based feature-attention module proposed by [37] in 2017. This module enhanced the network's ability to extract vital features of the object to be detected by modeling each feature channel, distinguishing the importance of different channels, and enhancing key features while suppressing irrelevant information. Incorporating this attention mechanism in practical applications can greatly improve model performance, enabling it to better capture the details and features in the image. Therefore, we utilize SeNet to process the pseudo-images obtained from PointPillars for feature learning and attention weights' calculation. The resulting weighted features are then mapped onto the original point cloud data to further enhance the accuracy and robustness of object detection. By combining the strengths of PointPillars with SeNet, we can generate a more expressive and generalizable feature representation.

In the preceding steps, we obtain the feature map U with dimensions $H \times W \times C$. Following the feature-map processing transformation, two separate operations are performed. Firstly, the new feature map U undergoes global average pooling, compressing the height and width dimensions into a vector of size $1 \times 1 \times C$ while preserving channel information, which we denote as z. To calculate the cth element of z, we use the formula

$$z_c = F_{sq}(u_c) = \frac{1}{H \times W} \sum_{i=1}^{H} \sum_{j=1}^{W} u_c(i,j) \tag{5}$$

where F_{sq} is the compression operation, u_c is the feature vector after feature extraction, and i and j are the positions of the cth element on the corresponding feature map. Through global average pooling, the input feature map undergoes compression, and output features z_c are obtained to learn feature weights for each channel. Then z_c are sequentially passed through the fully connected layer, the ReLU layer, and sigmoid layers to complete the excitation and achieve a comprehensive dependency capture between channels, calculated as

$$s = F_{ex}(z, W) - \sigma(g(z, W)) - \sigma(W_2 \delta(W_1 z)) \tag{6}$$

where W_1 and W_2 are the two weight matrices to be learned, $\sigma(\cdot)$ is the sigmoid activation function, $\delta(\cdot)$ is the ReLU activation function, and s is the feature matrix obtained after the excitation operation. Finally, the obtained feature matrix is weighted with the feature matrix u. The weighting formula is

$$\widetilde{x} = F_{scale}(u_c, s_c) \tag{7}$$

where $\widetilde{x}$ is the final feature matrix obtained by the dot product weighting operation of the feature matrix u_c and s_c.

3.2.2. RPN

Similarly to [13], this unit follows a strategy showcased in Figure 1. We employ the RPN (region proposal network) to process the feature map obtained from the previous step, aiming to enhance the detection capability of objects across various sizes with multi-scale receptive field. The unit consists of three convolutional layers, each serving a specific purpose. The initial layer within each block downsamples the feature map by half, accomplished through a convolution operation with a stride size of 2. Subsequently, a series of convolutions with a stride of 1 are applied, complemented by BN (batch normalization) and ReLU operations after each convolutional layer. These features from different scales are then concatenated together and used for the final detection.

3.3. Detection Head

We employ the single shot detector (SSD) [38] to achieve 3D object detection, which exhibits exceptional capabilities in terms of rapid detection speed and real-time performance. Moreover, it showcases its versatility by enabling various detection head configurations, allowing for adaptation to specific tasks.

3.4. Loss Function

We use the similar loss function as in [15]. We parameterize a 3D bounding box of the ground truth as $(x, y, z, w, l, h, \theta)$, where (x, y, z) are the center location, and (w, l, h) and θ represent the size and orientation angle of the bounding box, respectively. The regression residuals between the ground truth and the anchors are as follows

$$\Delta x = \frac{x^{gt} - x^a}{d^a}, \Delta y = \frac{y^{gt} - y^a}{d^a}, \Delta z = \frac{z^{gt} - z^a}{h^a} \tag{8}$$

$$\Delta w = \log \frac{w^{gt}}{w^a}, \ \Delta l = \log \frac{l^{gt}}{l^a}, \ \Delta h = \log \frac{h^{gt}}{h^a} \tag{9}$$

$$\Delta\theta = \sin(\theta^{gt} - \theta^a) \tag{10}$$

where x^{gt} denotes ground truth and x^a is the anchor box, with $d^a = \sqrt{(l^a)^2 + (w^a)^2}$. The location loss is denoted as

$$L_{loc} = \sum_{b \in (x,y,z,w,l,h,\theta)} SmoothL1(\Delta b) \tag{11}$$

For the classification loss, we use the focal loss [39]

$$L_{cls} = -a(1-p)^r \log p \tag{12}$$

where p represents the probability of an anchor. We use the settings of $r = 2$ and $a = 0.25$. Moreover, the heading direction loss L_{dir} is measured by a softmax classification loss on discretized directions. Adding up all the losses to the total loss of the whole network, the total loss function is defined as

$$L = \frac{1}{N_{pos}}(\beta_{loc} L_{loc} + \beta_{cls} L_{cls} + \beta_{dir} L_{dir}) \tag{13}$$

where N_{pos} is the number of positive anchors. β_{loc}, β_{cls} and β_{dir} are weighting coefficients for the localization loss, classification loss, and direction loss, respectively. We use the settings of $\beta_{loc} = 2$, $\beta_{cls} = 1$, $\beta_{dir} = 0.2$.

4. Experiment

4.1. Dataset

All experimental results are evaluated on KITTI's official evaluation test metrics, which include both bird's-eye view (BEV) and 3D. KITTI dataset are divided into easy, moderate, and hard categories according to object size, occlusion level, and truncation. The main metric of interest is average precision (AP) with intersection over union (IoU) thresholds, where a 3D bounding box overlap of 0.7 for cars is considered reasonable, whereas an overlap of 0.5 is required for pedestrians and cyclists. Notably, the official KITTI leaderboard is ranked based on performance on the moderate category, with performance being measured as the mean average precision (mAP) on KITTI validation. The experiments are conducted on the KITTI 3D object-detection benchmark dataset, which consists of 7481 training samples and 7518 test samples. The KITTI benchmark requires the detection of specific categories [40], including cars, pedestrians and cyclists. We also follow the widely used training–validation split, which comprises 3712 training samples and 3769 validation samples.

4.2. Implementation Details

Here, the detection range of the point cloud is [0, 70.4] m, [−40, 40] m, and [−3, 1] m along the X, Y, Z axes, and we set the X, Y, Z pillar resolution of (0.16, 0.16, 4) m. The maximum number of points per pillar is 32 and maximum number of pillars is 16,000. When performing feature encoding, the dimensions of the point-wise feature D and pillar-wise feature G after MLP are both 32. The proposed approach is based on the PyTorch framework, with all networks trained on the NVIDIA GTX3090 computing platform. The model is trained for 160 epochs with an initial learning rate of 2×10^{-3} and decrease by 0.8 every 15 epochs with Adam [41] optimizer.

4.3. Result

In this unit, we present the evaluation results of the proposed network using Ts-PFE and the SeNet backbone on the KITTI dataset. Our analysis comprises both quantitative and qualitative assessments, aimed at providing a comprehensive evaluation of the network's performance.

4.3.1. Quantitative Evaluation

All detection results are measured using the official KITTI evaluation detection metrics of both BEV and 3D detection. We use an IoU threshold of 0.7 for the car category and 0.5 for the pedestrian and cyclist categories to calculate an average precision.

Compared with other similar algorithms, the proposed algorithm achieves promising results. Tables 1 and 2 show the results of comparison under BEV and 3D settings. The proposed method achieves BEV mAP scores of 84.95%, 53.24%, and 64.86%, and 3D mAP scores of 76.09%, 47.31%, and 61.30% for the moderate level of car, pedestrian, and cyclist detection, respectively. In particular, the improvement over PointPillars is 1.1%, 3.78%, and 2.23% under a moderately difficult level of 3D detection for the car, pedestrian, and cyclist categories, respectively. This improvement can be attributed to the utilization of Ts-PFE, which effectively captures the relationship between individual pillars and the overall point cloud, as well as the use of SeNet backbone.

Table 1. Results on the KITTI test BEV detection benchmark (in %). The bold number indicates the best result in a table.

Model	Car			Pedestrian			Cyclist		
	Easy	Mod.	Hard	Easy	Mod.	Hard	Easy	Mod.	Hard
PointPillars [15]	88.35	86.10	79.83	58.66	50.23	47.19	79.14	62.25	56.00
SECOND [20]	88.07	79.37	77.95	55.10	46.27	44.76	73.67	56.04	48.78
VoxelNet [13]	89.35	79.26	77.39	46.13	40.74	38.11	66.70	54.76	50.55
TANet [42]	**91.58**	86.54	**81.19**	**60.58**	51.38	47.54	79.16	63.77	56.21
AVODFPN [43]	88.53	83.79	77.90	58.75	51.50	47.54	68.09	57.48	50.77
FPointNet [44]	88.70	84.00	75.33	58.09	50.22	47.20	75.38	61.96	54.68
HDNET [45]	89.14	**86.57**	78.32	N/A	N/A	N/A	N/A	N/A	N/A
PRGBNet [46]	91.39	85.73	80.68	38.07	29.32	26.94	73.09	57.59	51.78
Ours	89.62	84.95	79.53	59.50	**53.24**	**49.13**	**83.07**	**64.86**	**60.74**

Table 2. Results on the KITTI test 3D detection benchmark (in %). The bold number indicates the best result in a table.

Model	Car			Pedestrian			Cyclist		
	Easy	Mod.	Hard	Easy	Mod.	Hard	Easy	Mod.	Hard
PointPillars [15]	79.05	74.99	68.30	52.08	43.53	41.49	75.78	59.07	52.92
SECOND [20]	83.13	73.66	66.20	51.07	42.56	37.29	70.51	53.85	46.90
VoxelNet [13]	77.47	65.11	57.73	39.48	33.69	31.50	61.22	48.36	44.37
TANet [42]	**84.39**	75.94	68.82	53.72	44.34	40.49	75.70	59.44	52.53
AVODFPN [43]	81.94	71.88	66.38	50.80	42.81	40.88	64.00	52.18	46.61
FPointNet [44]	81.20	70.39	62.19	51.21	44.89	40.23	71.96	56.77	50.39
SATGCN [47]	83.20	76.04	71.17	44.63	37.37	34.92	75.24	**61.70**	55.32
PRGBNet [46]	83.99	73.49	68.56	34.77	26.40	24.03	67.05	52.15	46.78
Ours	83.52	**76.09**	**73.88**	**53.99**	**47.31**	**42.98**	**81.23**	61.30	**57.57**

4.3.2. Qualitative Evaluation

In order to qualitatively evaluate the performance of the network, we utilize it to predict the results on the validation set and visualize it under a 3D view, as shown in Figures 3 and 4.

Typical object-detection results are shown in Figure 3, indicating that the proposed model can accurately detect targets in the scene, even at longer distances or in instances of occlusion or overlap. Conversely, Figure 4 shows the individual failure examples. For example, it can be difficult to accurately identify objects that are not labeled in the data, such as trucks (as seen in Figure 4a). Additionally, certain instances produce missed detections (as shown in Figure 4b). The model may also mistakenly classify vertical objects such as poles, garbage cans, or trees as pedestrians or cyclists (as seen in Figure 4c). In some cases, there may be a slight over-segmentation, where the model detects one object as two (as seen in Figure 4d).

Figure 3. Qualitative analysis on the KITTI validation dataset. The results are only from LiDAR. For each sample, the upper part is the image and the lower part is a representative view of the corresponding point cloud with 3D bounding box.

Figure 4. Failure cases on the KITTI validation dataset. Same visualized setup from Figure 3 but focusing on several common failure examples. (**a**). It can be difficult to accurately identify objects that are not labeled in the data, such as trucks. (**b**). Certain instances produce missed detections. (**c**). The model may also mistakenly classify vertical objects such as poles, garbage cans, or trees as pedestrians or cyclists. (**d**). In some cases, there may be a slight over-segmentation, where the model detects one object as two.

4.4. Ablation Experiments

We conducted some ablation experiments to evaluate the importance and contribution of each component within the proposed network. These experiments are performed on the KITTL validation dataset, with PointPillars serving as the baseline.

4.4.1. SeNet Backbone Module Analysis

The results of "with SeNet" in Tables 3 and 4 demonstrate the effect of adding only the SeNet module. This module serves to focus on key features in the pseudo-images, which highlights important information while ignoring irrelevant information. The results in the tables indicate a discernible improvement in performance, with increases of 2.25% and 1.65% under moderate and hard levels of BEV detection for pedestrian category, as well as 0.86% and 3.19% improvement under moderate and hard levels of BEV detection for cyclist category. These findings firmly attest to the beneficial impact of the SeNet backbone module.

Table 3. Results on the KITTI test BEV detection benchmark (in %). The bold number indicates the best result in a table.

Model	Car			Pedestrian			Cyclist		
	Easy	Mod.	Hard	Easy	Mod.	Hard	Easy	Mod.	Hard
PointPillars [15]	88.35	86.10	**79.83**	58.66	50.23	49.19	79.14	62.25	56.00
withSeNet	**90.03**	86.35	79.83	58.21	52.48	48.84	79.37	63.11	59.19
withTsPFE	89.50	**86.43**	79.74	58.29	52.51	49.01	80.79	61.71	59.27
withboth	89.62	84.95	79.53	**59.50**	**53.24**	**49.13**	**83.07**	**64.86**	**60.74**

Table 4. Results on the KITTI test 3D detection benchmark (in %). The bold number indicates the best result in a table.

Model	Car			Pedestrian			Cyclist		
	Easy	Mod.	Hard	Easy	Mod.	Hard	Easy	Mod.	Hard
PointPillars [15]	79.05	74.99	68.30	52.08	43.53	41.49	75.78	59.07	52.92
withSeNet	**83.85**	**76.09**	69.06	52.02	46.97	42.54	76.28	59.14	55.51
withTsPFE	83.36	75.74	68.90	52.64	46.59	**42.99**	75.98	58.28	54.16
withboth	83.52	**76.09**	**73.88**	**53.99**	**47.31**	42.98	**81.23**	**61.30**	**57.57**

4.4.2. Ts-PFE Module Analysis

The results of "with Ts-PFE" in Tables 3 and 4 show the effect of integrating only the Ts-PFE module in the main architecture. By considering both inter- and intra-relational features among and in the pillars, the Ts-PFE module achieves noteworthy performance gains across various metrics. Specifically, the improvement is 4.31% under the easy level of 3D detection for the car category, and 3.06% and 1.5% under the moderate and hard level of 3D detection for pedestrian category, respectively. For the cyclist category, we observe an improvement of 1.24% under the hard level of 3D detection. These findings demonstrate the crucial role played by Ts-PFE in enhancing the network's performance.

4.4.3. Ts-PFE and SeNet Backbone Analysis

According to the findings presented in Tables 3 and 4 of "with both", the impact of incorporating both SeNet backbone and Ts-PFE is examined. Remarkably, the results reveal that the performance of the models is consistently improved with the addition of both modules, surpassing the baseline and outperforming the networks with individual module.

The preceding discussion presents ablation experiments conducted at the data level. Furthermore, it is feasible to evaluate the network's performance pre- and post-improvement through visualization, as shown in Figure 5. We utilize the accompanying figures to compare the visualization outputs, with and without the integration of the improved module. The baseline network fails to account for the relationship between pillars and the overall point cloud, which can lead to under-segmentation in case of overlap or occlusion and missed detection, as shown in the left part of Figure 5a–d. In contrast, due to the application of Ts-PFE in the proposed network, the relational features will be taken into account to reduce the occurrence of error cases. As observed in the right portion of Figure 5a–d, the improved network effectively detects occluded objects.

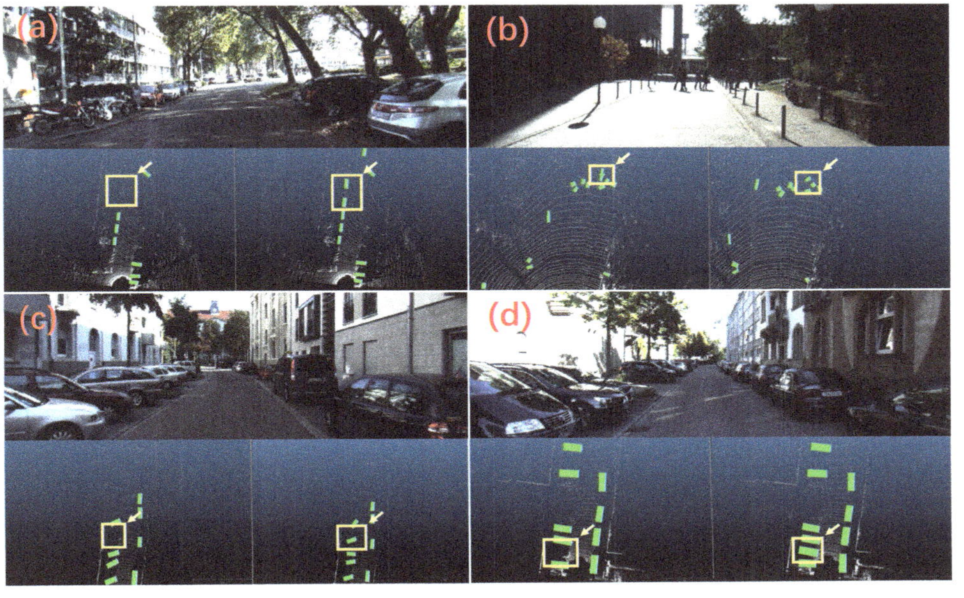

Figure 5. Qualitative improvement of the model with Ts-PFE module (without SeNet) over the baseline PointPillars for 3D detection. The results are shown under BEV in different scenes, with boxes and arrows indicating key differences for a better comparison. The left part of each scene is the result of the baseline, and the right part is the result of the proposed approach. (**a**,**c**,**d**) show improvements for the missed detection problem, and (**b**) shows improvement for the under-segmentation problem.

5. Conclusions

In conclusion, we have proposed an improved pillar-based 3D object-detection network that incorporates a two-stage pillar feature-encoding (Ts-PFE) module and a backbone with SeNet. The utilization of the Ts-PFE module as the feature-encoding network in pillar-based 3D object-detection network greatly enhances the network's capability to handle detection tasks under occlusion or overlapping scenes, while the backbone with SeNet makes the network more focused on key features. The experiments on the KITTI dataset show that the proposed method achieves superior detection accuracy compared to existing methods. The improvement of AP for car, pedestrian, and cyclist 3D detection are 1.1%, 3.78%, and 2.23% over the baseline. More importantly, the ablation experiments also show the significance of the proposed Ts-PFE module for performance improvement. Additionally, the results of qualitative evaluation show that the proposed approach improve the under-segmentation and missed detection problems in occluded or overlapping scenes. These findings demonstrate the potential of the proposed approach in improving pillar-based 3D object detection. However, it should be noted that the proposed method lacks the evaluation in different levels of occlusion and overlap; further validation experiments are needed in various occlusion scenarios. Additionally, our evaluation has been limited to a single dataset, and future work should involve testing on a broader range of datasets to assess the algorithm's generalizability.

Author Contributions: Conceptualization, H.X. and B.Y.; methodology, H.X. and B.Y.; software, H.X. and X.D.; validation, H.X. and W.W.; formal analysis, W.W.; investigation, H.X. and X.D.; data curation, H.X. and W.W.; writing—original draft preparation, H.X.; writing—review and editing, B.Y.; visualization, H.X.; supervision, B.Y. and X.D.; project administration, X.D. and H.Z.; funding acquisition, X.D. and B.Y. All authors have read and agreed to the published version of the manuscript.

Funding: This work was partially supported by the National Key Research and Development Project Monitoring and Prevention of Major Natural Disasters Special Program (Grant No. 2020YFC1512202); the National Natural Science Foundation of China (Grant No. 62273342); and the Youth Innovation Promotion Association of the Chinese Academy of Sciences (Grant No. Y2021115).

Institutional Review Board Statement: Not applicable.

Informed Consent Statement: Not applicable.

Data Availability Statement: All data generated or analyzed during this study are included in this published article.

Conflicts of Interest: The authors declare no conflict of interest.

References

1. Li, F.; Jin, W.; Fan, C.; Zou, L.; Chen, Q.; Li, X.; Jiang, H.; Liu, Y. PSANet: Pyramid splitting and aggregation network for 3D object detection in point cloud. *Sensors* **2020**, *21*, 136. [CrossRef]
2. Bai, Z.; Wu, G.; Barth, M.J.; Liu, Y.; Sisbot, E.A.; Oguchi, K. Pillargrid: Deep learning-based cooperative perception for 3d object detection from onboard-roadside lidar. In Proceedings of the 2022 IEEE 25th International Conference on Intelligent Transportation Systems (ITSC), Macau, China, 8–12 October 2022; pp. 1743–1749.
3. Wang, B.; Zhu, M.; Lu, Y.; Wang, J.; Gao, W.; Wei, H. Real-time 3D object detection from point cloud through foreground segmentation. *IEEE Access* **2021**, *9*, 84886–84898. [CrossRef]
4. He, C.; Zeng, H.; Huang, J.; Hua, X.S.; Zhang, L. Structure aware single-stage 3d object detection from point cloud. In Proceedings of the IEEE/CVF Conference on Computer Vision and Pattern Recognition, Seattle, WA, USA, 13–19 June 2020; pp. 11873–11882.
5. Wang, Q.; Chen, J.; Deng, J.; Zhang, X. 3D-CenterNet: 3D object detection network for point clouds with center estimation priority. *Pattern Recognit.* **2021**, *115*, 107884. [CrossRef]
6. Bello, S.A.; Yu, S.; Wang, C.; Adam, J.M.; Li, J. Deep learning on 3D point clouds. *Remote Sens.* **2020**, *12*, 1729. [CrossRef]
7. Yang, B.; Luo, W.; Urtasun, R. Pixor: Real-time 3d object detection from point clouds. In Proceedings of the IEEE Conference on Computer Vision and Pattern Recognition, Salt Lake City, UT, USA, 18–23 June 2018; pp. 7652–7660.
8. Alaba, S.Y.; Ball, J.E. A survey on deep-learning-based lidar 3d object detection for autonomous driving. *Sensors* **2022**, *22*, 9577. [CrossRef] [PubMed]
9. Liang, Z.; Zhang, Z.; Zhang, M.; Zhao, X.; Pu, S. Rangeioudet: Range image based real-time 3d object detector optimized by intersection over union. In Proceedings of the IEEE/CVF Conference on Computer Vision and Pattern Recognition, Nashville, TN, USA, 20–25 June 2021; pp. 7140–7149.
10. Fan, L.; Xiong, X.; Wang, F.; Wang, N.; Zhang, Z. Rangedet: In defense of range view for lidar-based 3d object detection. In Proceedings of the IEEE/CVF International Conference on Computer Vision, Montreal, QC, Canada, 10–17 October 2021; pp. 2918–2927.
11. Qi, C.R.; Su, H.; Mo, K.; Guibas, L.J. Pointnet: Deep learning on point sets for 3d classification and segmentation. In Proceedings of the IEEE Conference on Computer Vision and Pattern Recognition, Honolulu, HI, USA, 21–26 July 2017; pp. 652–660.
12. Qi, C.R.; Yi, L.; Su, H.; Guibas, L.J. Pointnet++: Deep hierarchical feature learning on point sets in a metric space. *Adv. Neural Inf. Process. Syst.* **2017**, *30*, 1–14.
13. Zhou, Y.; Tuzel, O. Voxelnet: End-to-end learning for point cloud based 3d object detection. In Proceedings of the IEEE Conference on Computer Vision and Pattern Recognition, Salt Lake City, UT, USA, 18–23 June 2018; pp. 4490–4499.
14. Xie, J.; Zheng, Z.; Gao, R.; Wang, W.; Zhu, S.C.; Wu, Y.N. Generative VoxelNet: Learning energy-based models for 3D shape synthesis and analysis. *IEEE Trans. Pattern Anal. Mach. Intell.* **2020**, *44*, 2468–2484. [CrossRef] [PubMed]
15. Lang, A.H.; Vora, S.; Caesar, H.; Zhou, L.; Yang, J.; Beijbom, O. Pointpillars: Fast encoders for object detection from point clouds. In Proceedings of the IEEE/CVF Conference on Computer Vision and Pattern Recognition, Long Beach, CA, USA, 15–20 June 2019; pp. 12697–12705.
16. Liang, N.; Sun, S.; Zhou, L.; Zhao, N.; Taha, M.F.; He, Y.; Qiu, Z. High-throughput instance segmentation and shape restoration of overlapping vegetable seeds based on sim2real method. *Measurement* **2023**, *207*, 112414. [CrossRef]
17. Wang, Y.; Jiang, Z.; Li, Y.; Hwang, J.N.; Xing, G.; Liu, H. RODNet: A real-time radar object detection network cross-supervised by camera-radar fused object 3D localization. *IEEE J. Sel. Top. Signal Process.* **2021**, *15*, 954–967. [CrossRef]
18. Qi, C.R.; Liu, W.; Wu, C.; Su, H.; Guibas, L.J. Frustum pointnets for 3d object detection from rgb-d data. In Proceedings of the IEEE Conference on Computer Vision and Pattern Recognition, Salt Lake City, UT, USA, 18–23 June 2018; pp. 918–927.
19. Fernandes, D.; Silva, A.; Névoa, R.; Simões, C.; Gonzalez, D.; Guevara, M.; Novais, P.; Monteiro, J.; Melo-Pinto, P. Point-cloud based 3D object detection and classification methods for self-driving applications: A survey and taxonomy. *Inf. Fusion* **2021**, *68*, 161–191. [CrossRef]
20. Yan, Y.; Mao, Y.; Li, B. Second: Sparsely embedded convolutional detection. *Sensors* **2018**, *18*, 3337. [CrossRef] [PubMed]
21. Yin, T.; Zhou, X.; Krahenbuhl, P. Center-based 3d object detection and tracking. In Proceedings of the IEEE/CVF Conference on Computer Vision and Pattern Recognition, Nashville, TN, USA, 20–25 June 2021; pp. 11784–11793.

22. Li, J.; Chen, B.M.; Lee, G.H. So-net: Self-organizing network for point cloud analysis. In Proceedings of the IEEE Conference on Computer Vision and Pattern Recognition, Salt Lake City, UT, USA, 18–23 June 2018; pp. 9397–9406.
23. Wang, S.; Lu, K.; Xue, J.; Zhao, Y. DA-Net: Density-Aware 3D Object Detection Network for Point Clouds. *IEEE Trans. Multimed.* **2023**. [CrossRef]
24. Li, C.; Gao, F.; Han, X.; Zhang, B. A New Density-Based Clustering Method Considering Spatial Distribution of Lidar Point Cloud for Object Detection of Autonomous Driving. *Electronics* **2021**, *10*, 2005. [CrossRef]
25. Wang, Z.; Fu, H.; Wang, L.; Xiao, L.; Dai, B. SCNet: Subdivision coding network for object detection based on 3D point cloud. *IEEE Access* **2019**, *7*, 120449–120462. [CrossRef]
26. Bhattacharyya, P.; Huang, C.; Czarnecki, K. Sa-det3d: Self-attention based context-aware 3d object detection. In Proceedings of the IEEE/CVF International Conference on Computer Vision, Nashville, TN, USA, 20–25 June 2021; pp. 3022–3031.
27. Shi, S.; Guo, C.; Jiang, L.; Wang, Z.; Shi, J.; Wang, X.; Li, H. Pv-rcnn: Point-voxel feature set abstraction for 3d object detection. In Proceedings of the IEEE/CVF Conference on Computer Vision and Pattern Recognition, Seattle, WA, USA, 13–19 June 2020; pp. 10529–10538.
28. Guo, M.H.; Xu, T.X.; Liu, J.J.; Liu, Z.N.; Jiang, P.T.; Mu, T.J.; Zhang, S.H.; Martin, R.R.; Cheng, M.M.; Hu, S.M. Attention mechanisms in computer vision: A survey. *Comput. Vis. Media* **2022**, *8*, 331–368. [CrossRef]
29. Wu, C.; Zhang, F.; Xia, J.; Xu, Y.; Li, G.; Xie, J.; Du, Z.; Liu, R. Building damage detection using U-Net with attention mechanism from pre-and post-disaster remote sensing datasets. *Remote Sens.* **2021**, *13*, 905. [CrossRef]
30. Zhai, Z.; Wang, Q.; Pan, Z.; Gao, Z.; Hu, W. Muti-Frame Point Cloud Feature Fusion Based on Attention Mechanisms for 3D Object Detection. *Sensors* **2022**, *22*, 7473. [CrossRef]
31. Wang, G.; Zhai, Q.; Liu, H. Cross self-attention network for 3D point cloud. *Knowl.-Based Syst.* **2022**, *247*, 108769. [CrossRef]
32. Han, J.; Zeng, L.; Du, L.; Ye, X.; Ding, W.; Feng, J. Modify Self-Attention via Skeleton Decomposition for Effective Point Cloud Transformer. In Proceedings of the AAAI Conference on Artificial Intelligence, Held Virtually, 22 February–1 March 2022; pp. 808–816.
33. Zhao, X.; Liu, Z.; Hu, R.; Huang, K. 3D object detection using scale invariant and feature reweighting networks. In Proceedings of the AAAI Conference on Artificial Intelligence, Honolulu, HI, USA, 27 January–1 February 2019; pp. 9267–9274.
34. Qiu, S.; Wu, Y.; Anwar, S.; Li, C. Investigating attention mechanism in 3d point cloud object detection. In Proceedings of the 2021 International Conference on 3D Vision (3DV), London, UK, 1–3 December 2021; pp. 403–412.
35. Li, X.; Liang, B.; Huang, J.; Peng, Y.; Yan, Y.; Li, J.; Shang, W.; Wei, W. Pillar-Based 3D Object Detection from Point Cloud with Multiattention Mechanism. *Wirel. Commun. Mob. Comput.* **2023**, *2023*, 5603123. [CrossRef]
36. Chen, S.; Miao, Z.; Chen, H.; Mukherjee, M.; Zhang, Y. Point-attention Net: A graph attention convolution network for point cloud segmentation. *Appl. Intell.* **2022**, *53*, 11344–11356. [CrossRef]
37. Hu, J.; Shen, L.; Sun, G. Squeeze-and-excitation networks. In Proceedings of the IEEE Conference on Computer Vision and Pattern Recognition, Salt Lake City, UT, USA, 18–23 June 2018; pp. 7132–7141.
38. Liu, W.; Anguelov, D.; Erhan, D.; Szegedy, C.; Reed, S.; Fu, C.Y.; Berg, A.C. Ssd: Single shot multibox detector. In Proceedings of the Computer Vision–ECCV 2016: 14th European Conference, Amsterdam, The Netherlands, 11–14 October 2016; pp. 21–37.
39. Lin, T.Y.; Goyal, P.; Girshick, R.; He, K.; Dollár, P. Focal loss for dense object detection. In Proceedings of the IEEE International Conference on Computer Vision, Venice, Italy, 22–29 October 2017; pp. 2980–2988.
40. Geiger, A.; Lenz, P.; Urtasun, R. Are we ready for autonomous driving? The kitti vision benchmark suite. In Proceedings of the 2012 IEEE Conference on Computer Vision and Pattern Recognition, Providence, RI, USA, 16–21 June 2012; pp. 3354–3361.
41. Kingma, D.P.; Ba, J. Adam: A method for stochastic optimization. *arXiv* **2014**, arXiv:1412.6980.
42. Liu, Z.; Zhao, X.; Huang, T.; Hu, R.; Zhou, Y.; Bai, X. Tanet: Robust 3d object detection from point clouds with triple attention. In Proceedings of the AAAI Conference on Artificial Intelligence, New York, NY, USA, 7–12 February 2020; pp. 11677–11684.
43. Brekke, A.; Vatsendvik, F.; Lindseth, F. Multimodal 3d object detection from simulated pretraining. In Proceedings of the Nordic Artificial Intelligence Research and Development: Third Symposium of the Norwegian AI Society, Trondheim, Norway, 27–28 May 2019; pp. 102–113.
44. Cao, P.; Chen, H.; Zhang, Y.; Wang, G. Multi-view frustum pointnet for object detection in autonomous driving. In Proceedings of the 2019 IEEE International Conference on Image Processing (ICIP), Taipei, Taiwan, 22–25 September 2019; pp. 3896–3899.
45. Yang, B.; Liang, M.; Urtasun, R. Hdnet: Exploiting hd maps for 3d object detection. In Proceedings of the Conference on Robot Learning, Zürich, Switzerland, 29–31 October 2018; pp. 146–155.
46. Desheng, X.; Youchun, X.; Feng, L; Shiju, P. Real-time Detection of 3D Objects Based on Multi-Sensor Information Fusion. *Autom. Eng.* **2022**, *44*, 3.
47. Wang, L.; Song, Z.; Zhang, X.; Wang, C.; Zhang, G.; Zhu, L.; Li, J.; Liu, H. SAT-GCN: Self-attention graph convolutional network-based 3D object detection for autonomous driving. *Knowl.-Based Syst.* **2023**, *259*, 110080. [CrossRef]

World Electric Vehicle Journal

MDPI

Article

Driving Decisions for Autonomous Vehicles in Intersection Environments: Deep Reinforcement Learning Approaches with Risk Assessment

Wangpengfei Yu, Yubin Qian *, Jiejie Xu, Hongtao Sun and Junxiang Wang

School of Mechanical and Automotive Engineering, Shanghai University of Engineering Science, Shanghai 201620, China
* Correspondence: qianyb@sues.edu.cn

Abstract: Intersection scenarios are one of the most complex and high-risk traffic scenarios. Therefore, it is important to propose a vehicle driving decision algorithm for intersection scenarios. Most of the related studies have focused on considering explicit collision risks while lacking consideration for potential driving risks. Therefore, this study proposes a deep-reinforcement-learning-based driving decision algorithm to address these problems. In this study, a non-deterministic vehicle driving risk assessment method is proposed for intersection scenarios and introduced into a learning-based intelligent driving decision algorithm. In addition, this study proposes an attention network based on state information. In this study, a typical intersection scenario was constructed using simulation software, and experiments were conducted. The experimental results show that the algorithm proposed in this paper can effectively derive a driving strategy with both driving efficiency and driving safety in the intersection driving scenario. It is also demonstrated that the attentional neural network designed in this study helps intelligent vehicles to perceive the surrounding environment more accurately, improves the performance of intelligent vehicles, as well as accelerates the convergence speed.

Keywords: intelligent vehicle; decision-making; driving safety; deep reinforcement learning; risk assessment

Citation: Yu, W.; Qian, Y.; Xu, J.; Sun, H.; Wang, J. Driving Decisions for Autonomous Vehicles in Intersection Environments: Deep Reinforcement Learning Approaches with Risk Assessment. *World Electr. Veh. J.* **2023**, *14*, 79. https://doi.org/10.3390/wevj14040079

Academic Editors: Biao Yu, Linglong Lin and Jiajia Chen

Received: 4 March 2023
Revised: 16 March 2023
Accepted: 20 March 2023
Published: 23 March 2023

1. Introduction

Due to the improvement of people's living standards and economic development, as well as growing urbanization and transportation needs, the overall trend of car ownership in various countries around the world is rising. According to the International Energy Agency (IEA) [1], global car ownership has grown from about 560 million in 2000 to about 1.32 billion in 2020, with China's car ownership already exceeding 300 million. With the growth of car ownership, society faces many challenges and problems. One of the major issues is the increasing frequency of traffic accidents. According to data released by the World Health Organization's Global Status Report on Road Safety 2018, more than 1.3 million people worldwide die each year due to traffic accidents, and traffic accidents are the leading cause of death among children and young people aged 5–29 years of age [2], and this number is increasing year by year. Among the various driving scenarios, intersections are one of the most frequent scenarios for traffic accidents due to the complex traffic environment. According to the German In-Depth Accident Study (GIDAS) and other organizations, 40% of road injury accidents occur at intersections [3]. Therefore, in this context, it is important to propose a vehicle driving decision algorithm applicable to intersection scenarios. The problem of improving traffic safety is complicated by the fact that drivers have different physical and psychological states that make them perceive and react to risks differently during the driving process. Automated driving essentially changes the closed-loop human–vehicle–road system, reducing or minimizing the driver's influence in the system and making it more efficient and safer. Therefore, in the current context, the method of replacing the driver with an autonomous driving decision algorithm has become a hot research topic.

There have been many studies on intersection driving problems [4]; for example, Li et al. [5] proposed a deep reinforcement learning-based driving decision framework to build an end-to-end decision framework by convolutional neural networks to derive driving strategies at intersections without traffic signals using traffic images as input. Seong et al. [6] proposed an attention-based deep reinforcement learning for driving decision method that uses local vehicle perception data as input to derive driving strategies at intersections without traffic signals. Many driving decision methods have attempted to use various vehicle perception data as input to derive driving strategies. However, driving strategies using vehicle perception data as input are difficult to apply to natural or more complex driving environments because vehicle perception data are affected by changes in the driving environment. Many researchers use vehicle-to-vehicle (V2V) and vehicle-to-infrastructure (V2I) technologies to derive driving strategies by sharing information about the state and environment of individual vehicles at intersections through communication [7]. V2V and V2I technologies can transmit traffic information from intersections to intelligent vehicles regardless of weather conditions. The research in this study assumes that each vehicle's status and environmental information at the intersection is transmitted to the intelligent vehicle through communication sharing to carry out the research work.

As shown in Figure 1, current intelligent driving decision methods can be categorized into three types according to the technical approach: rule-based decision methods, risk assessment-based decision methods, and learning-based decision methods. Among them, the rule-based decision method, as the most traditional and common method, can meet most of the regular driving scenarios but lacks flexibility and applicability to deal with unexpected situations. Therefore, based on this, researchers have proposed an intelligent driving decision method based on risk assessment. In recent years, with the rapid development of machine learning technology, researchers have begun to explore learning-based decision methods to solve the problem of intelligent driving decision-making. However, learning-based decision methods lack the consideration of uncertainty problems such as traffic rules, driving risks, etc. Therefore, this paper integrates risk-assessment-based and learning-based driving decision methods and proposes an intelligent driving decision algorithm based on deep reinforcement learning, aiming to derive an autonomous driving decision strategy with low expected risk and high driving efficiency to meet the driving challenges in intersection scenarios.

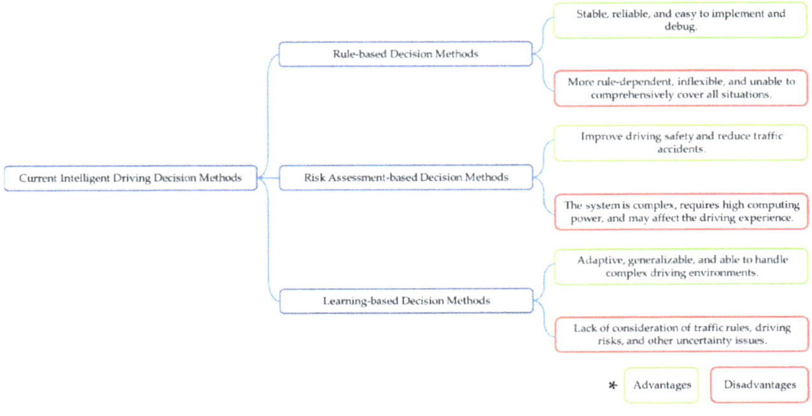

Figure 1. Advantages and disadvantages of different driving decision methods (* The green boxes represent advantages, and the red boxes represent disadvantages).

1.1. Rule-Based Decision Methods

The rule-based approach is the most traditional and common approach in driving decision-making. The rule-based decision-making approach builds a rule base through regular driving habits and traffic regulations. It develops corresponding driving strategies based on different driving situations. Furda et al. [8] proposed a method based on Petri nets and multi-criteria decision-making to solve the problem of real-time decision-making for autonomous driving. The proposed method obtains a set of feasible alternative driving decisions through Petri nets and then uses a multi-criteria decision-making method to select the best automatic driving decision from them. Chong et al. [9] proposed a rule-based neural network model to simulate the driver's driving behavior. The proposed method is based on a fuzzy rule-based neural network model to obtain rules for driver behavior decisions from the driver's vehicle trajectory. Li et al. [10] developed a decision model for automatic driving behavior in emergency situations based on the T-S fuzzy neural network. The rule-based driving decision method, although able to satisfy regular driving situations, cannot cope with unexpected unconsidered situations. Therefore, the rule-based driving decision method lacks a certain degree of flexibility and adaptability [11].

1.2. Risk-Assessment-Based Decision Methods

Decision-making methods based on risk assessment are generally used to determine driving strategies by assessing the risk profile of the current driving state. Currently, risk assessment is divided into two main categories: deterministic risk assessment and uncertainty risk assessment.

Deterministic risk assessment is usually based on multiple parameters in order to derive from obtaining a numerical value or different risk level regions to represent the risk magnitude. Hillenbrand J et al. [12] used TTC to assess the risk magnitude to make driving decisions by tracking information about vehicles in the area associated with the main vehicle lane change and calculating the time to collision (TTC) between them. Glaser et al. [13], based on the driving environment using TTC and headway time distance (THW) for risk assessment of the possible driving behavior of the primary vehicle, ranked the vehicles according to the risk assessment results to make the best driving strategy. Lee H et al. [14] similarly used the relative speed and position information between vehicles to calculate the risk coefficients, and based on this, the acceleration and speed change characteristics of the surrounding vehicles were taken into account in the risk assessment to help vehicles to make reasonable lane change decisions. Moreover, there are also many research scholars who establish risk assessment methods based on different risk factors from different perspectives [15–17].

Uncertainty risk assessment is usually based on mathematical probability models to represent the degree of risk. Many related studies of risk assessment methods assume that the state of the surroundings of the autonomous vehicle is constant, but the surrounding vehicle's driving state in a realistic environment is uncertain. It is this uncertainty that makes probabilistic model-based risk assessment methods important. Schubert R et al. [18] proposed a driving risk assessment method that considers sensor uncertainty, evaluates the risk of vehicle lane change behavior based on the Bayesian networks, and provides the driver with suggestions for lane change operations, thus improving the safety and reliability of vehicle driving. Kim B et al. [19] proposed a probabilistic threat assessment method based on road traffic information to predict and avoid possible collisions in multi-vehicle traffic. The proposed collision occurrence probability algorithm follows the basic idea of particle filtering and implements the numerical calculation of collision probability. Noh S et al. [20] proposed an automated driving decision framework for highway environments. The framework robustly assesses the potential risk of a collision for a given highway condition and determines the appropriate driving strategy for that situation. The proposed risk assessment method takes into account the uncertainty of the input data and infers the potential crash risk of the driving environment based on the Bayesian networks. In addition to this, there are probabilistic model-based risk assessment methods using Markov models, Gaussian processes, and deep learning [21,22].

1.3. Learning-Based Decision Methods

With the rapid development of machine learning in recent years, researchers have started to experiment with learning-based approaches to solve the problem of intelligent vehicle driving decisions. These approaches are further divided into imitation learning approaches and reinforcement learning approaches depending on the learning objectives. In the area of imitation-based learning research, Xu et al. [23] conducted a large-scale study on driving behavior learning and tested it using the BDDV dataset with image segmentation as an additional task of the network. The study achieved good accuracy in action mapping. However, in real driving situations, there are multiple solutions. For example, when crossing an intersection, there may be multiple ways of driving a vehicle. Codevilla et al. [24] proposed a conditional imitation learning approach, which inputs higher-level decision commands as conditions into the imitation learning framework to derive driving strategies for autonomous vehicles. It has been tested in both simulation and realistic environments with good results. However, imitation learning-based approaches usually rely on a large amount of data for training, which is costly to collect on the one hand. On the other hand, the collected data are usually related to the subjective judgment of the driver, and therefore, optimal driving decisions are not always obtained.

In recent years, reinforcement learning has been increasingly used for autonomous driving behavior decision-making. Mircheveska et al. [25] proposed a reinforcement learning method applying random forests for autonomous driving in highway scenarios. Mukadam et al. [26] proposed a deep Q-learning-based method to solve the problem of automatic vehicle lane changing in a multi-lane, multi-vehicle environment. The proposed method takes the vehicle state and the surrounding environment state as network inputs and the output is a score of five driving behaviors. Finally, the performance and generalization capability of the algorithm were verified in a SUMO simulation environment. Hu et al. [27] further proposed a training method based on a multi-intelligent body framework, aiming to obtain a more diverse training environment and to train and validate it in a lane merging task. The method extends the driving environment to more complex traffic scenarios by introducing multiple intelligences as agents of other vehicles, thus improving the diversity and realism of the training data. Bouton et al. [28] proposed a reinforcement learning method based on probabilistic guarantees to constrain the action selection of intelligence by using a desired probability specification represented by linear temporal logic (LTL) to achieve a more realistic training environment at autonomous driving at intersections involving multiple participants. Most of the current reinforcement learning is still explored in more idealized lane scenarios, lacking consideration of traffic rules, driving risks, and other uncertainty issues. Further research is still needed on how to introduce reinforcement learning into more complex traffic environments and realistic driving scenarios.

1.4. Contribution

This paper aims to design a risk-aware intelligent driving decision algorithm based on deep reinforcement learning methods to derive an autonomous driving decision strategy with low expected risk and high driving efficiency in intersection scenarios. To this end, this paper integrates risk-based assessment and learning-based driving decision methods. It introduces a non-deterministic risk assessment method for vehicle driving in the study of learning-based driving decision methods to address the lack of consideration of uncertainty issues such as driving risk in learning-based driving decision methods. The research in this paper constructs an intersection driving scenario in RoadRunner simulation software and conducts experiments to verify the effectiveness of the proposed method. The main work of this research paper is as follows.

1. Using a driving simulation platform, a typical intersection driving scenario is established and used for training and testing of reinforcement learning for autonomous driving;

2. Based on the Bayesian probability theory, this paper proposes a vehicle driving risk assessment method for intersection driving scenarios and incorporates the method into a learning-based driving decision method;
3. To improve driving safety, this paper proposes a driving policy learning algorithm based on state-action value distributed deep q-networks and introduces an attention network based on state information;
4. This paper is divided into six main sections. Section 1 is an introduction that describes the research background of this paper and the current state of research on rule-based, risk assessment, and learning-driving decision methods. Section 2 introduces the intersection driving scenario constructed by this paper's research using a driving simulation platform. Section 3 introduces the vehicle driving risk assessment method based on the Bayesian probability theory for the intersection scenario proposed in this paper. Section 4 presents a driving strategy learning algorithm based on state-action value distributed deep q-networks proposed in this paper, in which the relevant state space, action space, reward function, and neural network are systematically analyzed and designed. Section 5 is the experimental design and result analysis of this paper. Section 6 is the summary and outlook of this paper. This chapter summarizes the research results and shortcomings and looks toward future work.

2. Simulation Environment

In this paper, we study vehicle driving strategies in intersection scenarios through deep reinforcement learning methods. As a self-learning method, reinforcement learning requires an intelligent body to interact continuously with the environment in order to continuously learn strategies and achieve specific goals. In this paper, we simulated the vehicle driving environment using the RoadRunner simulation software in MathWorks. This software can simulate various real-world traffic scenarios and also supports developers in building custom driving scenarios and interaction behaviors by writing scripts to help developers test and optimize autonomous driving algorithms and promote the progress of autonomous driving technologies. As one of the most complex traffic environments and accident-prone traffic scenarios, intersections also present complex and diverse causes of traffic accidents, but according to relevant studies, driver error is one of the most important causes. According to statistics, about 96% of traffic accidents are caused by driver's misoperation, such as misunderstanding of traffic signs, negligence, and violation of traffic rules [29]. Therefore, considering the impact of driver misoperation with other vehicles in the scenario on the intelligent vehicle, the intersection scenario under study is assumed to be an unprotected intersection in this paper.

2.1. Experimental Scene Construction

The complexity of an intersection scenario comes from its dense traffic flow and complex traffic conditions. Therefore, the intersection simulation design requires careful consideration of various factors. In order to improve the simulation effect of the experiment and enhance the generalization ability of the intelligent driving strategy obtained from the training, the intersection scenario designed in this paper is a cross-shaped intersection connecting eight roads from different driving directions, and each road contains two parallel sub-lanes in the same direction, as shown in Figure 2. As shown in Figure 2, this paper sets the starting and ending points of intelligent vehicles according to three path scenarios of the left turn, straight ahead, and right turn. Other vehicles around are randomly generated in the lane at a random speed and pass through the intersection. The intelligent vehicle needs to learn to pass the intersection safely and efficiently in the designed intersection scenario.

This study proposes a hierarchical framework for vehicle motion behavior control for the intersection scenario. The intelligent drive model (IDM) controls the surrounding vehicles in the intersection scenario. However, this model only considers the interaction of vehicles within the same lane. Therefore, this paper studies the need to pay special attention to avoid lateral collisions with surrounding vehicles when constructing intersection scenarios. We used the following simplification strategy in building the scenario to address this issue: by using a fixed speed model, each surrounding vehicle predicts the position of its neighboring vehicle in the next 3 s, and when a risk of collision with the neighboring vehicle is detected, it decides whether to yield or not based on the priority of the road and the braking performance until the risk of collision is eliminated. Yielding is essential to traffic law in that it helps ensure safe traffic flow and prevents accidents. The lower-level model also designs speed and steering controllers that enable surrounding vehicles to travel according to the target speed given by the upper-level control model and the target lane.

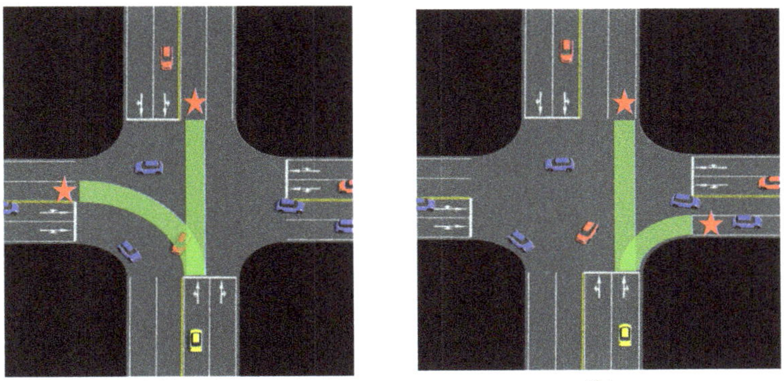

(a) (b)

Figure 2. Intersection simulation environment: (**a**) left turn and straight ahead; (**b**) right turn and straight ahead.

A hierarchical framework is used for the control of the motion behavior of intelligent vehicles, with the upper layer being the driving policy managed by the deep reinforcement learning model and the lower layer being responsible for controlling the controllers for longitudinal and lateral motion, respectively, which together with the surrounding vehicles form the entire control system.

2.2. Vehicle Behavior Control

In this study, the driving motions of the intelligent vehicle and its surrounding vehicles are managed by a hierarchical control framework. The IDM model in the upper layer of the surrounding vehicle framework manages the longitudinal driving behavior.

In the environment constructed in this paper, the surrounding vehicle uses a simple model that matches the actual driving behavior, which influences the acceleration and steering of the vehicle. Among them, the IDM model, as a typical representative of the microscopic model, is used to achieve collision avoidance following driving. Therefore, the longitudinal behavior of the surrounding vehicles is controlled by the IDM model, whose longitudinal acceleration is given by the IDM model [30].

$$\dot{v} = a[1 - (\frac{v}{v_0})^{\delta} - (\frac{d^*}{d})^2] \quad (1)$$

In the equation, v and $\dot{v}$ represent the travel speed and acceleration of the surrounding vehicle, respectively; a is the maximum acceleration of the surrounding vehicle; d denotes the relative distance between the surrounding vehicle and the vehicle in front of it; δ is the acceleration index; v_0 is the desired target speed; and d^* is the desired target relative distance. Where d^* is influenced by the vehicle ahead.

$$d^* = d_0 + Tv + \frac{v\Delta v}{2\sqrt{ab}} \quad (2)$$

where d_0 represents the safe distance from the vehicle ahead; T represents the driver's reaction time; Δv represents the relative speed difference with the vehicle ahead; and b represents the maximum deceleration rate of the vehicle.

In this paper, the relative velocities and distances in the IDM model are predefined to induce velocities and accelerations within each time step. The default parameters are set as follows: maximum acceleration $a = 6\ \text{m/s}^2$; maximum deceleration $b = -5\ \text{m/s}^2$; acceleration index $\delta = 4$; relative safety distance $d_0 = 10$ m from the vehicle in front; and driver's reaction time $T = 1.5$ s.

In the environment constructed in this study, the intelligent driving vehicle motion control system consists of a decision layer and a control layer. The decision layer uses deep reinforcement learning algorithms to generate driving decisions based on the input environmental information to determine the driving behavior of the intelligent vehicle (including accelerating, decelerating, and maintaining the current state). The control layer uses the relevant speed controller to realize the intelligent vehicle driving according to the target speed given by the decision layer. The following mathematical expression can represent its relevant mathematical model.

$$v_{target} = v + i\Delta v \tag{3}$$

where v_{target} represents the target driving speed that the intelligent vehicle needs to achieve through acceleration or braking behavior; v represents the current driving speed of the intelligent vehicle; i represents the speed variation coefficient, $i = 1$ when the decision layer determines the driving behavior as acceleration, $i = -1$ when deceleration, otherwise, $i = 0$; Δv represents the amount of speed variation under intelligent vehicle control, which is related to the policy frequency of the algorithm to influence the accuracy of intelligent vehicle control.

The above is the upper framework of the motion behavior control framework for intelligent vehicles and surrounding vehicles in the environment constructed in this paper; the upper framework determines the driving behavior, and the lower layer controls the vehicle driving through the motion controller according to the determined driving behavior.

2.3. Vehicle Motion Control

In the environment constructed in this paper, the framework for controlling the motion behavior of intelligent vehicles and surrounding vehicles is divided into two frameworks: the upper and lower. Among them, the lower framework is the vehicle motion controller, which is responsible for controlling the vehicle motion according to the driving behavior and target driving speed determined by the upper framework and the target lane. The motion controller is a longitudinal controller, where the longitudinal controller uses the control strategy of a proportional–integral–differential controller with the following mathematical expressions.

$$a = K_p \times (v_t - v) + K_i \times \left(\int (v_t - v)dt\right) + K_d \times \frac{d(v_t - v)}{dt} \tag{4}$$

where a represents the vehicle's acceleration; v represents the vehicle's travel speed; v_t is the reference speed; and K_p, K_i, and K_d are the proportional, integral, and differential gains of the controller, respectively.

In the experimental scenario constructed in this paper, the intelligent vehicle and the surrounding vehicles realize the driving control of the vehicle through the designed two-layer control framework. In this experimental environment, the intelligent vehicle will master how to achieve safe and efficient driving in the designed scenario by continuously trying and learning.

2.4. Experimental Scene Setting

In order to increase the diversity of cases in the experimental scenario and enhance the simulation effect and the generalization of the derived driving strategies, multiple other surrounding vehicles are set up by the scenario characteristics when constructing the experimental scenario in this paper; these surrounding vehicles follow the hierarchical vehicle motion behavior control framework proposed above to drive in the scenario.

The intersection scenario has a high traffic flow density and a more complex traffic situation scenario. The intersection scenario designed in this paper includes a cross-shaped intersection and several other surrounding vehicles. The designed intersection connects eight roads from different driving directions, and each road contains two parallel sub-lanes in the same direction, roughly ranging from a 50 m $\times$ 50 m square area. Specifically, the intersection scenario designed in this paper includes the following.

1. Intelligent vehicles are generated from the starting point at the beginning of the simulation at a speed of 7 m/s and travel along the road toward the set end point;
2. In the intersection scenario designed in this paper, two to three surrounding vehicles are set up at each road bordering the intersection. The peripheral vehicles are generated and driven at random locations on the road with a random speed in the interval of [0 m/s, 9 m/s];
3. Considering the characteristics of the intersection scenario and the driving task that the intelligent vehicle needs to complete, this paper sets the termination conditions for each round of the intelligent vehicle training process in the intersection scenario as follows: the intelligent vehicle collides, the intelligent vehicle reaches the endpoint, or the driving time reaches 30 s.

3. Vehicle Driving Risk Assessment Methods for Intersection Scenarios

Intersection scenarios are among the most complex and high-risk traffic scenarios. Therefore, it is challenging to propose an evaluation method that can robustly and correctly assess vehicle driving risk. In this paper, we assume that the intersection scenario under study is an unprotected intersection and design a vehicle driving risk assessment method for the intersection scenario based on the Bayesian probability theory, which has good robustness and applicability.

3.1. Scene Analysis

Intersection scenarios are more complex than other scenarios. However, the structured characteristics of the road network environment, road markings, and traffic regulations make it possible to predict the movement patterns of vehicles within intersection scenarios in advance. For example, by identifying the vehicle's position on a digital map, the vehicle's driving motive can be inferred. Therefore, this research paper determines a limited future driving path of a vehicle based on its location, intersection geometry, and topological features by projecting the vehicle in the intersection scene onto a digital map, as shown in Figure 3. In the figure, a finite set of future paths of the vehicle is given as it passes through the intersection, and the proposed method uses this finite set to identify potential threats.

When an autonomous vehicle passes through an intersection, it predicts the future path for all vehicles within the scenario. Furthermore, to achieve the main task of safely crossing the intersection, the autonomous vehicle pays special attention to the vehicles with intersection points with its predicted path and refers to them as relevant vehicles. In this process, the autonomous vehicle identifies the locations of the intersection points on the path of the relevant vehicles. It pays attention to the kinematic information of the relevant vehicles. All remaining vehicles are subsequently defined as irrelevant vehicles, as they do not threaten the autonomous vehicle, so they are not evaluated to improve the efficiency of the computation. Finally, the information on the relevant vehicles is used by the autonomous vehicle to evaluate the possibility of a collision when crossing the intersection in the current motion state.

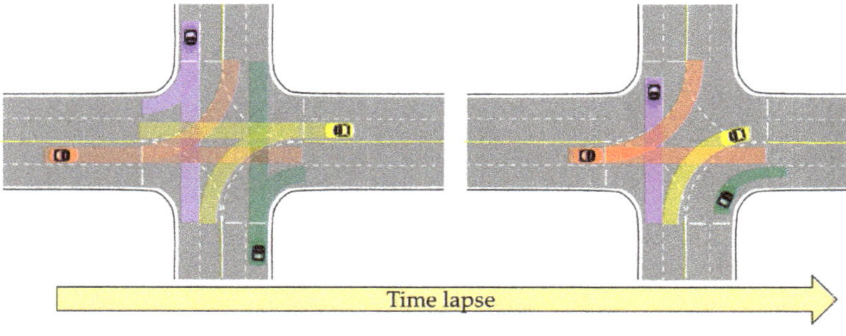

Figure 3. Projecting the vehicle onto the digital map.

The potential threat identification schematic in the intersection scenario is shown in Figure 4. In the figure, yellow, orange, and blue vehicles represent intelligent, related, and unrelated vehicles, respectively. The intelligent vehicle and related vehicles are represented by a and r_i, respectively, where i denotes the serial number of the related vehicle. The red area in the figure is the potential collision area, which indicates the area where the relevant vehicle's possible future path intersects with the intelligent vehicle's future path. In the study, the potential collision regions are denoted by c_j, where j denotes the serial number of the potential collision domain by distance on the future path of the intelligent vehicle. In addition, each potential collision region c_j has a safety line and an end line corresponding to the intelligent driving vehicle and the related vehicles. In the risk assessment process, the probability of a two-vehicle collision is evaluated using the time when the intelligent vehicle and the related vehicle drive into the safety zone and out of the end line. v indicates the driving speed of the corresponding vehicle.

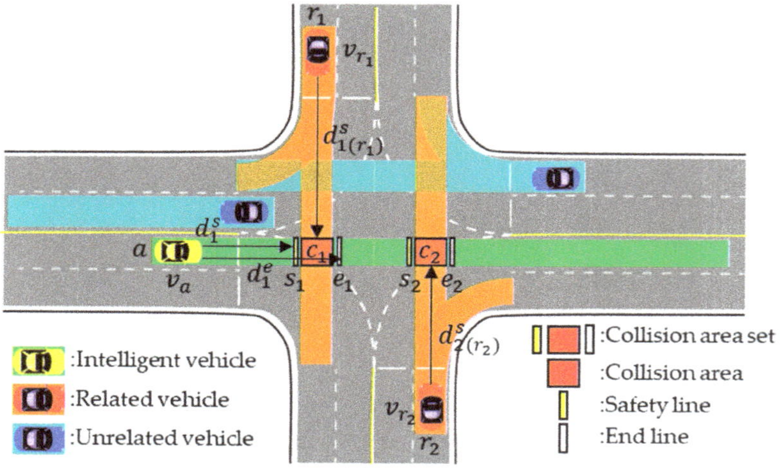

Figure 4. Potential threat identification.

3.2. Vehicle Driving Risk Evaluation Metrics for Intersection Scenarios

For the intersection scenario, this study uses time to enter (TTE) to evaluate the driving condition of intelligent vehicles through the intersection scenario. The TTE indicates the time that the vehicle continues to travel in its current state of motion from its current position until it enters the collision zone. The calculation process is as follows.

$$t_j^{TTE} = \frac{\sqrt{2a_{AV}d_j + v_{AV}{}^2} - v_{AV}}{a_{AV}} \tag{5}$$

In the above equation, t_j^{TTE} denotes the TTE value of the intelligent vehicle relative to the potential collision region c_j; a_{AV} denotes the acceleration of the intelligent vehicle; d_j denotes the distance of the intelligent vehicle from its current position along the road curvature to the safety line corresponding to the potential collision region c; and v_{AV} denotes the travel speed of the intelligent vehicle. However, the existence of an intersection region between the future path of the vehicle in question and the future path of the intelligent driving vehicle does not necessarily mean that a collision will occur between them. Therefore, assessing the driving risk in the current traffic environment requires using the kinematic information of the intelligent vehicle and the related vehicle to determine the possibility of collision in the corresponding potential collision region. In this study, we use the temporal overlap of intelligent and related vehicles through the potential collision region to determine whether they will collide in the corresponding potential collision region. The mathematical expression of this temporal overlap degree is as follows.

$$\begin{cases} t_j^s = \frac{\sqrt{2ad_j^s + v^2} - v}{a} \\ t_j^e = \frac{\sqrt{2a(d_j^e + l) + v^2} - v}{a} \end{cases} \tag{6}$$

In the above formula, t_j^s indicates the time when the vehicle drives into the safety line corresponding to the potential collision area c_j; t_j^e indicates the time when the vehicle drives out of the end line corresponding to the potential collision area c_j; d_j^s is the distance when the vehicle drives from the current position along the road curvature to the safety line corresponding to the potential collision area c_j; d_j^e is the distance when the vehicle drives from the current position along the road curvature to the end line corresponding to the potential collision area c_j; v is the current driving speed of the vehicle; a is the acceleration of the vehicle; and l is the body length of the vehicle.

By substituting the information of the map, the intelligent vehicle, and the related vehicles into Equation (6), we can obtain the time intervals $[t_j^{s_{AV}}, t_j^{e_{AV}}]$ and $[t_j^{s_{SV}}, t_j^{e_{SV}}]$ that the intelligent vehicle and the related vehicles pass in the potential collision region c_j. By analyzing the overlap of these two-time intervals, we can determine whether a collision occurs between the intelligent vehicle and the vehicle of interest in the potential collision region c_j. According to the different temporal arrangements of $t_j^{s_{AV}}$, $t_j^{e_{AV}}$, $t_j^{s_{SV}}$ and $t_j^{e_{SV}}$, we can classify them into six cases, as shown in Figure 5. Except for the cases of ⑤ and ⑥, there is an overlap of time intervals between the intelligent vehicle and the related vehicle in passing the potential collision area c_j, and the intelligent vehicle needs to make reasonable driving decisions to avoid the collision. In the cases of ⑤ and ⑥, there is no time interval overlap between the intelligent vehicle and the related vehicle in passing the potential collision area c_j, and the intelligent vehicle can pass safely.

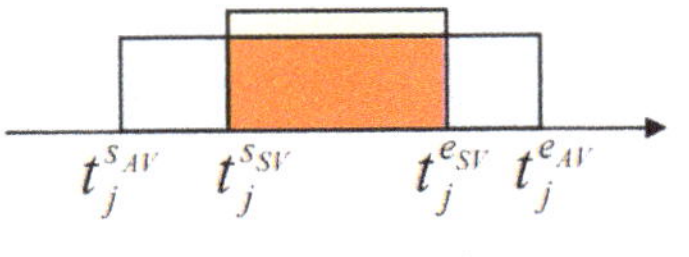

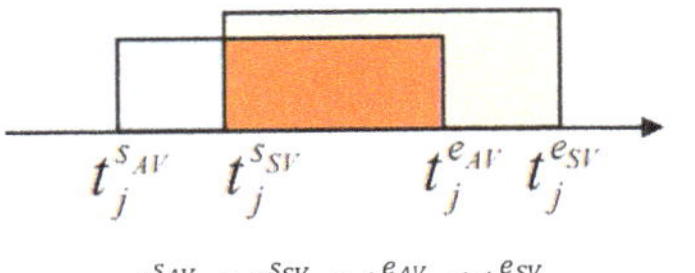

Figure 5. *Cont.*

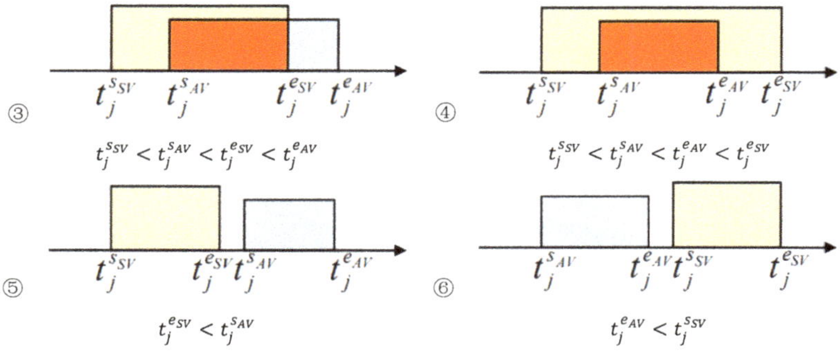

Figure 5. Classification of TTE sequence arrangement in the collision area.

The time overlap between the intelligent vehicle passing through each potential collision area and the related vehicle passing through the potential collision area is calculated and analyzed to determine whether a collision will occur in each potential collision area for the intelligent vehicle. In the case of a collision, the driving risk of the intelligent vehicle regarding the potential collision region c_j is evaluated based on the value of t_j^{TTE}.

3.3. Risk Probability Reasoning

In the intersection scenario, this paper assesses vehicle driving risk by defining three risk levels associated with t_j^{TTE} values. These three risk levels represent the risk levels associated with the likelihood of a collision, as shown below.

$$z_j \in Z = \{Dangerous, Attentive, Safe\} = \{D, A, S\} \tag{7}$$

where z_j is defined as a random variable that represents the risk level of the intelligent vehicle in the potential collision area c_j with relevant vehicles. In order to determine the boundaries of these three risk levels, two thresholds t_d and t_a are defined, which represent the TTE value thresholds for the "dangerous" and "attentive" risk levels, respectively. According to the international safety standards [31,32] and vehicle tracking performance [33], T_D and T_A are defined as 4 s and 7 s, respectively. In addition, to consider the uncertainty in the driving scenario, we introduce the uncertainty measure σ_t when constructing the likelihood function. t_j^{TTE} is defined as follows:

$$\begin{aligned}
p(t_j^{TEE}|z_j = D) &\triangleq \begin{cases} \exp(-\frac{(t_j^{TEE}-t_d)^2}{2\sigma_t{}^2}), \text{for } t_j^{TEE} > t_d \\ 1, \text{otherwise} \end{cases} \\
p(t_j^{TEE}|z_j = A) &\triangleq \begin{cases} \exp(-\frac{(t_j^{TEE}-t_d)^2}{2\sigma_t{}^2}), \text{for } t_j^{TEE} < t_d \\ \exp(-\frac{(t_j^{TEE}-t_a)^2}{2\sigma_t{}^2}), \text{for } t_j^{TEE} > t_a \\ 1, \text{otherwise} \end{cases} \\
p(t_j^{TEE}|z_j = S) &\triangleq \begin{cases} \exp(-\frac{(t_j^{TEE}-t_a)^2}{2\sigma_t{}^2}), \text{for } t_j^{TEE} < t_a \\ 1, \text{otherwise} \end{cases}
\end{aligned} \tag{8}$$

In this paper, based on the assumption that the prior probability of vehicle risk level $P(z_j)$ follows a uniform distribution, the Bayesian probability theory is used to determine the probability distribution of each risk level of an intelligent vehicle for a potential collision area c_j under a given traffic situation, which is calculated as follows.

$$P(z_j|t_j^{TEE}) = \frac{p(t_j^{TEE}|z_j) \cdot p(z_j)}{\sum\limits_{z_j \in Z} p(z_j) \cdot p(t_j^{TEE}|z_j)} \tag{9}$$

This paper uses a numerical approach to measure the degree of the risk level. Precisely, values 2, 1, and 0 denote the different risk levels. Therefore, this paper defines the risk level value ε, which is defined as follows.

$$\varepsilon \in Z = \{Dangerous, Attentive, Safe\} = \{2, 1, 0\} \tag{10}$$

Based on the calculated probability distribution of each risk level of the intelligent driving vehicle regarding the potential collision area c_j and the expected risk level value ε_{c_j} of the intelligent driving vehicle regarding the potential collision area c_j can be calculated at the current moment, and the calculation process is as follows:

$$\varepsilon_{c_j} = \sum_{\varepsilon \in \{2,1,0\}} \varepsilon \cdot P(z_j|t_j^{TEE}) \tag{11}$$

In an intersection scenario, the threat to the intelligent vehicle may come from multiple possible future paths of multiple related vehicles. For example, in Figure 4, the related vehicle r_1 may have multiple future paths before entering the intersection. However, only the straight-ahead path intersects with the planned path of the intelligent vehicle. Therefore, the possibility of the future trajectory of the relevant vehicle also needs to be considered when assessing the driving risk of the intelligent vehicle. In addition, the intelligent vehicle will pass through each potential collision area in turn when passing through an intersection. In this process, the level of attention of the intelligent vehicle to each potential collision area is not equal. In order to measure the overall expected risk level value of the intelligent vehicle in the intersection scenario, the relevant vehicle trajectory likelihood weight w_j and the recession factor γ are introduced, as follows:

$$\varepsilon_{risk} = \frac{\sum\limits_{j=1}^{n} \gamma^{j-1} w_j \varepsilon_{c_j}}{\sum\limits_{j=1}^{n} \gamma^{j-1} w_j \varepsilon_{\max}} \tag{12}$$

where ε_{risk} represents the overall expected risk level value of the intelligent vehicle in the intersection scenario. w_j represents the future trajectory possibility of the relevant vehicle corresponding to the potential collision region c_j. In this paper, we assume that multiple future trajectories of the relevant vehicle follow a uniform distribution. $\gamma \in [0, 1]$ is the attenuation factor used to balance the potential collision region that is about to be passed and the potential collision region that will be passed in the future—the importance of the potential collision area. When γ is close to 0, the risk assessment method will pay more attention to the potential collision areas that are about to pass by; when γ is close to 1, the risk assessment method will pay more attention to the potential collision areas that are about to pass by in the future.

4. Intelligent Driving Decision Algorithm Based on Deep Reinforcement Learning

4.1. Deep Q-Network Learning Based on State-Action Value Distribution

Q-learning-based algorithms suffer from the problem of overestimation due to valuation errors, which can lead to a significant bias in the final obtained algorithmic model and result in degraded algorithm performance. To solve this problem, H. V. Hasselt et al. proposed the theory of double Q learning and applied it to the Deep Q-Network (DQN) learning algorithm [34]. The Double Deep Q-Network (DDQN) learning algorithm solves the overestimation problem by decoupling the selection of the target Q-value action and the computation of the target Q-value. Although the DDQN algorithm improves the algorithm performance by optimizing the computation of the target Q-value, there are still exceptional cases during the learning process, such as when an intelligent body chooses an action that may be accompanied by a significant risk to obtain a greater reward return. This situation is perilous and is manifested in the intelligent vehicle driving strategy learning process, as the intelligent vehicle may improve driving efficiency by driving dangerously. Therefore, safety should be considered along with reward maximization in the intelligent vehicle policy learning process. To this end, this paper uses a state-action value distributed deep Q-network-based learning algorithm distributional DQN learning algorithm through which optimal driving strategies for intelligent vehicles are derived.

The distributional DQN learning algorithm aims to improve the estimation of the distribution of possible rewards, making learning more stable when combined with a neural network. Specifically, the distributional DQN algorithm transforms the estimation of Q-values into a probability distribution so that the intelligent body can choose the best action based on the probability of Q-values for different actions. This change not only allows the intelligence to effectively perceive the risks present in the environment and improve the exploration of the environment, it also allows the intelligence to avoid risks when deciding on actions effectively, and the intelligence is more inclined to choose the action with a better worst-case scenario rather than just choosing the action with a larger Q-value. The basic structure of the distributional DQN algorithm used in this paper is shown in Figure 6.

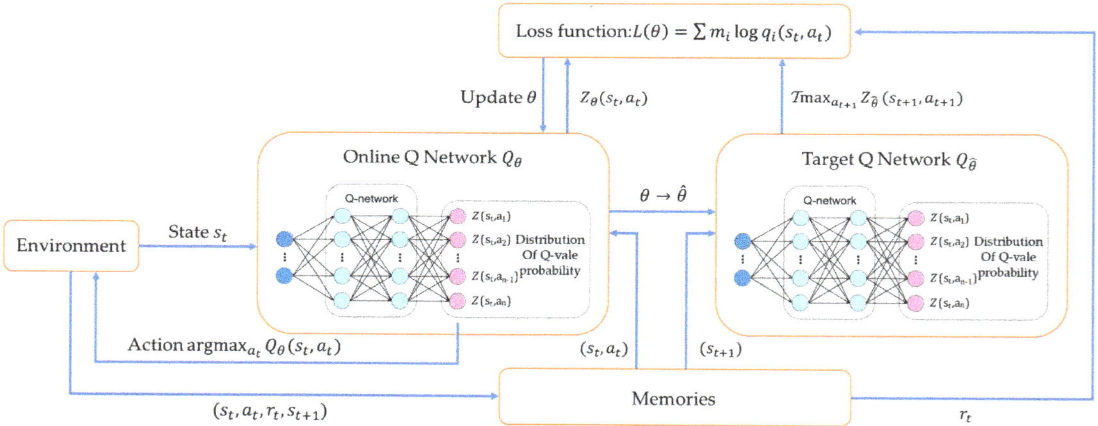

Figure 6. Framework of the distributional DQN algorithm.

In the distributional DQN algorithm used in this paper, the neural network's output is the probability distribution of Q-values for each action, as shown in Figure 7. In the figure, the x-axis represents the normalized Q-values, and the y-axis represents the probability that the intelligence achieves the corresponding Q-value after taking action. To model the Q-value distribution, we use a discrete distribution represented by parameters $N \in \mathbb{N}$ and $V_{max}, V_{min} \in \mathbb{R}$. In parameterizing the distribution, the range of Q-value taking $[V_{min}, V_{max}]$ is discretized into N branches and divided into $N-1$ equal branching sets, and each is represented as follows:

$$\{z_i = V_{\min} + i\Delta z; 0 \le i < N, \Delta z = \frac{V_{\max} - V_{\min}}{N-1}\} \tag{13}$$

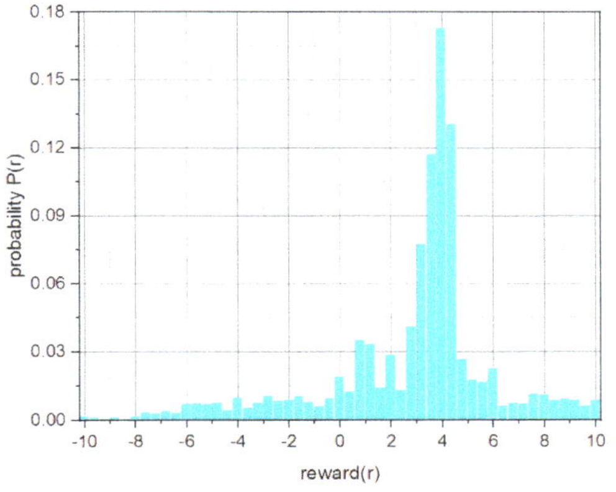

Figure 7. Example output of expected future earnings in the distributional DQN.

The branch set is fixed, and the neural network outputs the probability that each action Q-value takes the current branch value. Therefore, the Q-value of each action is calculated as follows:

$$Q(s_{t+1}, a) = \sum_{i=0}^{N-1} z_i p_i(s_{t+1}, a) \tag{14}$$

The neural network of the distributional DQN algorithm takes the predicted Q-value distribution as the output. Accordingly, the training target should be obtained as a distribution as well. Therefore, the Bellman update for each branch is calculated as follows:

$$\mathcal{T} z_i = r_t + \gamma z_i \tag{15}$$

After the Bellman update calculation, there may be cases where the distribution takes values outside the range of the original distribution. In order to obtain the target probability distribution $(m_j, i = \{0, \cdots, N-1\})$ for each branch, the distribution $\mathcal{T} z_i$ needs to be projected onto the branch of $\mathcal{T} z_i$. The specific procedure is as follows:

$$\Phi \mathcal{T} z_i = \sum_{j=0}^{N-1} \left[1 - \frac{\left| [\mathcal{T} z_j]_{V_{\min}}^{V_{\max}} - z_i \right|}{\Delta z} \right]_0^1 p_j \tag{16}$$

In the distributional DQN algorithm, since both the estimation network and the target network use the distribution as the output, this paper uses a measure of the similarity between the two distributions as the loss. Therefore, the loss function of the Distributional DQN algorithm can be implemented by calculating the cross-entropy term of the Kullback–Leibler divergence. Precisely, the loss function is calculated as follows:

$$L = -\sum_{i=0}^{N-1} m_i \log p_i(s_t, a_t) \tag{17}$$

The distributional DQN algorithm has the same structure as the DQN algorithm, including empirical recall and a separate target network, and uses a sampling process with an ϵ-greedy strategy. However, the neural network output of the distributional DQN algorithm is no longer a Q-value function for each action but a probability distribution of Q-values for each action. This makes the action selection of the distributional DQN algorithm based on the ϵ-greedy strategy of expected Q-values. In the DQN algorithm, the Q-network is trained with the mean squared deviation as the loss function by finding the action corresponding to the maximum Q-value in the estimation network. However, the distributional DQN algorithm trains the network with KL divergence as the loss function by finding the action that corresponds to the best probability distribution of Q values in the estimation network. This change allows intelligence to use the probability distribution of Q-values to provide more information in action decision-making. In specific risk scenarios, the intelligence is more inclined to choose the action with less variance or a better worst-case scenario rather than just choosing the action with a larger Q value. Such a strategy allows intelligence to deal with uncertainty more robustly and improve its performance in risky environments.

4.2. Structure of the Intelligent Driving Decision Algorithm

In this paper, we propose an intelligent vehicle driving decision learning algorithm based on the deep reinforcement learning method by considering the vehicle driving risk approach and the Distributional DQN learning algorithm proposed above. The algorithm consists of three components: a perception layer, a decision layer, and a control layer. The perception layer fuses information from the intelligent vehicle and the surrounding environment as input and constructs it as a deep reinforcement learning state space. The decision layer uses a deep neural network to output driving decision commands. Finally, the control layer implements the longitudinal control of the intelligent vehicle using a PID model to control the vehicle according to the decision commands outputted by the decision layer. Combining the above descriptions, the structure of the intelligent driving decision algorithm in this paper is shown in Figure 8.

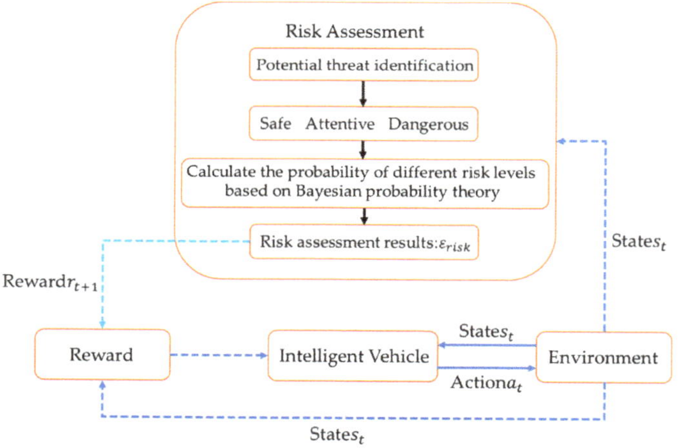

Figure 8. Intelligent driving decision algorithm framework.

4.3. Markov Decision Process for Intelligent Driving

In this study, the driving decision problem of an intelligent vehicle is studied by describing it as a Markov decision process. In this process, the intelligent vehicle decides its action at the next time step based on the current state and is rewarded accordingly. For the driving decision problem in the straight ahead and intersection scenarios, this paper uses the tuple $(S, \mathcal{A}, R, \gamma)$ to describe its Markov decision process. where $s \in S$ is the intelligent vehicle state space, $a \in \mathcal{A}$ is the action space, R is a model on immediate rewards $r(s, a, s')$, and $\gamma \in [0, 1]$ is a discount factor for delayed rewards, which is used to reduce the weight of distant rewards when balancing the importance of current and future rewards.

4.4. State Space

This paper proposes a deep-reinforcement-learning-based algorithm for intelligent vehicle driving decision-making. The algorithm describes the intelligent vehicle driving decision problem as a Markov decision process and represents it by a quadratic group $(S, \mathcal{A}, R, \gamma)$. In the intelligent vehicle driving decision algorithm proposed in this paper, the state information represents the environmental information perceived by the intelligent vehicle and the changes caused by its actions. Therefore, the state space designed in this study contains the location information, driving state information, and environmental information of the intelligent vehicle and other vehicles around it. Among them, the joint state information of the autonomous vehicle s_0 and the location and driving state information of the surrounding N other vehicles are represented as follows:

$$s_v = (s_k)_{k\in[0,N]}, s_k = \left[x_k\ y_k\ v_k^x\ v_k^y\ a_k^x\ a_k^y\ yaw_k\right]^T \tag{18}$$

where x and y represent the position of the vehicle along the lateral and longitudinal axes of space, respectively; v^x and v^y represent the velocity of the vehicle along the lateral and longitudinal axes of space, respectively; a^x and a^y represent the acceleration of the vehicle along the lateral and longitudinal axes of space, respectively; and yaw represents the direction of vehicle travel.

Using such a representation minimizes the traffic information in the information representation environment. However, this representation has two problems. First, the dimensionality of its information vector varies with the number of vehicles present in the environment, which is detrimental to the approximation of an input function that expects a constant size. Second, driving decisions trained using this form are affected by the order in which the information about other states around the scene is arranged. This leads to a lack of some generalizability of the derived driving strategies. Therefore, when designing the state space, the filtered state information must be processed to ensure its formal and logical uniformity to improve the universality of the driving decisions derived in this paper.

In order to ensure the uniformity of the designed state space form and to guarantee the constant length of the output state information vector, this paper preprocesses the filtered state information. Specifically, as shown in Figure 9, the map information is gridded, and the vehicle state information in the environment is represented in the grid as a tensor. This processing not only ensures the stability of the state space structure but also makes the intelligent vehicle better perceive the relative position relationships of other vehicles around. At the same time, to learn a more general driving strategy, the designed state space must ensure that the state information has logical uniformity. Therefore, this paper takes the center of the intelligent vehicle as the origin and defines the state information of other surrounding vehicles as the position information and motion state information relative to the intelligent vehicle. The specific state information definition is shown in Table 1.

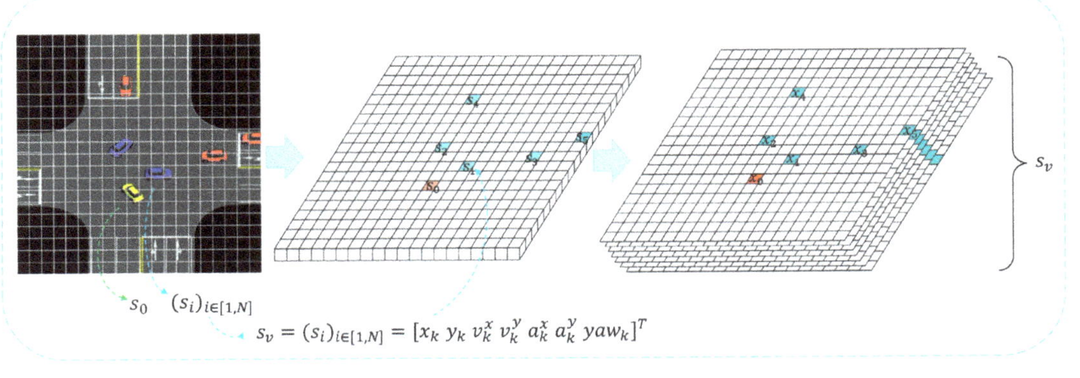

Figure 9. Vehicle status information pre-processing process.

Table 1. Definition of vehicle status information.

Status Information	Scope (Unit)	Description
x_k	[−200, 200] (m)	Relative distance along the longitudinal axis to the intelligent vehicle
y_k	[−200, 200] (m)	Relative distance along the transverse axis to the intelligent vehicle
v_k^x	[−40, 40] (m/s)	Relative intelligent vehicle travel speed along the longitudinal axis
v_k^y	[−40, 40] (m/s)	Relative intelligent vehicle travel speed along the transverse axis
a_k^x	[−8, 8] (m/s^2)	Acceleration of travel along the longitudinal axis relative to the intelligent vehicle
a_k^y	[−8, 8] (m/s^2)	Acceleration of travel along the transverse axis relative to the intelligent vehicle
yaw_k	[−π, π] (rad)	Vehicle direction of travel

The road information in the state space follows the same formal and logical consistency described before. Specifically, the process is shown in Figure 10 by representing the road information in the form of 0 and 1 in the grid to ensure the uniformity and logical uniformity of the road information in the state space.

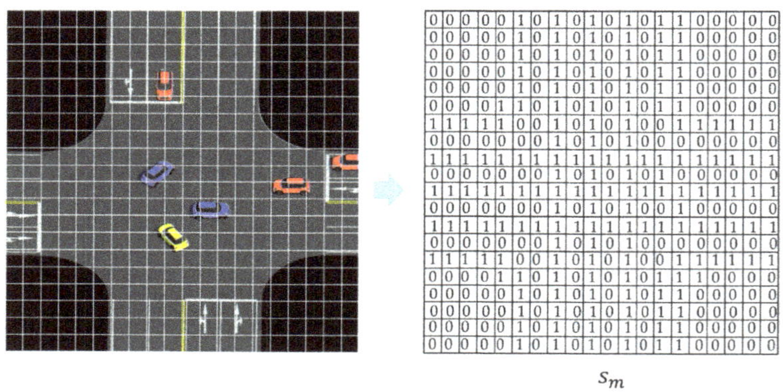

Figure 10. Map status information pre-processing process.

In summary, the state space in the study of this paper is the set of the joint state information of the position information, motion state information, and road information of the vehicles in the environment, and the state information of the environment at the moment is represented as follows (Figure 11):

$$s_t = \{s_m, s_v\} \tag{19}$$

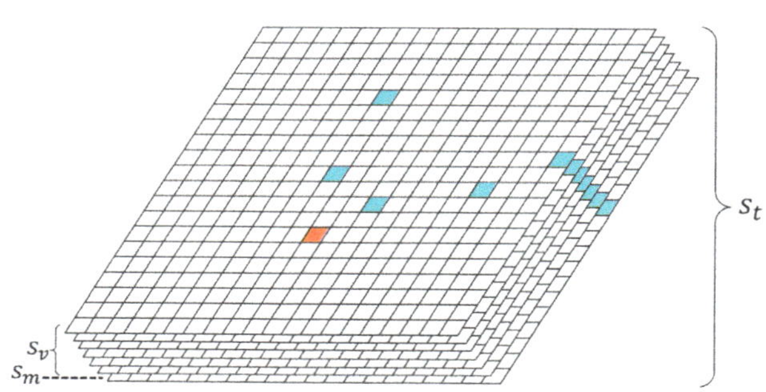

Figure 11. Environment status information.

4.5. Action Space

As an intelligent vehicle passes through an intersection, the intelligent vehicle can adjust the throttle and brake operations to accelerate, decelerate, or maintain a constant speed for safety, depending on the surrounding traffic environment. Therefore, the action space of the intersection driving decision learning algorithm is defined as three speed-related operations: acceleration, deceleration, and constant speed driving.

4.6. Reward Functions

Reinforcement learning aims to maximize the designed reward function by seeking a strategy. Therefore, an appropriate reward function must be designed for an intelligent body that accomplishes a specific task. In this paper, the intelligent vehicle must drive safely and efficiently in an intersection scenario. To solve the reward sparsity problem, we decompose the target task of the intelligent vehicle into multiple sub-goal tasks with appropriate rewards or penalties to ensure that the intelligent vehicle can receive timely feedback at each time step. Therefore, in this paper, the goal task is decomposed into the following subgoal tasks: avoid the collision, drive at high speed, and try to ensure minimum driving risk. Based on this sub-goal task, we define the primary reward function as follows:

$$r_{total} = \lambda_1 r_{collision} + \lambda_2 r_{risk} + \lambda_3 r_{velocity} \tag{20}$$

where $r_{collision}$ denotes the penalty for a collision of an intelligent vehicle; r_{risk} is the reward for the driving risk of an intelligent vehicle based on the current traffic environment assessment; $r_{velocity}$ is the reward for evaluating the driving efficiency of an intelligent vehicle; λ_1, λ_2, and λ_3 denote the relative weight coefficients of the reward functions $r_{collision}$, r_{risk}, and $r_{velocity}$ in the total immediate reward, respectively.

The penalty function $r_{collision}$ for a collision of an intelligent vehicle is defined as follows:

$$r_{collision} = \begin{cases} 1 & \text{if ego vehicle collides} \\ 0 & \text{otherwise} \end{cases} \tag{21}$$

In order to ensure the safety of intelligent driving vehicles and avoid possible collisions, intelligent vehicles should assess the risk level of their driving state according to the surrounding traffic environment and adjust their driving strategies accordingly. Therefore, this paper designs a reward function r_{risk} based on the risk assessment of the current driving state of the intelligent vehicle based on the vehicle driving risk assessment method based on the Bayesian probability theory proposed in Section 3, defined as follows:

$$r_{risk} = 1 - \frac{\varepsilon_{risk}}{\varepsilon_{max}} \tag{22}$$

where ε_{risk} denotes the intelligent vehicle's expected driving risk level value at the current time step and ε_{max} denotes the defined maximum risk level value.

In this paper, it is proposed that excessive conservatism should be avoided in the driving safety of intelligent vehicles because it may not only negatively affect the efficiency of driving but also cause traffic accidents, delays, and gridlock [35]. Therefore, to maximize driving efficiency while ensuring safety, this paper designs the reward function $r_{velocity}$ on driving efficiency, defined as follows:

$$r_{velocity} = \begin{cases} 0 & \text{if } v_x < v_{min} \\ 1 & \text{if } v_x > v_{max} \\ \frac{v_x - v_{min}}{v_{max} - v_{min}} & \text{otherwise} \end{cases} \tag{23}$$

where v_x represents the driving speed of the intelligent vehicle in the road direction; v_{max} and v_{min} represent the maximum desired speed and the minimum desired speed of the intelligent vehicle in the driving scenario studied in this paper, which is set to 9 m/s and 7 m/s, respectively, in this paper.

The total immediate reward r_{total} designed in this paper is the weighted sum of the reward term $\lambda_1 r_{collision}$ for collision avoidance, the reward term $\lambda_2 r_{risk}$ for driving risk, and the reward term $\lambda_3 r_{velocity}$ for driving efficiency. Where λ_1, λ_2, and λ_3 are the weighting coefficients of the three reward terms in the total immediate reward. According to the above, $r_{collision}$ takes values between $\{0, 1\}$, r_{risk} takes values in the range of $[0, 1]$, and $r_{velocity}$ takes values in the range of [0, 1], so the total immediate reward r_{total} takes values in the range of $[\lambda_1, \lambda_2 + \lambda_3]$. In deep reinforcement learning, it is beneficial to use normalized rewards [36]. Therefore, in this paper, the total immediate reward r_total is normalized from the range of values $[\lambda_1, \lambda_2 + \lambda_3]$ to the range [0, 1] to normalize the total reward value within each time step as shown below:

$$r = \frac{r_{total} - \lambda_1}{\lambda_2 + \lambda_3 - \lambda_1} \tag{24}$$

In setting the three reward weight coefficients λ_1, λ_2, and λ_3, this paper avoids using negative values as much as possible to avoid the situation in which the intelligent vehicle chooses to pre-emptively end a training set by colliding with other vehicles around it to avoid receiving large negative rewards. In this paper, the weight coefficient of the reward function for the collision of the intelligent vehicle is set to -1. In addition, this paper believes that intelligent vehicles should pay more attention to driving risk than driving efficiency during driving, so the weight coefficients of the reward functions for driving risk and driving efficiency are set to 0.3 and 0.2, respectively.

4.7. Neural Network Structure

The model in this study consists of a value function network containing an online Q-value network and a target Q-value network with identical structures. The network's input comes from the state space, and the output is the distribution of state-action values. This paper argues that intelligent vehicles should focus on the vehicles associated with their driving rather than all the vehicles around them during driving. Therefore, this paper proposes an attention network based on state information to realize this idea, as shown in Figure 12. The network introduces a particular attention module consisting of two parts: a channel attention module and a spatial attention module.

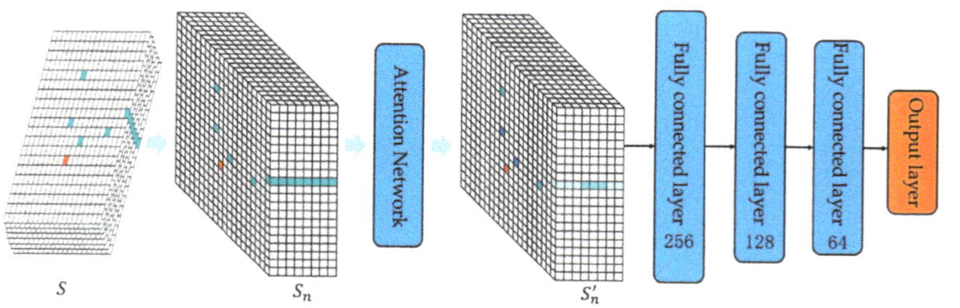

Figure 12. Structure of the attention network based on state information.

The attention network designed in this paper uses the state space $S \in \mathbb{R}^{C \times H \times W}$ as input, and the state information in each dimension of the state space is normalized and processed by the channel attention module $M_c \in \mathbb{R}^{C \times 1 \times 1}$ and the spatial attention module $M_s \in \mathbb{R}^{1 \times H \times W}$, as shown in Figure 13. The attentional process of the designed attentional network can be summarized as follows.

$$S \rightarrow S_n \tag{25}$$

$$S'_n = M_c(S_n) \otimes S_n \otimes M_s(S_n) \tag{26}$$

where $\otimes$ denotes element-by-element multiplication. S_n is the normalized state space. In this study, the state information of each dimension of the state space $S \in \mathbb{R}^{C \times H \times W}$ is normalized by batch normalization. $M_c(S_n)$ is the output of S_n after the channel attention module M_c; $M_s(S_n)$ is the final output of S_n after the spatial attention module M_s.

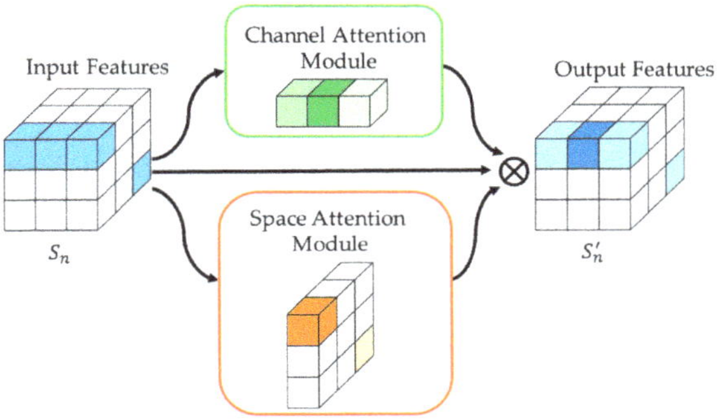

Figure 13. Attention network.

The channel attention module implements the channel attention mechanism by analyzing the relationships between feature channels. Each information dimension in the state space can be considered a channel, and each channel can be regarded as a feature extractor. The channel attention module uses the attention mechanism to focus on meaningful information dimensions, thus improving the model's focus on essential features and enhancing its expressive power. Woo et al. argue that average and maximum pooling, which collect information about different object features, leads to more accurate channel attention [37]. Therefore, in this paper, both averaging and maximum pooling methods are used with reference to the above research work. The computation process of the two attention modules is shown in Figure 14, and the two are computed as follows:

$$\begin{aligned} M_c(S_n) &= \sigma(MLP(MaxPool(S_n)) + MLP(AvgPool(S_n))) \\ &= \sigma(W_1(W_0(S_{n\,max}^{c})) + W_1(W_0(S_{n\,avg}^{c}))) \end{aligned} \tag{27}$$

$$\begin{aligned} M_s(S_n) &= \sigma(f^{3\times3}([AvgPool(S_n); MaxPool(S_n)])) \\ &= \sigma(f^{3\times3}([S_{navg}^{s}; S_{nmax}^{s}])) \end{aligned} \tag{28}$$

where $f^{3\times3}$ denotes a convolution operation with a convolution kernel size of 3×3.

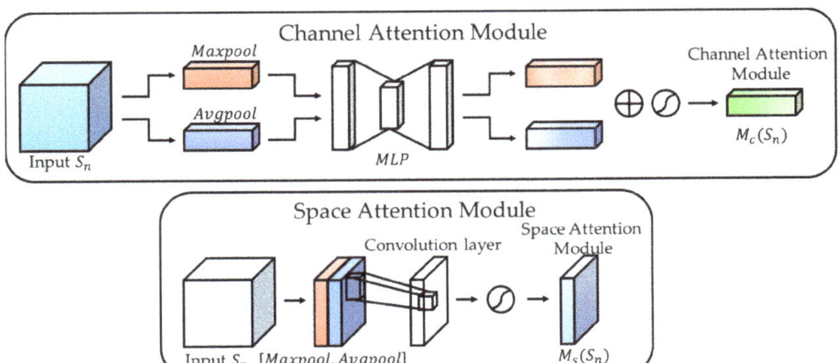

Figure 14. Attention modules.

The attention network designed in this paper uses the channel and spatial attention modules to analyze the state information in the current driving scenario to determine important information and the surrounding vehicles relevant to their driving. This attention network helps intelligent driving vehicles understand their surroundings more accurately and make more effective driving decisions.

5. Results

The algorithms studied in this paper are based on the implementation of PyTorch. The learning framework and the models are trained using CUDA acceleration techniques. CUDA is a parallel computing platform and programming model from NVIDIA that can be used to accelerate high-performance computing applications based on NVIDIA GPUs. The CUDA platform provides higher parallel computing performance than traditional CPUs by distributing parallel computing work to the thousands of compute cores in the GPU to achieve acceleration. All experiments in this chapter are conducted on the same experimental platform. In this paper, the network is trained using the Adam optimizer, and the network parameters are updated with the learning rate η. All hyperparameters are detailed in Table 2.

Table 2. Network training hyperparameters.

Hyperparameter	Value
Discount factor γ	0.99
Learning rate η	0.001
Experience replay memory size M_{replay}	500,000
Number of empirically collected samples M_{batch}	128
Target network update frequency F	10
Number of discrete branch points of Q-value distribution N	51
Q-value range $[V_{min}, V_{max}]$	[−10, 10]

In this paper, based on the constructed experimental environment and the pre-set experimental platform, the algorithm is trained and evaluated through the network structure and parameter settings.

5.1. Algorithm Performance Analysis

In this section, intelligent vehicles are trained to learn through the intersection scenario designed in this study using the distributional DQN algorithm, DDQN algorithm, and DQN algorithm, and with all of them using state information-based attention networks. This paper uses the intersection passing rate and the standardized reward of the intelligent vehicle in the scenario as quantitative metrics to verify the effectiveness and superior performance of the driving decision algorithms in this study. Among them, the intersection passing rate is defined as follows:

$$throughput\ rate = \sum_{n}^{n-9} \frac{I_n(throughput = true)}{10} \tag{29}$$

Figure 15 shows the comparison of different quantitative metrics during the training process for the distributional DQN algorithm, DDQN algorithm, and DQN algorithm. The vertical axis indicates the specific values of the quantified metrics, and the horizontal axis indicates the number of episodes in the training process.

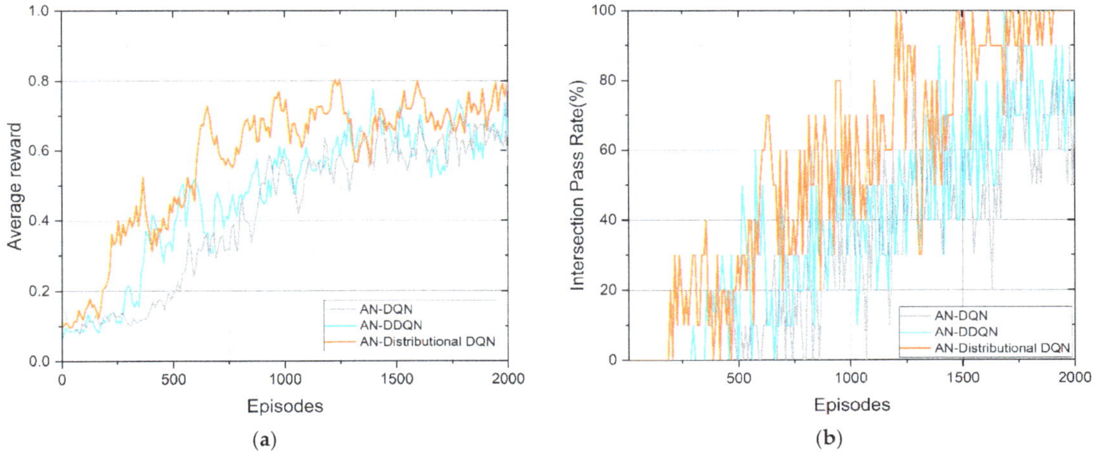

Figure 15. Intersection scenario training results: (**a**) average reward variation for the three methods during training and (**b**) intersection pass rate variation for the three methods during training.

As shown in Figure 15, the intelligent vehicle trained with reinforcement learning showed significant performance improvement in the intersection scenario. In the aspect of reward value, the intelligent vehicle showed an apparent upward trend during the training process, and all of them reached a stable state after a period of training; meanwhile, in the aspect of the intersection passing rate, the intelligent vehicle showed a trend of gradual improvement during the training process, which indicates that with the training process, the intelligent vehicle gradually mastered the driving skills of passing the intersection as the training process progressed.

As can be seen from the curves in Figure 15 regarding the intersection passing rate, the collision rate curves of the intelligent vehicles using reinforcement learning show a sudden drop during the rise. The analysis reveals that during the reinforcement learning training process, the intelligent vehicle cannot fully explore the state-space aspect, so it will prioritize the decision to increase the driving speed without colliding with the surrounding vehicles to obtain a higher reward for driving efficiency. In this process, in order to obtain higher driving efficiency, the intelligent vehicle will lead to frequent collisions. After this, the intelligent vehicle learns the driving measures it should take when encountering obstacles by fully exploring the environment. After experiencing the penalty of collisions, the intelligent vehicle learns the driving skill of avoiding obstacles, which increases the reward value. The above description shows that the reward function designed in this study plays an influential guiding role in the training process of driving measures of intelligent vehicles.

The experimental analysis conducted for the three algorithms in the intersection scenario shows that the distributional DQN algorithm can converge to the optimal policy relatively quickly. Specifically, the algorithm converged to the optimal policy at about 970 training rounds, much faster than the other two. Regarding the intersection passing rate, both algorithms used in this study show a trend of gradually increasing the success rate of intelligent vehicles passing the intersection. Specifically, after 1910 rounds, the intersection passing rate of the intelligent vehicles trained with the distributional DQN algorithm was stable at about 100%. In contrast, the intersection passing rate of the intelligent vehicles trained with the DDQN algorithm was slightly lower during the training process, stabilizing at around 80%. In comparison, the intersection passing rate of the intelligent vehicles trained with the DQN algorithm was around 70%. These experimental results show that the distributional DQN algorithm performs better in complex intersection scenarios and can achieve the optimal strategy in a shorter time and a higher intersection passing rate.

Under the experimental conditions designed in this section, the intelligent vehicles are trained for 2000 episodes. To further verify the effectiveness of their strategies, we create an independent test set with 100 episodes and the same environment settings as the training process. The evaluation of the test set allows us to compare the differences in policy effectiveness of the intelligent vehicles trained by different algorithms in the intersection scenario in this section of the experiments. Table 3 shows the differences between the average driving performance and intersection passing rate scenarios of the intelligent vehicles using the distributional DQN algorithm, DDQN algorithm, and DQN algorithm proposed in this paper during the testing process, indicating the relative rate of change. Specifically, the average passing rate of the intelligent vehicle using the distributional DQN algorithm proposed in this paper is 98% during the test, which is a 22% improvement compared to the intelligent vehicle with the DQN algorithm. In addition, in terms of driving efficiency, the average driving speed of the intelligent vehicle using the distributional DQN algorithm proposed in this paper is 8.71 m/s during the test, which is 10.39% higher relative to the intelligent vehicle with the DQN algorithm. The above results reveal that the driving strategy derived using the distributional DQN algorithm proposed in this paper has higher driving efficiency and intersection passing success rate than the DDQN and DQN algorithms.

Table 3. Test results of using different algorithms in the intersection scenario.

Algorithm	Passing Rate (%)	Relative Rate of Change Δ (%)	Average Speed (m/s)	Relative Rate of Change Δ (%)
DQN	77	-	7.89	-
DDQN	86	9 ↑	8.27	4.82 ↑
Distributional DQN	98	22 ↑	8.71	10.39 ↑

Combining the above, the distributional DQN algorithm proposed in this paper outperforms the DDQN algorithm and the DQN algorithm, enabling the trained intelligent vehicles to master the driving strategies in intersection scenarios faster and better. This research proposes a deep Q-network learning method based on state-action value distribution to derive the best driving strategy for intelligent vehicles in intersection scenarios, considering driving safety. The algorithm proposed in this paper reconstructs the neural network to output each state-action's value probability distribution. This change allows the algorithm to predict expected returns more accurately than traditional deep Q-network learning algorithms that rely on individual scalar value predictions and improves learning stability. As an example, Figure 16 shows the process of finding an action through the state-action value distribution for each state of the intelligent vehicle trained by the Distributional DQN algorithm in an intersection scenario, where Ac, Dc, and Keep denote "accelerate," "decelerate" and "keep current state", respectively. In this traffic situation, the intelligent vehicle selects actions in the following order: Keep, Dc, Ac, and Ac. This further validates that the proposed algorithm can help the intelligent vehicle avoid risky maneuvers and make optimal driving decisions in scenarios where there are many uncertainties and risks.

The experimental results show that using deep reinforcement learning algorithms in intersection scenarios enables intelligent vehicles to learn practical and intelligent driving decisions. This approach of continuously exploring driving skills using deep reinforcement learning algorithms using a reward function enables vehicles to learn more diverse driving skills to deal with complex and random traffic environments and improve as the scenario changes continuously.

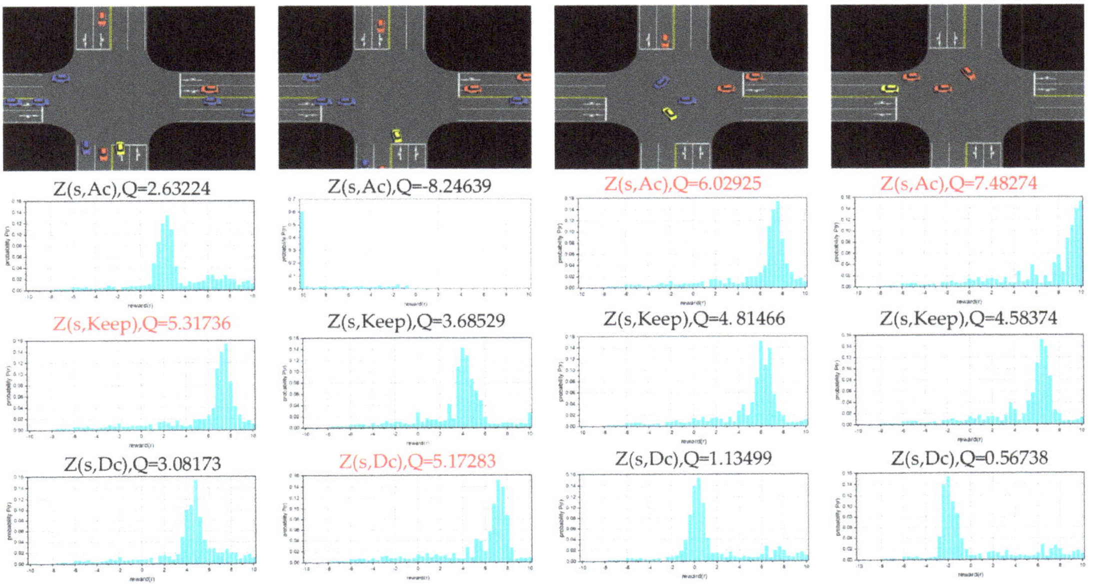

Figure 16. Intelligent vehicles make driving decisions through state-action value distribution.

5.2. Risk Assessment Method Analysis

The vehicle driving risk assessment method proposed in this paper applies continuous risk assessment values to the reward function to avoid behaviors that may lead the intelligent vehicle into higher-risk situations during training. This allows the intelligent vehicle to become aware of driving behaviors that may lead the vehicle into dangerous situations by penalizing them to different degrees when making driving decisions. However, even without using risk assessment methods, training an intelligent vehicle with a reward function based only on collision penalties allows it to learn to avoid collisions after some time. However, the driving strategy derived from this algorithm tends to put the intelligent vehicle into a potentially dangerous driving state. Figure 17 shows the driving risk states per second for successful test sets of intelligent vehicles using the driving risk assessment method and the deep reinforcement learning driving decision algorithm without the driving risk assessment method. It can be found that the intelligent vehicles with driving strategies derived using the risk assessment method maintain a lower risk state during driving, while the intelligent vehicles without driving strategies derived using the risk assessment method are always at a higher risk state. In summary, applying risk assessment methods can more accurately reflect the potential risks of intelligent vehicles in driving scenarios and help intelligent vehicles find the best driving strategy in the current driving environment.

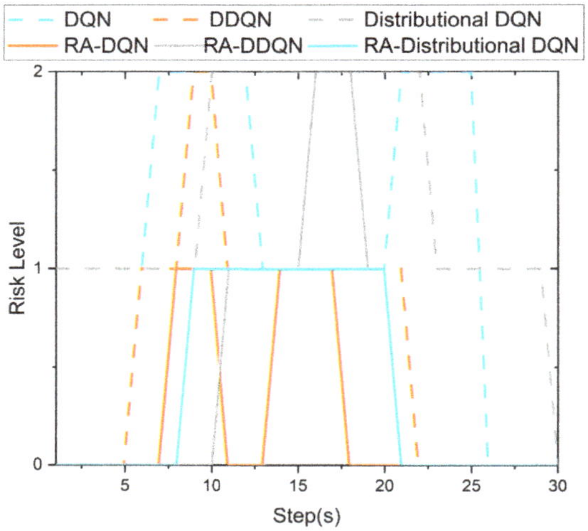

Figure 17. Intersection scenario test results: RA denotes that the algorithm uses a risk assessment method; 0 denotes that the risk level is safe; 1 denotes that the risk level is attentive; and 2 denotes that the risk level is dangerous.

5.3. Neural Network Structure Analysis

In this paper, to demonstrate the superiority of the state-information-based attention network proposed in this paper, we designed experiments using the Distributional DQN algorithm and a multilayer perceptron (MLP) neural network for reinforcement learning training in an intersection scenario. In this section, the attention network and MLP network were trained using the distributional DQN algorithm for reinforcement learning, and a comparison of their training results is shown in Figure 18. In Figure 18, the horizontal axis represents the number of episodes during training and the vertical axis represents the round reward.

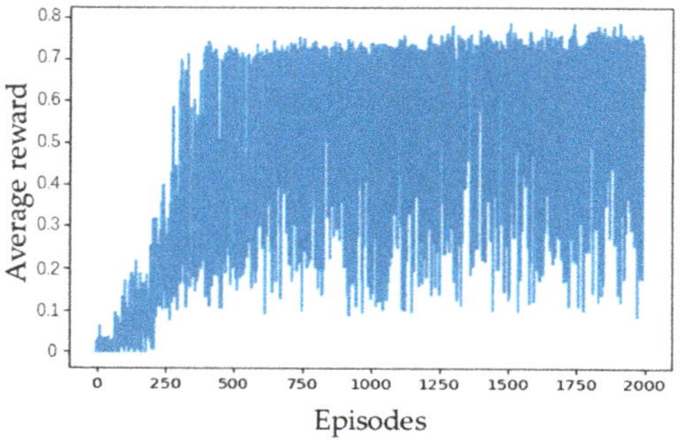

(**a**)

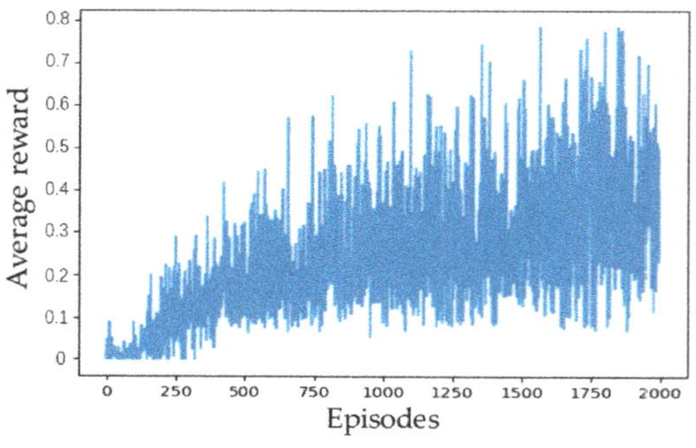

(**b**)

Figure 18. Training results of algorithms using different network structures: (**a**) using the attention network and (**b**) not using the attention network.

The state-information-based attention network proposed in this paper contains a channel attention module and a spatial attention module, which focus on the practical information in state space and the spatial location of the practical information, respectively. The channel attention module can help the intelligent vehicle identify critical state information in the current driving scenario, such as relative speed information. In contrast, the spatial attention module can help the intelligent car identify which vehicles around it are of importance to understand the surrounding environment. By analyzing the experimental results of different network structures, we find that intelligent vehicles using the attention network algorithm perform better than those using the MLP network algorithm.

The specific performance of the intelligent vehicle during the experiment also proves that the intelligent vehicle with an attention network based on state information can complete the driving task more safely in the complex intersection scenario. The specific performance during the experiments is shown in Figure 18. In this paper, we use green-width lines to connect the intelligent vehicle with the surrounding vehicles, and the width of the connecting lines is proportional to the corresponding attention weights. The list on the right side shows the attention weight ratio for each dimension of state information output by the attention network. In Figure 19a, the intelligent vehicle has not yet entered the intersection, the surrounding vehicles 1, 2, 3, and 4 are far away, and the destinations of these vehicles are not yet precise. Therefore, the intelligent vehicle shows equal attention to different dimensions of state information and these vehicles. In addition, for vehicles that do not pose a threat, intelligent vehicles do not assign attention. In Figure 19b, the direction of travel of vehicles 1 and 2, whose original destinations are uncertain, changes, showing that vehicle 1 turns left and vehicle 2 goes straight, while vehicles 3 and 4 remain far away from the intelligent vehicles and their destinations are unclear. Therefore, the intelligent vehicle pays more attention to vehicles 1 and 2 than vehicles 3 and 4. As the intelligent vehicle drove into the intersection, the attention level of the attention network for map information increased. As the relative distance between vehicles 1 and 2 and the intelligent vehicles on the x-axis gradually decreases, the attention network starts to pay more attention to the state information in the x-axis direction.

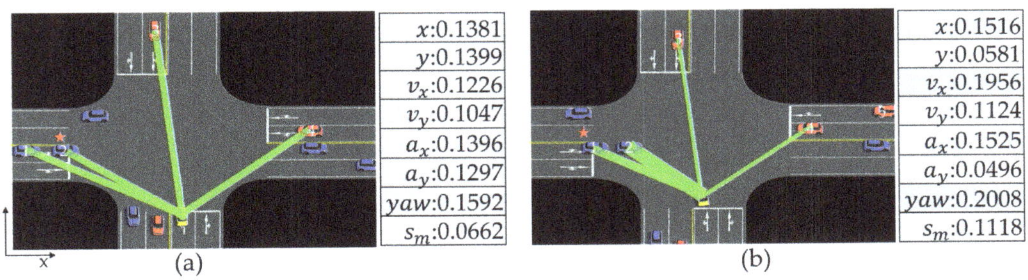

Figure 19. Changes in attention distribution during driving in the intelligent vehicle. (**a**) The intelligent vehicle has not yet entered the intersection.; (**b**) The intelligent vehicle has entered the intersection.

In summary, this section demonstrates the superiority of attention networks based on state information through experimental results. Introducing the attention mechanism makes the algorithm training process more stable, it converges faster, and it improves intelligence performance.

6. Conclusions

In this paper, a non-deterministic vehicle driving risk assessment method is proposed for the intersection scenario and introduced into the learning-based intelligent driving decision algorithm to derive an automated driving decision strategy with low expected risk and high driving efficiency. According to the analysis of the experimental results, the proposed algorithm can effectively derive a driving strategy with both driving efficiency and safety in intersection driving scenarios. The designed risk assessment method can improve driving safety while ensuring the driving efficiency of the intelligent vehicle. In contrast, the designed attention neural network helps the intelligent vehicle perceive the surrounding environment more accurately and identify important information related to its driving and the surrounding vehicles, thus improving the training stability of the algorithm, enhancing the performance of the intelligent body, and accelerating the convergence speed. The research in this paper is still an exploration of an ideal lane environment, where various other driving scenarios exist in the actual traffic environment. Therefore, introducing the idea of transfer learning to transfer knowledge or skills learned in a single scenario to other environments, or transferring knowledge gained in virtual environments to real scenarios, will be the direction of continued research in future work.

Author Contributions: Conceptualization, W.Y. and Y.Q.; methodology, W.Y.; software, W.Y.; validation, W.Y., Y.Q. and H.S.; formal analysis, W.Y.; investigation, J.W.; resources, J.X.; data curation, J.X.; writing—original draft preparation, W.Y.; writing—review and editing, Y.Q.; visualization, H.S.; supervision, J.X.; project administration, Y.Q.; funding acquisition, Y.Q. All authors have read and agreed to the published version of the manuscript.

Funding: This research received no external funding.

Data Availability Statement: Not applicable.

Conflicts of Interest: The authors declare no conflict of interest.

References

1. Bigra, E.M.; Connelly, E.; Gorner, M.; Lowans, C.; Paoli, L.; Tattini, J.; Teter, J.; LeCroy, C.; MacDonnell, O.; Welch, D.; et al. *Global EV Outlook 2021 | Accelerating Ambitions Despite the Pandemic*; IEA: Paris, France, 2021.
2. World Health Organization. *Global Status Report on Road Safety: Summary*; World Health Organization: Geneva, Switzerland, 2018.
3. GIDAS: German In-Depth Accident Study. Available online: http://www.gidas.org/ (accessed on 1 February 2018).
4. Namazi, E.; Li, J.; Lu, C. Intelligent intersection management systems considering autonomous vehicles: A systematic literature review. *IEEE Access* **2019**, *7*, 91946–91965. [CrossRef]
5. Li, G.; Li, S.; Li, S.; Qin, Y.; Cao, D.; Qu, X.; Cheng, B. Deep reinforcement learning enabled decision-making for autonomous driving at intersections. *Automot. Innov.* **2020**, *3*, 374–385. [CrossRef]
6. Seong, H.; Jung, C.; Lee, S.; Shim, D.H. Learning to drive at unsignalized intersections using attention-based deep reinforcement learning. In Proceedings of the 2021 IEEE International Intelligent Transportation Systems Conference (ITSC), Indianapolis, IN, USA, 19–22 September 2021; pp. 559–566.
7. Jurgen, R.K. *V2V/V2I Communications for Improved road Safety and Efficiency*; SAE International: Warrendale, PA, USA, 2012; Volume 154.
8. Furda, A.; Vlacic, L. Enabling safe autonomous driving in real-world city traffic using multiple criteria decision making. *IEEE Intell. Transp. Syst. Mag.* **2011**, *3*, 4–17. [CrossRef]
9. Chong, L.; Abbas, M.M.; Flintsch, A.M.; Higgs, B. A rule-based neural network approach to model driver naturalistic behavior in traffic. *Transp. Res. Part C Emerg. Technol.* **2013**, *32*, 207–223. [CrossRef]
10. Li, S.; Zhang, J.; Wang, S.; Li, P.; Liao, Y. Ethical and legal dilemma of autonomous vehicles: Study on driving decision-making model under the emergency situations of red light-running behaviors. *Electronics* **2018**, *7*, 264. [CrossRef]
11. Bai, Z.; Shangguan, W.; Cai, B.; Chai, L. Deep reinforcement learning based high-level driving behavior decision-making model in heterogeneous traffic. In Proceedings of the 2019 Chinese Control Conference (CCC), Guangzhou, China, 27–30 July 2019; pp. 8600–8605.
12. Hillenbrand, J.; Spieker, A.M.; Kroschel, K. A multilevel collision mitigation approach—Its situation assessment, decision making, and performance tradeoffs. *IEEE Trans. Intell. Transp. Syst.* **2006**, *7*, 528–540. [CrossRef]
13. Glaser, S.; Vanholme, B.; Mammar, S.; Gruyer, D.; Nouveliere, L. Maneuver-based trajectory planning for highly autonomous vehicles on real road with traffic and driver interaction. *IEEE Trans. Intell. Transp. Syst.* **2010**, *11*, 589–606. [CrossRef]
14. Lee, H.; Kang, C.M.; Kim, W.; Choi, W.Y.; Chung, C.C. Predictive risk assessment using cooperation concept for collision avoidance of side crash in autonomous lane change systems. In Proceedings of the 2017 17th International Conference on Control, Automation and Systems (ICCAS), Jeju, Republic of Korea, 18–21 October 2017; pp. 47–52.
15. Mar, J.; Lin, H.-T. The car-following and lane-changing collision prevention system based on the cascaded fuzzy inference system. *IEEE Trans. Veh. Technol.* **2005**, *54*, 910–924. [CrossRef]
16. Wang, C.; Stamatiadis, N. Surrogate safety measure for simulation-based conflict study. *Transp. Res. Rec.* **2013**, *2386*, 72–80. [CrossRef]
17. Li, Y.; Lu, J.; Xu, K. Crash risk prediction model of lane-change behavior on approaching intersections. *Discret. Dyn. Nat. Soc.* **2017**, *2017*, 7328562. [CrossRef]
18. Schubert, R.; Schulze, K.; Wanielik, G. Situation assessment for automatic lane-change maneuvers. *IEEE Trans. Intell. Transp. Syst.* **2010**, *11*, 607–616. [CrossRef]
19. Kim, B.; Park, K.; Yi, K. Probabilistic threat assessment with environment description and rule-based multi-traffic prediction for integrated risk management system. *IEEE Intell. Transp. Syst. Mag.* **2017**, *9*, 8–22. [CrossRef]

20. Noh, S.; An, K. Decision-making framework for automated driving in highway environments. *IEEE Trans. Intell. Transp. Syst.* **2017**, *19*, 58–71. [CrossRef]
21. Laugier, C.; Paromtchik, I.E.; Perrollaz, M.; Yong, M.; Yoder, J.-D.; Tay, C.; Mekhnacha, K.; Nègre, A. Probabilistic analysis of dynamic scenes and collision risks assessment to improve driving safety. *IEEE Intell. Transp. Syst. Mag.* **2011**, *3*, 4–19. [CrossRef]
22. Lambert, A.; Gruyer, D.; Saint Pierre, G. A fast monte carlo algorithm for collision probability estimation. In Proceedings of the 2008 10th International Conference on Control, Automation, Robotics and Vision, Hanoi, Vietnam, 17–20 December 2008; pp. 406–411.
23. Xu, H.; Gao, Y.; Yu, F.; Darrell, T. End-to-end learning of driving models from large-scale video datasets. In Proceedings of the IEEE Conference on Computer Vision and Pattern Recognition, Honolulu, HI, USA, 21–26 July 2017; pp. 2174–2182.
24. Codevilla, F.; Müller, M.; López, A.; Koltun, V.; Dosovitskiy, A. End-to-end driving via conditional imitation learning. In Proceedings of the 2018 IEEE International Conference on Robotics and Automation (ICRA), Brisbane, QSL, Australia, 21–25 May 2018; pp. 4693–4700.
25. Mirchevska, B.; Blum, M.; Louis, L.; Boedecker, J.; Werling, M. Reinforcement learning for autonomous maneuvering in highway scenarios. In Proceedings of the Workshop for Driving Assistance Systems and Autonomous Driving, Yokohama, Japan, 16–19 October 2017; pp. 32–41.
26. Mukadam, M.; Cosgun, A.; Nakhaei, A.; Fujimura, K. Tactical decision making for lane changing with deep reinforcement learning. In Proceedings of the NIPS Workshop on Machine Learning for Intelligent Transportation Systems, Long Beach, CA, USA, 9 December 2017.
27. Hu, Y.; Nakhaei, A.; Tomizuka, M.; Fujimura, K. Interaction-aware decision making with adaptive strategies under merging scenarios. In Proceedings of the 2019 IEEE/RSJ International Conference on Intelligent Robots and Systems (IROS), Macau, China, 3–8 November 2019; pp. 151–158.
28. Bouton, M.; Nakhaei, A.; Fujimura, K.; Kochenderfer, M.J. Cooperation-aware reinforcement learning for merging in dense traffic. In Proceedings of the 2019 IEEE Intelligent Transportation Systems Conference (ITSC), Auckland, New Zealand, 27–30 October 2019; pp. 3441–3447.
29. Choi, E.H. *Crash Factors in Intersection-Related Crashes: An On-Scene Perspective*; Technical Report No. DOT HS 811 366; U.S. Department of Transportation, National Highway Traffic Safety Administration (NHTSA): Washington, DC, USA, 2010.
30. Zhou, M.; Qu, X.; Jin, S. On the impact of cooperative autonomous vehicles in improving freeway merging: A modified intelligent driver model-based approach. *IEEE Trans. Intell. Transp. Syst.* **2016**, *18*, 1422–1428. [CrossRef]
31. *ISO Std. 22 179:2009*; Intelligent Transport Systems—Full Speed Range Adaptive Cruise Control (FSRA) Systems—Performance Requirements and Test Procedures. ISO: Geneva, Switzerland, 2009.
32. *ISO Std. 26 684:2015*; Intelligent Transport Systems (ITS)—Cooperative Intersection Signal Information and Violation Warning Systems (CIWS)—Performance Requirements and Test Procedures. ISO: Geneva, Switzerland, 2015.
33. Na, K.; Byun, J.; Roh, M.; Seo, B. RoadPlot-DATMO: Moving object tracking and track fusion system using multiple sensors. In Proceedings of the 2015 International Conference on Connected Vehicles and EXPO (ICCVE), Shenzhen, China, 19–23 October 2015; pp. 142–143.
34. Van Hasselt, H.; Guez, A.; Silver, D. Deep reinforcement learning with double q-learning. In Proceedings of the AAAI Conference on Artificial Intelligence, Phoenix, AZ, USA, 12–17 February 2016.
35. Zhan, W.; Liu, C.; Chan, C.-Y.; Tomizuka, M. A non-conservatively defensive strategy for urban autonomous driving. In Proceedings of the 2016 IEEE 19th International Conference on Intelligent Transportation Systems (ITSC), Rio de Janeiro, Brazil, 1–4 November 2016; pp. 459–464.
36. Min, K.; Kim, H.; Huh, K. Deep distributional reinforcement learning based high-level driving policy determination. *IEEE Trans. Intell. Veh.* **2019**, *4*, 416–424. [CrossRef]
37. Woo, S.; Park, J.; Lee, J.-Y.; Kweon, I.S. Cbam: Convolutional block attention module. In Proceedings of the European Conference on Computer Vision (ECCV), Munich, Germany, 8–14 September 2018; pp. 3–19.

World Electric Vehicle Journal

MDPI

Article

Accelerated and Refined Lane-Level Route-Planning Method Based on a New Road Network Model for Autonomous Vehicle Navigation

Ke He, Haitao Ding *, Nan Xu and Konghui Guo

State Key Laboratory of Automotive Simulation and Control, Jilin University, Changchun 130025, China; heke20@mails.jlu.edu.cn (K.H.)
* Correspondence: dinght@jlu.edu.cn

Abstract: Lane-level route planning is a critical issue for a lane-level navigation system for autonomous vehicles. Current route-planning methods mainly focus on the road level and applying them directly to search for lane-level routes results in a reduction in search efficiency. In addition, previously developed lane-level methods lack consideration for vehicle characteristics and adaptability to multiple road network structures. To solve this issue, this study proposes an accelerated and refined lane-level route-planning algorithm based on a new lane-level road network model. First, five sub-layers are designed to refine the internal structure of the divided road and intersection areas so that the model can express multiple variations in road network structures. Then, a multi-level route-planning algorithm is designed for sequential planning at the road level, lane group level, lane section level, and lane level to reduce the search space and significantly improve routing efficiency. Last, an optimal lane determination algorithm considering traffic rules, vehicle characteristics, and optimization objectives is developed at the lane level to find the optimal lanes on roads with different configurations, including those with a constant or variable number of lanes while satisfying traffic rules and vehicle characteristics. Tests were performed on simulated road networks and a real road network. The results demonstrate the algorithm's better adaptability to changing road network structures and vehicle characteristics compared with past hierarchical route planning, and its higher efficiency compared with direct route planning, past hierarchical route planning, and the Apollo route-planning method, which can better support autonomous vehicle navigation.

Keywords: lane-level; road network model; route planning

Citation: He, K.; Ding, H.; Xu, N.; Guo, K. Accelerated and Refined Lane-Level Route-Planning Method Based on a New Road Network Model for Autonomous Vehicle Navigation. *World Electr. Veh. J.* **2023**, *14*, 98. https://doi.org/10.3390/wevj14040098

Academic Editors: Biao Yu, Linglong Lin, Jiajia Chen and Ghanim A. Putrus

Received: 8 January 2023
Revised: 5 March 2023
Accepted: 4 April 2023
Published: 6 April 2023

1. Introduction

Navigation systems based on digital maps can help drivers or intelligent vehicles choose the optimal route given an origin and a destination, and map and route planning are two major parts of a navigation system [1,2]. The convenience and efficiency of navigation technology are apparent, especially in complicated road and traffic conditions. Traditional navigation systems affect driving behavior by notifying a driver using a display and sound, which requires the driver to process the guidance information further and choose a trajectory in real time. However, for autonomous vehicles, the navigation system can provide more detailed guidance if it reaches the lane level, which can significantly relieve the computational burden and reduce the risk of failure of real-time perception and decision systems [3]. Lane-level localization is the basis for lane-level navigation. The global navigation satellite system (GNSS) [4] is used in navigation systems; it can reach the centimeter level in open areas with real-time kinematic (RTK) [5] technology. When the signal is blocked, high positioning accuracy can be attained by camera-based [6] or LiDAR-based [7] feature-matching technology. More details about localization are not discussed here, but can be found in the literature.

The map is a foundation for route planning. Traditional route planning only outputs road-level routes and falls short of providing more delicate route guidance since it is based on road-level maps, which lack the context of lane information. To meet the needs of lane-level navigation for autonomous driving, route planning should output routes accurately to the lane level, which relies on lane-level maps enhanced with lane-level details of the environment, compared with road-level maps [8].

Many researchers have focused on the lane-level road network model, the mathematical expression used to abstract and simplify an actual lane-level map. On the one hand, various road network geometric models have been proposed, including polylines [9], arc splines [10], clothoids [11,12], Cubic Hermite Spline [13], B-spline [14], and piecewise polynomial curves [15]. They have achieved centimeter-level mapping accuracy. On the other hand, the logical representation of lane-level road networks is developing rapidly. In 2007, the DARPA Urban Challenge Road Network Definition File (RNDF) and Mission Definition File (MDF) were used to define map and mission data [16]. Each road segment in the map data contained multiple lanes represented by numerous waypoints, and some waypoints were marked as entries or exits to build connections between the roads. Nevertheless, the model and information in the map files are too concise to express a complex road environment. Bétaille et al. [12] established a one-way road and lane model for each road and represented it with a centerline but ignored the intersection model, so the model was not adapted for lane-level route planning. Jiang K et al. [3] designed a multi-layer map model for lane-level route planning. However, the model could not express possible changes in the number of lanes on the road (for example, when a road changes from three to four lanes), which is a typical road network structure. Several map standards, such as the Navigation Data Standard (NDS) [17] and OpenDRIVE [18], could be employed for lane-level route planning. However, the standards store detailed map information, most of which is not required for a given route-planning task. The direct application of these standards to lane-level route planning lacks pertinence, flexibility, and efficiency.

Route planning in road networks is usually simplified to the shortest path problem on a graph. According to whether the edge weights in the road network change with time, route planning can be divided into two major categories: static route planning and dynamic route planning. In static route planning, the edge weights do not change with time, and the key to the problem at this point is how to search for the shortest path on the graph structure of the road network. The shortest path algorithms applied to static route planning include Dijkstra's algorithm [19], Floyd's algorithm [20], the A* algorithm [21], the ant colony algorithm [22], and so on. The Dijkstra algorithm and Floyd algorithm can find the global optimal route, but the number of iterations of the algorithm is high. The A* algorithm uses a heuristic function to optimize the Dijkstra algorithm, but the specific degree of optimization is closely related to the selection of the heuristic function. The ant colony algorithm takes its inspiration from the natural biological environment and is also known as the bionic intelligence algorithm. In addition, some scholars have improved the above algorithms, for example, Xu et al. [23]. used a bidirectional search strategy to improve the traditional A* algorithm as a way to speed up the efficiency of the search. Lee et al. [24]. used a genetic algorithm to improve the ant colony algorithm and proposed the IAACO algorithm, which improved the convergence performance and optimal-solution-finding ability of the algorithm. Jiang et al. [25] proposed an intelligent optimization algorithm of adaptive ant colony and particle swarm optimization to improve efficiency. Lan et al. [26] proposed an improved algorithm based on the A* algorithm fused with the ant colony algorithm to improve the convergence speed compared with the ant colony algorithm. In dynamic route planning problems, some studies consider real-time traffic conditions and model the uncertain edge weights [27–29]. The above algorithms are all considered at the road level. When they are applied directly to the lane level, the efficiency is greatly reduced due to the large increase in the number of vertices and edges, especially in the large-scale road network, which makes it challenging to meet the real-time requirements for autonomous driving. Heuristic rules and hierarchy structures have been exploited to realize efficient routing in large networks [30–32]. However, they focus on road-level maps and do not reach the lane level. On the other hand, more factors should be considered to plan the optimal lane-level routes, such as whether the routes involving lane changing and turning operations meet vehicle characteristics and traffic regulations and how to determine the best lane-level route on a road with a variable number of lanes. Jiang K et al. proposed a seven-layer map structure and developed a hierarchical route-searching algorithm to accelerate the planning process [3]. However, for one thing, the algorithm lacked the adaptability to determine the optimal lanes on a road with variable lane numbers; for another, the algorithm did not consider minimum allowable distance constraints for vehicle turns and lane changes, which may lead to unreasonable results.

To sum up, an applicable road network model is needed as a foundation for lane-level route planning. Specifically, this model should include the necessary traffic elements to demonstrate lane-level details and express various road network structures while also being simple enough for practical applications. Then, based on this map model, it is important a lane-level route-planning method be provided that can improve the efficiency of existing search algorithms in searching for lane-level routes and can find the optimal lane sequence in various road network structures while satisfying traffic rules and vehicle characteristics.

The goal of this study is to design a route planning algorithm that is compatible with existing graph search methods based on the designed multi-layer road network model, which can find optimal lanes in changing road network structures and further improve the existing methods' efficiency when applied to lane-level route planning while also satisfying traffic rules and vehicle characteristics. This study's primary research content is summarized in Figure 1. The main contributions are listed as follows:

1. A new road network model for lane-level route planning is proposed. The model is divided into the road and intersection areas containing five sub-layers. The model can express multiple road network structures and can facilitate more flexible, fine-grained, and efficient route planning.
2. Based on the proposed road network model, an accelerated and refined lane-level route-planning method is proposed. First, a multi-level route-planning algorithm is designed for sequential planning at the road level, lane group level, lane section level, and lane level, reducing the number of nodes and edges traversed and improving routing efficiency. Then, an optimal lane determination algorithm considering traffic rules, vehicle characteristics, and optimization objectives is developed to find the optimal lanes at the lane level. The entire route-planning algorithm can remarkably improve existing search algorithms' efficiency in searching lane-level routes and acquire optimal lanes satisfying traffic rules and vehicle characteristics on roads with different structures, including those with a constant or variable number of lanes.

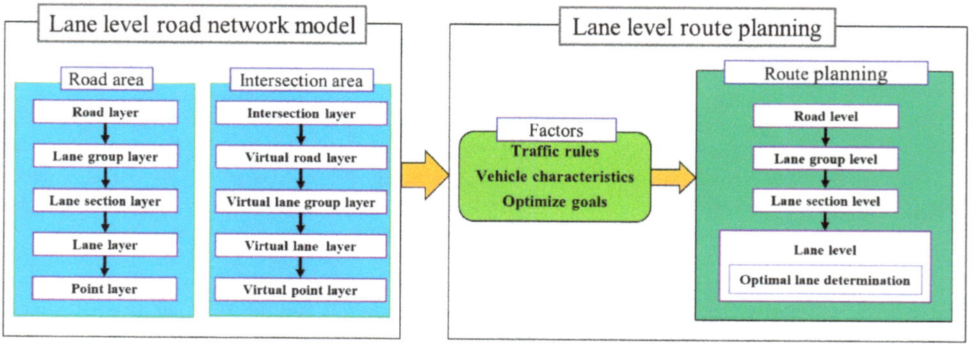

Figure 1. The main research content of this study.

The rest of this paper is organized as follows. Section 2 introduces the proposed road network model. Section 3 details the designed lane-level route-planning algorithm based on the model. Experiments are performed on simulated road networks and an actual road network to validate the route-planning algorithm, and the results and discussion are presented in Section 4. Finally, conclusions are given in Section 5.

2. Lane-Level Road Network Modeling

A new lane-level road network model, which is the foundation of the lane-level route planning elaborated upon later, is created in this section. The whole model consists of road and intersection areas containing five sub-layers, which are used to refine the structures of real roads and intersections. This multi-sub-layer modeling can express multiple variations in the road network structure and be adapted to the multi-level route-planning algorithm proposed in a later section.

The road network is expressed as follows:

$$G = (R, I) \tag{1}$$

where G represents the complete road network, R describes the road area consisting of a set of roads $\{r_m\}_{m=1}^{N_r}$, and I is the intersection area comprising a group of intersections $\{i_m\}_{m=1}^{N_i}$.

2.1. Road Area

2.1.1. Road Layer

This article defines a set of lanes in the same direction on one road as a lane group. When the lanes on the road have two opposite directions, the road has two lane groups; if there is only a single direction, there is only one lane group. One road r is

$$r = (LG, q_r, f, b) \tag{2}$$

where LG is a set of lane groups $\{lg_m\}_{m=1}^{N_{lg}}$, q_r represents common road attributes such as length, class, and type, and f and b indicate the intersection entering and leaving this road.

2.1.2. Lane Group Layer

One lane group in a road can be expressed as

$$lg = (LS, isequal) \tag{3}$$

where LS is a set of lane sections $\{ls_m\}_{m=1}^{N_{ls}}$, and one lane section is defined as a unit for which the number of lanes is constant. $isequal$ can be 1 or -1, where 1 means that the lane group's direction is the same as the forward direction defined by the road, and -1 is the opposite.

2.1.3. Lane Section Layer

One lane section in a lane group is

$$ls = \{l_m\}_{m=1}^{N_l} \tag{4}$$

where $\{l_m\}_{m=1}^{N_l}$ is a set of lanes.

2.1.4. Lane Layer

One lane in a lane section can be characterized as

$$l = (seq, pre, suc, P_{geo}, P_{attr}, q) \tag{5}$$

where seq denotes the sequence number of l in ls; starting from the lane closest to the road centerline, the serial number increases from 1 in order. pre and suc define the intersection or lane connecting the start and end of l. P_{geo} and P_{attr} are point sets defined in the point layer. q contains lane attributes such as the width, length and speed limit.

2.1.5. Point Layer

The point layer consists of the geometric control point set P_{geo} and the attribute control point set P_{attr} on each lane. P_{geo} comprises the points characterizing the lane geometry. Each point p_{geo} in P_{geo} can be expressed as

$$p_{geo} = (n, x, y, z) \tag{6}$$

where n is the serial number of p_{geo} on the entire lane and x, y, z indicate the X, Y, and Z coordinates in the ENU coordinate system.

P_{attr} is a set of points on the lane markings reflecting lane property changes, which can be represented as

$$P_{attr} = (side, \{p_{attr}{}^{m}\}_{m=1}^{N_{P_{attr}}}) \tag{7}$$

where *side* can be "left" or "right", indicating that the lane marking is located on the left or right side of the lane.

Each point p_{attr} in P_{attr} can be denoted as

$$p_{attr} = (x, y, z, s, h, type, lc) \tag{8}$$

where x, y, z indicate the coordinates in the ENU coordinate system, s is the length from the start point to this point along the lane, h is the lateral distance of the point relative to the lane centerline, and $type$ indicates the point type which can take the values "start", "end", "lanechange", or "typechange". "start" and "end" represent the start and end of a lane marking, respectively. "lanechange" or "typechange" indicate that p_{attr} is a demarcation point where the left or right lane marking type changes or the lane type changes. lc is true if the interval of the lane marking following this point in the lane direction can allow a lane change; otherwise, it is false.

A road is taken as an example to illustrate the model in the road area, as shown in Figure 2; the road whose forward direction points to the right comprises two lane groups. lg_1 and lg_2 both have two lane sections, ls_1 and ls_2. ls_1 contains three lanes and ls_2 contains four lanes. The left and right lane markings of l_3 in ls_2 in lg_2 are used to illustrate the attribute control points: according to (8), the right lane marking has two attribute control points, the start and end points; since the left lane marking type changes, it has one more control point than the right lane marking, which marks the place where the lane marking type changes. For each lane marking, combining the value of lc in (8) of attribute control points, the lane changeable interval can be known, facilitating the refined lane-level route-planning algorithm.

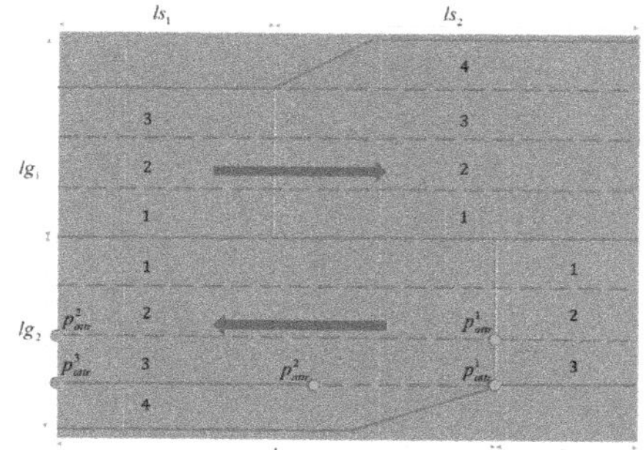

Figure 2. A road used to illustrate the model. The marked numbers are the sequence numbers of lanes.

2.2. Intersection Area

2.2.1. Intersection Layer

An intersection i is defined as

$$i = (VR, sig, P_{center}) \tag{9}$$

where VR represents a set of virtual roads $\{vr_m\}_{m=1}^{N_{vr}}$ that connect roads entering and leaving i, sig indicates the intersection attributes, such as the number and position of traffic lights and the junction type, and P_{center} includes the coordinates of the intersection center point.

2.2.2. Virtual Road Layer

One virtual road vr can be indicated as

$$vr = (r_f, r_r, w, VLG) \tag{10}$$

where r_f, r_r represent the roads entering or exiting the intersection and w describes the passage method, including going straight, turning left, turning right, and making a U-turn.

2.2.3. Virtual Lane Group Layer

VLG in (10) contains a set of virtual lane groups $\{vlg_m\}_{m=1}^{N_{vlg}}$.

One virtual lane group, vlg in VLG, comprises all virtual lanes connecting to the same lane entering the intersection:

$$vlg = \left(seq, \{vl_m\}_{m=1}^{N_{vl}}\right) \tag{11}$$

where seq denotes the sequence number of the lane entering the intersection and $\{vl_m\}_{m=1}^{N_{vl}}$ is a set of virtual lanes.

2.2.4. Virtual Lane Layer

One virtual lane vl can be denoted as

$$vl = (l_f, l_r, P_{vl}) \tag{12}$$

where l_f, l_r represent the lanes entering or exiting the intersection.

2.2.5. Virtual Point Layer

P_{vl} in (12) represents the control points approximating the possible turning trajectory of a vehicle traveling from lane l_f to l_r. Each point p_{vl} in P_{vl} can be expressed as

$$p_{vl} = (n, x, y, z) \tag{13}$$

where n is the serial number of p_{vl} on the entire virtual lane and x, y, z indicate the X, Y, and Z coordinates in the ENU coordinate system.

An intersection shown in Figure 3 illustrates the model in the intersection area. The figure depicts an intersection where four roads join. Each road has two lane groups represented by broad arrows on both sides of the road centerline, the direction of which indicates the lane group's driving direction. Each lane group has only one lane section containing three lanes. The virtual roads with Road 3 as the entrance road are displayed. Only one lane entering the intersection is linked with vr_1, vr_2, or vr_3, so there is only one virtual lane group for them. vr_4, connecting Road 3 with Road 2, has two entering lanes; namely, it has two virtual lane groups. The first virtual lane group vlg_1 links with the middle lane of the lane section on Road 3. The second virtual lane group vlg_2 merges with the rightmost lane in Road 3: vl_1, vl_2, and vl_3, marked in Figure 3, are three virtual lanes in vlg_2.

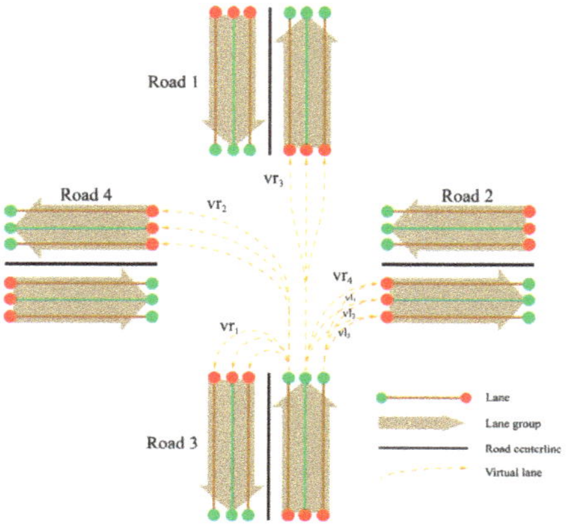

Figure 3. An intersection used to illustrate the model.

2.3. Comparison with Other Map Models

In light of the preceding, the designed road area accounts for lane groups with different driving directions. The lane group considers different lane sections to represent changes in the lane number on real roads, and the change in lane type can be reflected in attribute control points. The designed intersection area contains virtual roads consisting of virtual lane groups with many virtual lanes, hierarchically representing the connections within the intersection. Therefore, the model can express multiple road network structures. The proposed model's differences and advantages compared to the existing models are shown in Table 1.

Table 1. Comparison of the proposed model with other map models.

Existing Map Models	The Proposed Model (Referred to Here as A)'s Differences and Advantages Compared to Those of the Existing Model (Referred to Here as B)
RNDF	A can express possible changes in the number of lanes on one road, but B cannot.
The model proposed by Bétaille et al. [12]	A contains the intersection model, but B ignores it, which is a basic requirement for lane-level route planning.
The model proposed by Jiang K et al. [3]	A can express possible changes in the number of lanes on one road, but B cannot.
OpenDRIVE	Compared with B, A defines lane groups and the point layer in the road area. In addition, the designed intersection area contains five sub-layers that hierarchically represent the connections within the intersection.

3. Lane-Level Route Planning

This section proposes an accelerated and refined lane-level route-planning method based on the established road network model. First, a multi-level route-planning algorithm is proposed and implemented in turn at the road, lane group, lane section, and lane level; the number of nodes and edges traversed can be decreased significantly to enhance routing efficiency. Then, at the lane level, an optimal lane determination algorithm considering traffic rules, vehicle characteristics, and optimization goals is designed to find the optimal lanes on roads with different configurations, including those with a constant or variable number of lanes. The route-planning method at each level is discussed in detail.

3.1. Route Planning at the Road Level

Route planning at the road level is based on the road layer and the virtual road layer, generating a feasible route with a road or an intersection as a unit. A road-level directed graph G_R can be expressed as follows:

$$G_R = (V_R, E_R) \tag{14}$$

where V_R and E_R are a set of nodes comprising the intersections and edges that include roads.

When the starting point O and end point D are given in G_R, many existing graph search algorithms, such as Dijkstra and A*, can be used at this level. Finally, the result V_R contains an intersection sequence to pass from the start to the end:

$$V_R = (O, i_1, \ldots, i_k, \ldots, i_n, D) \tag{15}$$

where i_k is the kth intersection in the sequence and n is the number of intersections.

Then, the roads between the intersection sequence can be written as

$$E_R = (r_0, \ldots, r_k, \ldots, r_n) \tag{16}$$

where r_0 and r_n are the roads where point O and point D are located, respectively, and r_k is the road between i_k and i_{k+1}.

3.2. Route Planning at the Lane Group Level

Route planning at the lane group level is based on the lane group layer and the virtual lane group layer.

According to the attribute *is_equal* in (3), the road sequence in E_R can be refined into a result based on lane groups:

$$E_{LG} = (lg_{O-1}, \ldots, lg_{(k-1)-k}, \ldots, lg_{n-D}) \tag{17}$$

where lg_{O-1} represents the lane group between the origin O and the first intersection, lg_{n-D} represents the lane group between the nth intersection and the destination D, $lg_{(k-1)-k}$ indicates the lane group between the $(k-1)th$ intersection and the kth intersection, and $k = 2, 3, \ldots, n$.

According to (9) and (10), the intersection sequence in V_R can be described based on virtual lane groups:

$$V_{VLG} = (VLG_1, \ldots, VLG_k, \ldots, VLG_n) \tag{18}$$

where VLG_k indicates the set of virtual lane groups in the kth intersection, and $k = 1, 2, \ldots, n$.

Figure 4 displays the route planning from the road level to the lane group level; the black lines mark the planned route. Suppose the route-planning result at the road level is the road and intersection sequence from r_k to r_{k+3}; at the lane group level, it can be refined to the lane group sequence from $lg_{k-(k+1)}$ to $lg_{(k+3)-(k+4)}$.

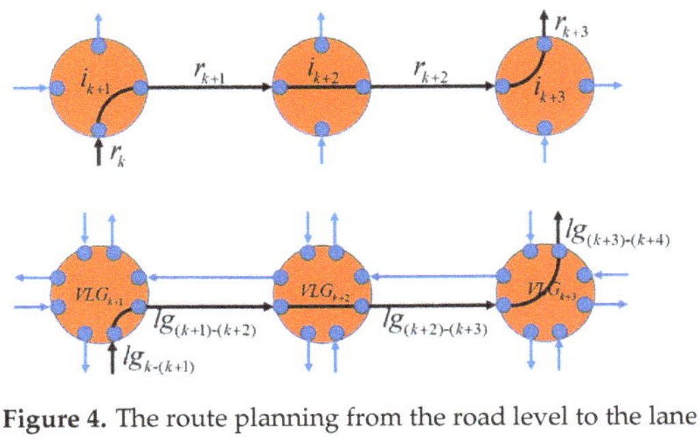

Figure 4. The route planning from the road level to the lane group level.

3.3. Route Planning at the Lane Section Level

Route planning at the lane section level uses the correspondence between the lane group layer and the lane section layer. According to (3) and the route-planning result at the lane group level, the lane sections in the lane group $lg_{(k-1)-k}$ can be written as

$$LS_{(k-1)-k} = (ls_1, \ldots, ls_k, \ldots, ls_n) \tag{19}$$

where ls_k is the kth lane section in $lg_{(k-1)-k}$ and n is the number of lane sections. ls_1 connects the previous intersection's exit and ls_n connects the entrance to the next intersection.

Then, the whole lane section sequence is

$$V_{LS} = (LS_{O-1}, \ldots, LS_{(k-1)-k}, \ldots, LS_{n-D}) \tag{20}$$

where LS_{O-1} represents the lane sections between the origin O and the first intersection, LS_{n-D} represents the lane sections between the nth intersection and the destination D, $LS_{(k-1)-k}$ indicates the lane sections between the $(k-1)th$ intersection and the kth intersection, and $k = 2, 3, \ldots, n$.

3.4. Route Planning at the Lane Level

Route planning at the lane level aims to find the optimal lanes in a sequence of lane sections in one lane group. Figure 5 displays route planning from the lane group level to the lane section level, and then to the lane level. Taking the lane group $lg_{(k+1)-(k-2)}$ as an example, it can be refined to the lane section sequence from ls_1 to ls_2 at the lane section level; at the lane level, the routing result is from l_1 to l_2. The route planning from the lane section level to the lane level adopts the optimal lane determination algorithm illustrated in Figure 6, which is designed as follows:

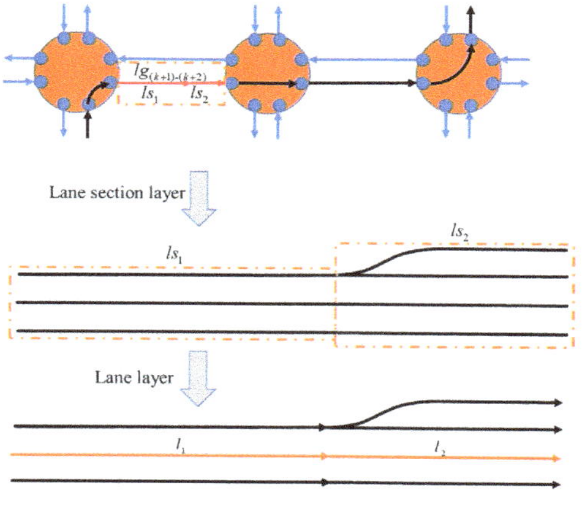

Figure 5. The route planning from the lane group level to the lane section level, and then to the lane level.

Step i: Referring to the correspondence between the lane section layer and the lane layer and the route-planning result at the lane section level, the lanes to be selected are determined.

Step ii: According to the traffic rules indicated by (12), infeasible lanes are screened out, and feasible lanes in both ls_n in $LS_{(k-1)-k}$ entering i_K and ls_1 in $LS_{k-(k+1)}$ leaving i_K are determined. In other words, for each planned lane group, the lanes conforming to traffic rules in the first and last lane sections are determined.

Step iii: Vehicle characteristics are considered for the elimination of unreasonable lanes entering and leaving intersections. For example, a vehicle may not make a U-turn from the leftmost lane to the adjacent reverse lane considering the minimum turning radius of the vehicle.

Step iv: After excluding lanes that do not meet traffic rules and vehicle turning characteristics, a directed graph is constructed. The lane group displayed in Figure 7 is taken as an example. Since the lane marking between every two adjacent lanes is dashed, lane changes can occur between every two adjacent lanes. Figure 8 provides an abstraction of the directed graph; nodes represent lanes, and edges represent the transitions between every two lanes. Each node's cost is the travel cost of driving along a single lane, and the cost of each edge perpendicular to lanes, such as that between A and B, is the travel cost of lane changing. These costs' computing methods, which are not the focus of this study, refer to the method in Ref. [3]. The cost of each edge parallel to lanes, such as that between A and D, is zero.

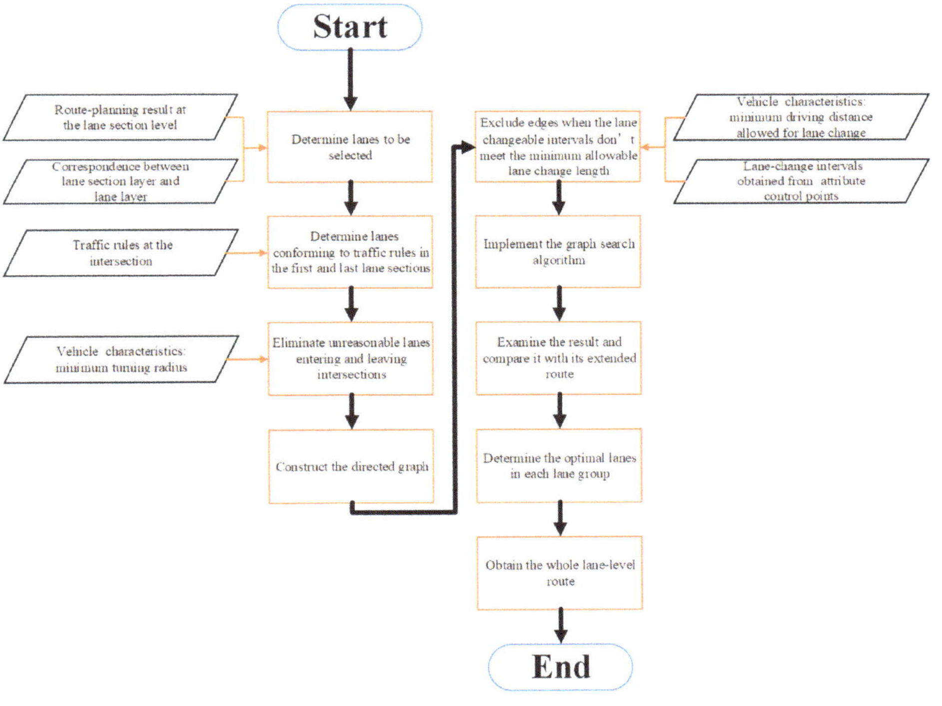

Figure 6. The flowchart of the optimal lane determination algorithm.

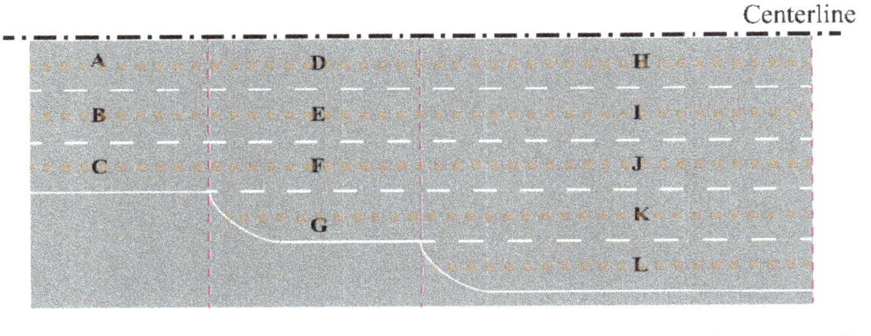

Figure 7. A lane group is shown, which comprises three lane sections that contain three, four, and five lanes, respectively. Each lane is identified by a capital letter from A to L.

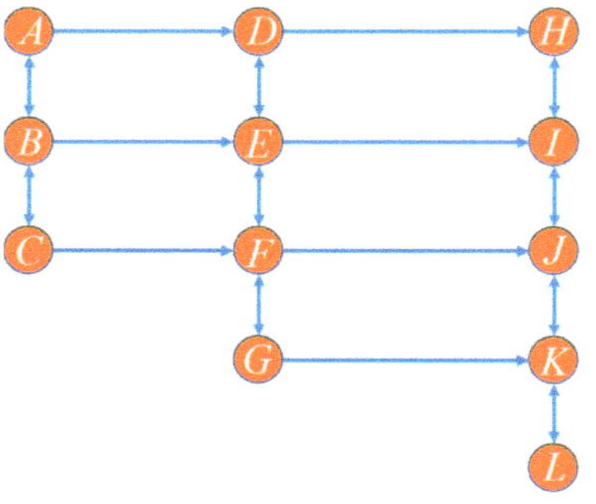

Figure 8. The directed graph abstracted from the lane group shown in Figure 7. The arrows indicate the travel directions. Two-way arrows represent mutual access; for example, the two-way arrow between D and E indicates that the vehicle can move from D to E and E to D.

Step v: The lane change intervals can be obtained according to attribute control points in the point layer. It is necessary to consider whether the lane change interval meets the minimum allowable length in combination with vehicle characteristic constraints. For example, if the lane change interval on the right lane marking of lane A is not enough for lane changing, the edge between A and B is eliminated.

Step vi: According to the predetermined first lane in ls_1 and last lane in ls_n in Step ii and Step iii and the constructed directed graph in Step iv and Step v, a graph search algorithm, such as Dijkstra or A*, is implemented.

Step vii: The result in Step vi needs to be examined because the route can be extended at the last lane section, and the extended route could not be searched by Step vi. For example, if the result is route A-B-E-I, the route can be extended to route A-B-E-I-H-I, which means changing lanes from I to H, and then back to I in the last lane section. The costs of driving along lane I and driving from lane I to lane H and then back to I are compared. If the cost of the extended route is lower than that of the original route, the extended route replaces the result.

Step viii: After Step i–vii, the optimal lanes in each lane group are determined. The optimal lane sequence in $lg_{(k-1)-k}$ can be expressed as follows:

$$L_{(k-1)-k} = (l_1, \ldots, l_k, \ldots, l_n) \tag{21}$$

where l_k is the kth lane in order and n is the number of lanes.

Step ix: The whole lane-level route can be obtained as

$$V_L = (L_{O-1}, \ldots, L_{(k-1)-k}, \ldots, L_{n-D}) \tag{22}$$

where L_{O-1} represents the lane sequence between the origin O and the first intersection, L_{n-D} is the lane sequence between i_n and the destination D, $L_{(k-1)-k}$ indicates the lane sequence between i_{k-1} and i_k, and $k = 2, 3, \ldots, n$.

4. Results and Discussion

A road network was first constructed to verify the lane-level routing algorithm using RoadRunner [33]. This interactive editor allows us to design 3D scenes for simulating and testing automated driving systems. Then, the OpenDrive file exported from RoadRunner was imported into SCANeR studio [34] to build an ultra-realistic virtual driving scene. SCANeR studio is a comprehensive software suite dedicated to automotive and transport simulation that provides all the tools and models necessary to create an ultra-realistic virtual world.

Figure 9 displays a constructed road network consisting of 16 intersections and 24 roads. All roads were bidirectional and comprised two lane groups; each lane group had one lane section containing three lanes. A green circle shows the host vehicle in the simulation. All of the road network parameters are given in Table 2. A partially enlarged intersection and corresponding virtual lanes in the intersection are displayed in Figure 10.

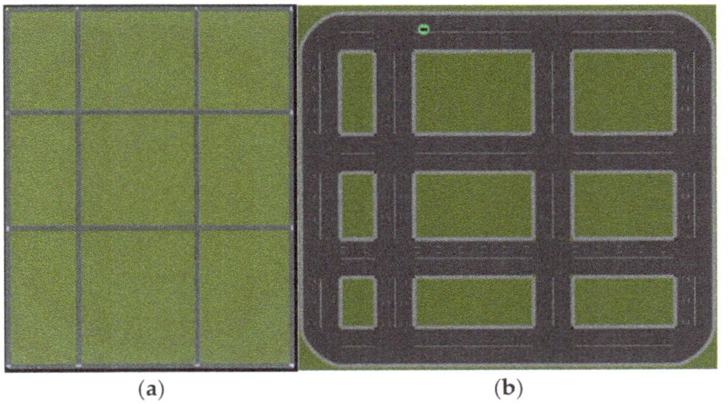

(a) (b)

Figure 9. The road network and the corresponding driving scene. (**a**) The road network constructed by RoadRunner. (**b**) The driving scene created using SCANeR studio.

Table 2. Road network parameter settings.

Parameter	Value
The number of lanes in one lane section	3
The number of lane sections in one lane group	1
The number of lane groups in one road	2
Lane width	3.5 m
Intersection width	24 m
The average speed of each road: v_{road} (km/h)	Randomly chosen from $\{80, 60, 40\}$
The average speed of each lane: v_{lane} (km/h)	Inner: $v_{road} + 20$; middle: v_{road}; outer: $v_{road} - 20$
The traffic rules designed for the lanes	Inner: left; middle: forward; outer: right

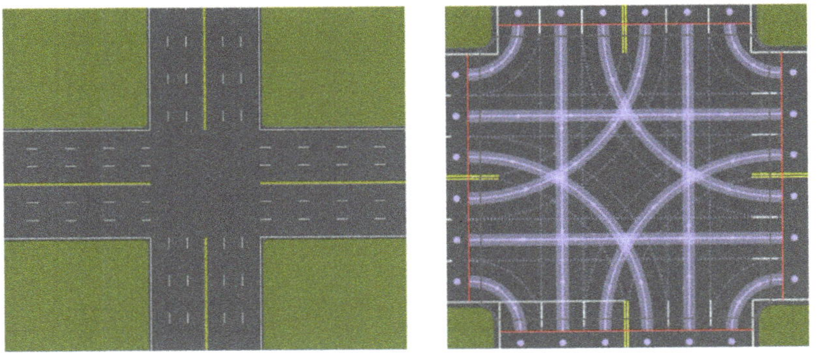

(a) (b)

Figure 10. An intersection and corresponding virtual lanes. (**a**) An intersection and four connecting roads are shown. (**b**) The virtual lanes in the intersection shown in (**a**) are displayed.

Figure 11 presents an abstraction of the simulated road network and the planned route from A to B. There are 16 vertices representing intersections and 24 edges representing roads. The average road speed is marked next to the road edge. The value on the coordinate axis shows the length of each edge. Taking the shortest total time as the goal, the route from vertex A on the lower left to vertex B on the upper right was planned using the algorithm proposed, and the red lines indicate the planned route. The average speed of the roads through the planned route is high, and the route meets the shortest time goal at the road level. The circles on the right side depict the lane-level details inside the indicated junctions: four roads enter the junction; the arrows represent the lanes' driving directions, the dotted lines represent the roads' centerlines, and the red lines show the planned route in the junction.

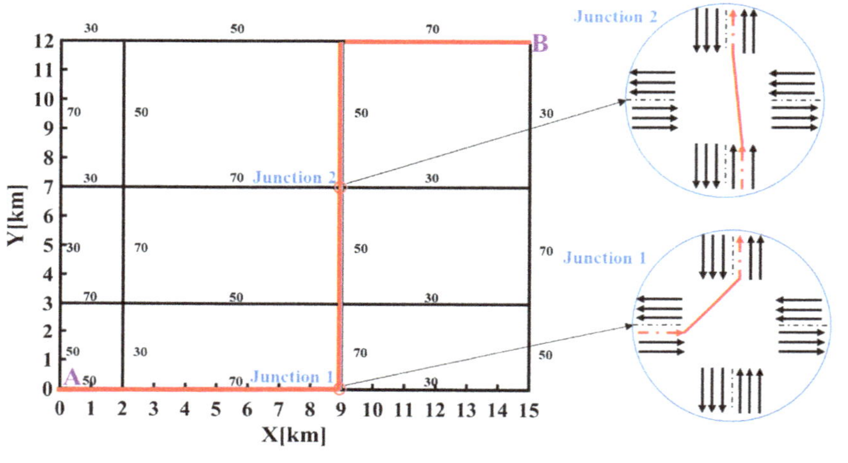

Figure 11. Abstraction of the simulated road network. The red lines show the planned route. A and B are the start and end of the route.

Two junctions are chosen to illustrate the lane-level routing results. At Junction 1, since a vehicle on the road left of the intersection must turn left when entering the intersection, it will travel in the inner lane, which meets the traffic regulations; on the other hand, after leaving the intersection, the inner lane with a high average speed is chosen, which satisfies the goal of achieving the shortest travel time. Similarly, at Junction 2, the vehicle will travel straight from the middle lane to the inner lane. This result reveals that the proposed algorithm based on the map model successfully chose the correct lanes on roads with a constant number of lanes to meet both the traffic rules and the routing goal; namely, the algorithm can find the optimal lane-level route in a road network.

Another test is shown in Figure 12, where the route from lane C to lane D is planned. To illustrate the main route-planning problem in this paper, it is assumed that the minimum turning radius, which can be estimated according to vehicle characteristics, is 8 m. The planned route can be obtained using the method in Ref. [3], and this study, it is represented by the green or red lines. We can see that the route planned by Ref. [3] goes from lane C to lane D directly; however, the minimum turning radius is larger than the two-lane width, such that the vehicle cannot directly U-turn from C to D; hence, the result is not feasible. In contrast, the route developed using the method in this study first includes a U-turn from C to E and then involves a change of lanes from E to D; although this route has a higher cost, it fits the vehicle turning characteristics. Therefore, the algorithm in this study can obtain a reasonable route considering vehicle turning characteristics, meeting the needs of practical applications.

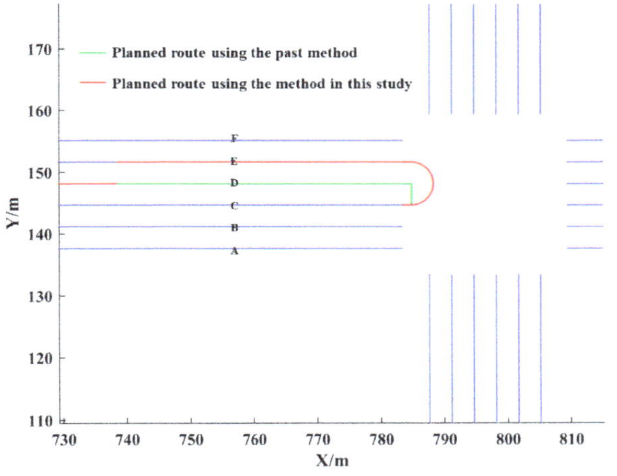

Figure 12. The constructed road network includes four roads and an intersection. Each road has six lanes, each with a width of 3.5 m. The lanes on the left road are represented by A–F. The average speed on all six lanes is set to be the same. Lane A, B, and C enter the intersection, and Lane D, E, and F exit. The traffic rules on Lane C allow for left turns and U-turns. The lane-level route from Lane C to Lane D is demonstrated. The green route is obtained using the method in Ref. [3].

Another road network, displayed in Figure 13, was constructed to verify the optimal lane determination algorithm at the lane level. The minimum length for lane changing, considering vehicle characteristics, was set to 10 m, so the interval CD was not available for a lane change; the interval AB and EF were allowed. The traffic rules designed on Lane 3–5 entering the intersection were left turn, forward and right turn, and forward and right turn, respectively. The average speeds in Lane 1–5 were 50 km/h, 30 km/h, 30 km/h, 60 km/h, and 30 km/h, respectively.

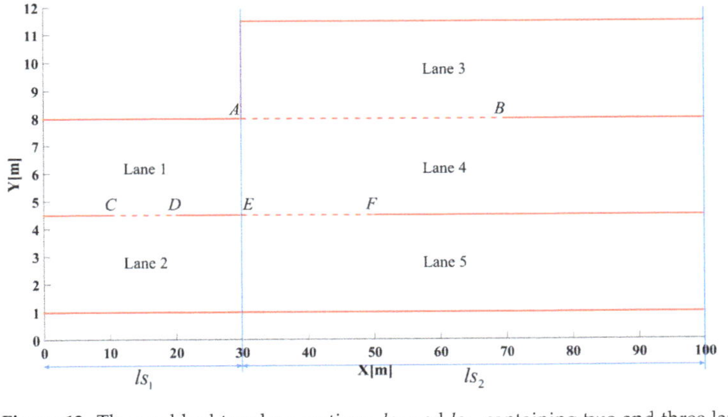

Figure 13. The road had two lane sections, ls_1 and ls_2, containing two and three lanes numbered from 1 to 5. A–F are attribute control points stored in the point layer, and the dotted line between the two points indicates that lane changes are allowed.

First, the lanes to be selected were determined according to Steps i–iii of the optimal lane determination algorithm at the lane level; the lanes chosen from ls_1 were Lane 1 and Lane 2, and the lanes chosen from ls_2 were Lane 4 and Lane 5. In contrast, Lane 3 was excluded due to the traffic rule. According to Steps iv–v, the directed graph was created as shown in Figure 14. Figure 15 illustrates all possible routes. After Step vi, Route 1 (Lane 1–Lane 4) in Figure 15a was chosen. Then, according to Step vii, Route 2 (Lane 1–Lane 4–Lane 3–Lane 4) in Figure 15a, which was the extended route of Route 1, was compared with Route 1. Finally, Route 1, with a lower cost than Route 2, was chosen as the optimal lane-level route. At the same time, the costs of all possible routes were calculated. By comparison, the cost of Route 1 was the lowest, which verified the correctness of the optimal lane determination algorithm.

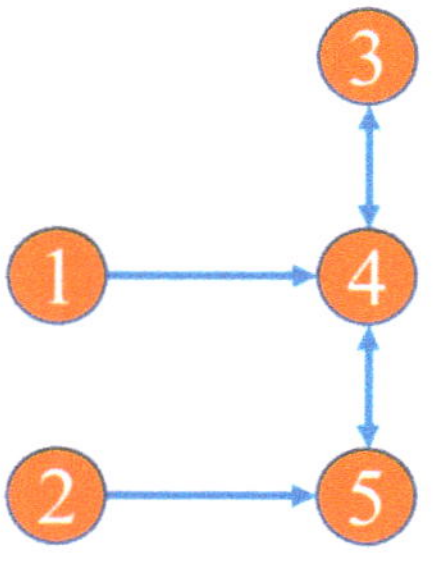

Figure 14. The directed graph of Figure 13.

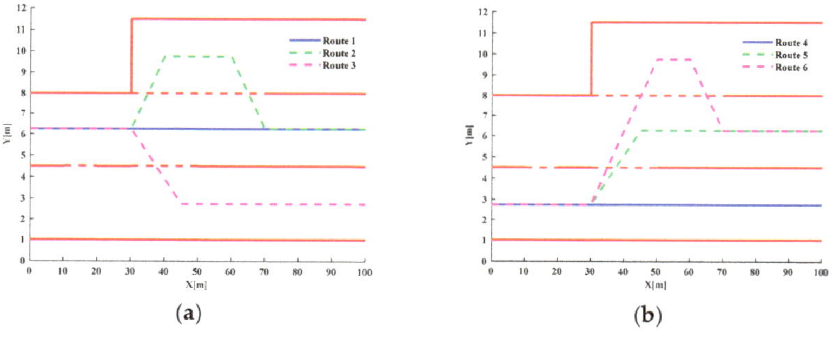

Figure 15. (**a**). Three possible routes with Lane 1 as the start lane. (**b**). Three possible routes with Lane 2 as the start lane.

This test demonstrates that the optimal lane determination algorithm could find the optimal lanes while satisfying traffic rules and vehicle characteristics on roads with a variable number of lanes. In contrast, the algorithm in Ref. [3] subdivides the road into a constant number of lanes and lacks the adaptability to this road structure; however, this road structure may result in the creation of an increased number of possible routes that need to be considered before one is chosen.

Whether the road has a constant or variable number of lanes, the directed graph can be constructed at the lane level. For example, if the lane number is constant at three, a simple directed graph consisting of three vertices and two edges can be constructed; if the road has a variable number of lanes, a more complex directed graph can be constructed. Then, the existing graph search algorithm can be applied to find the optimal route. The algorithm in Ref. [3] did not consider the lane sections as defined in this study, it could not adapt to the changing internal structure of the road. In addition, the lane change intervals are not considered in Ref. [3], meaning that the produced route may not be feasible. The tests, which assign different average speeds to different road and lanes, are mainly conducted to verify that the proposed algorithm can adapt to road networks with different structures while meeting the traffic rules and vehicle characteristics. In addition, the goal for the shortest time is satisfied under relative static conditions. The variable traffic conditions, which may influence the edges' costs, are not considered here; however, some existing algorithms considering the real-time traffic conditions can be applied at the road level and lane level and improve the practicability in the actual traffic conditions, which needs further research in the future.

From the previous section, it can be inferred that the suggested route-planning algorithm can determine the optimal lanes on roads with different configurations, including those with a constant or variable number of lanes, while satisfying the need for the shortest time, as well as the requirements for traffic regulations and vehicle characteristics, which allows the planned route to match the actual vehicle navigation needs.

Next, the increasingly intricate road networks depicted in Figure 16 were built to assess the algorithm's efficiency improvement in road networks of different sizes. To confirm the improved efficiency of our proposed algorithm, we compared the routing time of different routing algorithms, including direct lane-level route planning using the existing graph search algorithm, the route-planning method proposed in Ref. [3] and the route-planning algorithm proposed in this study, in the road networks detailed in Figures 9 and 16.

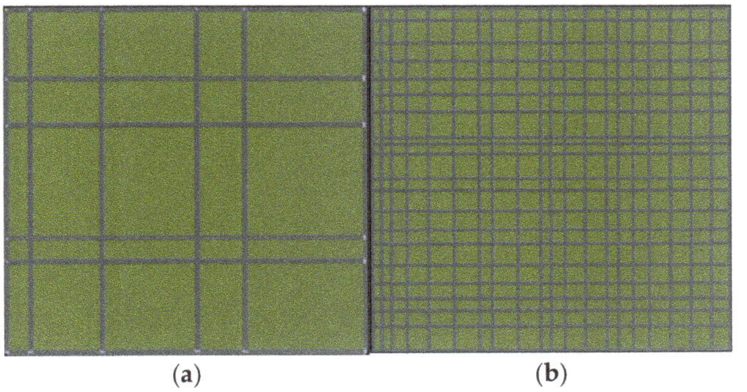

Figure 16. Two road networks were constructed, and the length of each road was randomly set. (**a**) A road network containing 6 roads in horizontal and vertical directions. (**b**) A road network containing 21 roads in horizontal and vertical directions.

Direct lane-level route planning searches the lane-level route directly on the road network. Figure 17 illustrates a road network comprising one road and an intersection. Following the directed graph abstraction of the road network (Figure 18), the direct routing can perform the existing graph search algorithm to search the lane-level route.

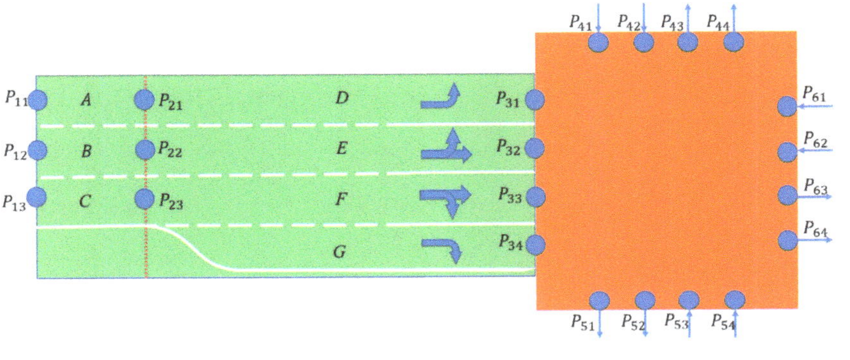

Figure 17. A road (green) with one lane group comprising two lane sections entering an intersection (orange). The lanes are treated as edges represented as A–G, the start and end nodes of which are expressed with P.

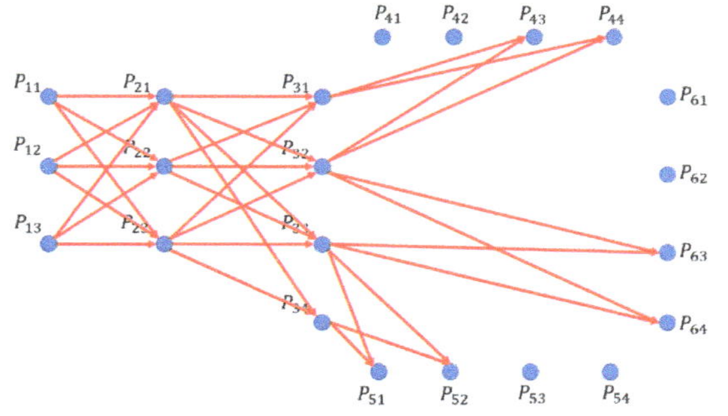

Figure 18. The abstracted directed graph. Blue dots represent nodes, and red lines represent edges.

We implemented the algorithms in C++, maintaining consistency in the hardware setup and data structure and varying the algorithms themselves. To verify the improvement in efficiency in the proposed algorithm over the existing graph search algorithm, the A* algorithm was taken as an example. Table 3 compares different routing methods, and the time saved is relative to the time spent using direct route planning based on A*. Both the algorithm in this study and the process in Ref. [3] can improve the efficiency compared with direct route planning; however, the algorithm in this study has higher efficiency for both small or large networks, and the improvement in efficiency becomes more remarkable as the scale of the road network grows.

Table 3. Time costs of different routing methods in three simulated road networks with different scales.

Number of Lanes	Routing Method	Time (μs)	Time Saved (%)
288	Direct lane-level route planning	61	
	Lane-level route planning proposed in Ref. [3]	38	37.7
	Lane-level route planning in this study	20	67.2
705	Direct lane-level route planning	383	
	Lane-level route planning proposed in Ref. [3]	77	79.9
	Lane-level route planning in this study	35	90.1
10,088	Direct lane-level route planning	7832	
	Lane-level route planning proposed in Ref. [3]	535	93.2
	Lane-level route planning in this study	304	96.1

In this test, the A* algorithm was utilized at the road level in the proposed algorithm, which is consistent with Ref. [3], so the efficiency difference between this study and Ref. [3] is brought about by different hierarchical structures. The proposed algorithm directly eliminates the candidate lanes with different directions through the defined lane groups, which further reduces the search space compared to the method described in Ref. [3], which direct maps from roads to lanes, and thus speeds up the search speed. This proposed method is generally applicable to various road networks and can effectively reduce the search space of various algorithms, so the algorithm proposed in this paper can effectively improve the efficiency of existing algorithms when used for lane-level route planning. Although the test did not consider the changing traffic, which mainly affects the edges' cost, the search space can also be reduced by the proposed algorithm in real-time changing traffic conditions, so it can be inferred that the existing algorithms' efficiency can also be improved in dynamic traffic conditions. In conclusion, this simulation test validated the proposed method as having better efficiency in supporting lane-level navigation than the direct and past hierarchical routing methods.

The test was also conducted on the physical road network at Changchun, China, to validate the effectiveness of route-planning algorithm. High-precision point cloud maps were initially retrieved using the probe vehicle presented in Figure 19. Since LiDAR delivers a high-resolution 360-degree environment, the entire road geometry can be captured. Based on the results of the collected LiDAR data, a high-precision point cloud map of the test site was built first, as shown in Figure 20a, and then a lane-level electronic map was established, as shown in Figure 20b.

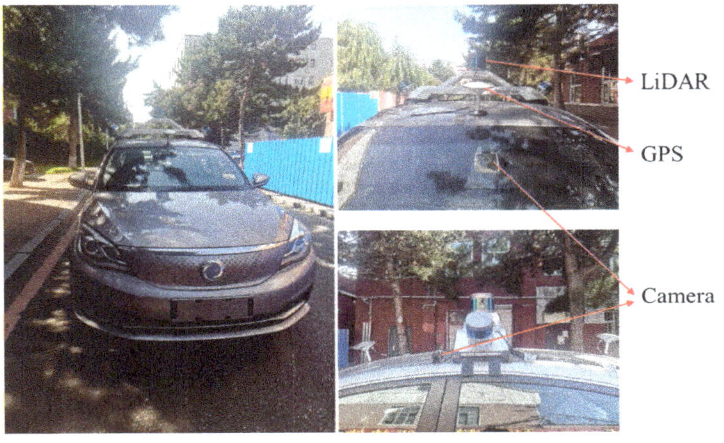

Figure 19. Vehicle for recording mapping data, equipped with RTK-GPS, LiDAR, and cameras.

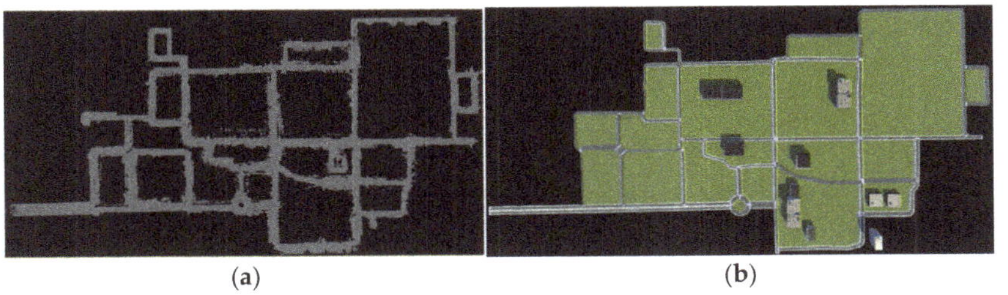

(**a**) (**b**)

Figure 20. (**a**) Point cloud map of the test site at Changchun, China. (**b**) Lane-level electronic map of the test site at Changchun, China.

The red lines in Figure 21 demonstrate the routing outcome. It is clear that the proposed method picked the correct lanes to arrive at the destination, demonstrating that the method can obtain reasonable routes on the physical road network. Table 4 illustrates the time costs of the Apollo routing method [35] and the proposed algorithm. We can see that the proposed algorithm improved routing efficiency by 78.0% compared with the Apollo method, so the proposed algorithm can better meet the real-time requirements of autonomous driving.

Table 4. Time costs of different routing methods in a real road network.

Number of Lanes	Routing Method	Time (μs)	Time Saved (%)
323	Routing algorithm of Apollo	132	
	Routing algorithm in this study	29	78.0

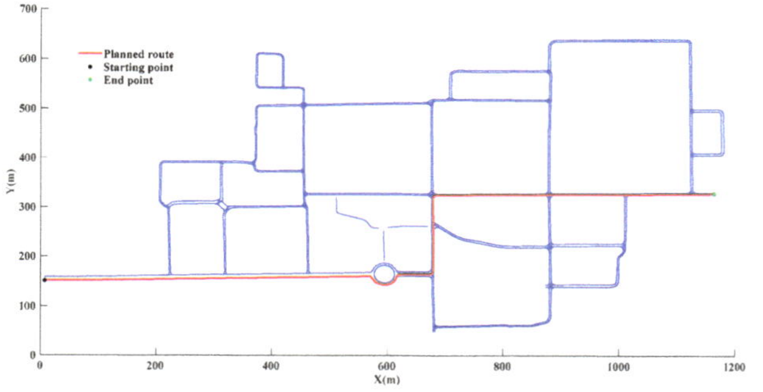

Figure 21. The blue lines represent the lane centerlines in the map demonstrated in Figure 20, and the red lines highlight the planned route.

5. Conclusions

In this study, which was aimed at providing better support for navigation systems in autonomous vehicles, an efficient lane-level route-planning algorithm based on a new lane-level road network model is proposed, which is manifested in the following aspects:

1. A new road network model for lane-level route planning is proposed. The whole model is divided into the road area and the intersection area, containing five sub-layers, refining the structures of real roads and intersections. On the one hand, this multi-sub-layer modeling can express variations in the road network structure; on the other hand, it facilitates the proposed multi-layer route-planning algorithm, improving the overall routing efficiency.
2. Based on the proposed road network model, an accelerated and refined lane-level route-planning method is proposed. First, a multi-layer route-planning algorithm is designed to plan sequentially at the road level, lane group level, lane section level, and lane level, resulting in remarkable improvements in routing efficiency, especially in large-scale road networks. Then, an optimal lane determination algorithm is developed at the lane level to find the optimal lanes while satisfying traffic rules and vehicle characteristics. The entire route-planning method can be seen as a framework compatible with existing algorithms, which means the existing graph search methods can be applied at the road and lane level to take advantage of the progress of existing research, and the routing efficiency can be dramatically improved compared to that of direct route planning using the graph search methods at the lane level. In addition, the proposed algorithm can better adapt to the changing road network structure and yield optimal lanes on roads with different configurations, including those with a constant or variable number of lanes. Tests were performed on simulated road networks and an actual road network. The results demonstrate the effectiveness, broader adaptability, and higher efficiency of the proposed algorithm compared with direct route planning, past hierarchical route planning, and the Apollo route-planning method. In particular, the efficiency can be improved by up to 96.1% compared with direct route planning using the graph search methods at the lane level. This study is complementary to existing studies and better supports autonomous vehicle navigation.

This proposed algorithm can effectively improve the efficiency of existing algorithms when used for lane-level route planning by reducing the search space. However, it is more theoretical than practical due to the lack of consideration of the influence of real driving conditions. Although it can be inferred that the search space can also be reduced by the proposed algorithm in real-time changing traffic conditions, to increase the utility of the method in practical navigation applications, further research will consider the traffic conditions, weather, and other factors that influence the actual travel costs in order to obtain the optimal route in practice. At the same time, considering these factors increases the complexity of routing, and a balance between routing complexity and efficiency will be considered in the future.

Author Contributions: Conceptualization, H.D.; Data curation, K.H.; Funding acquisition, H.D.; Investigation, K.H.; Methodology, K.H.; Project administration, N.X. and K.G.; Resources, H.D. and N.X.; Software, K.H.; Supervision, K.G.; Validation, K.H.; Writing—original draft, K.H.; Writing—review and editing, K.H. All authors have read and agreed to the published version of the manuscript.

Funding: This research was funded by the National Natural Science Foundation of China (grant number U1864206).

Data Availability Statement: Not applicable.

Conflicts of Interest: The authors declare no conflict of interest.

References

1. Groves, P.D. *Principles of GNSS, Inertial, and Multisensor Integrated Navigation Systems*, 2nd ed.; Artech House: London, UK, 2013; 776p, ISBN 978-1-60807-005-3.
2. Brügger, A.; Richter, K.-F.; Fabrikant, S.I. How does navigation system behavior influence human behavior? *Cogn. Res. Princ. Implic.* **2019**, *4*, 1. [CrossRef] [PubMed]
3. Jiang, K.; Yang, D.; Liu, C.; Zhang, T.; Xiao, Z. A flexible multi-layer map model designed for lane-level route planning in autonomous vehicles. *Engineering* **2019**, *5*, 305–318. [CrossRef]
4. Gyagenda, N.; Hatilima, J.V.; Roth, H.; Zhmud, V. A review of GNSS-independent UAV navigation techniques. *Robot. Auton. Syst.* **2022**, *152*, 104069. [CrossRef]
5. Zhang, J.; Wen, W.; Huang, F.; Wang, Y.; Chen, X.; Hsu, L.-T. GNSS-RTK Adaptively Integrated with LiDAR/IMU Odometry for Continuously Global Positioning in Urban Canyons. *Appl. Sci.* **2022**, *12*, 5193. [CrossRef]
6. Upadhyay, V.; Balakrishnan, M. Monocular Localization Using Invariant Image Feature Matching to Assist Navigation. In *Computers Helping People with Special Needs, Proceedings of the 18th International Conference, ICCHP-AAATE 2022, Lecco, Italy, 11–15 July 2022*; Proceedings, Part I; Springer International Publishing: Cham, Switzerland, 2022; pp. 178–186.
7. Zhang, Y.; Wang, L.; Jiang, X.; Zeng, Y.; Dai, Y. An efficient LiDAR-based localization method for self-driving cars in dynamic environments. *Robotica* **2022**, *40*, 38–55. [CrossRef]
8. Zheng, L.; Li, B.; Yang, B.; Song, H.; Lu, Z. Lane-level road network generation techniques for lane-level maps of autonomous vehicles: A survey. *Sustainability* **2019**, *11*, 4511. [CrossRef]
9. Sivaraman, S.; Trivedi, M.M. Dynamic probabilistic drivability maps for lane change and merge driver assistance. *IEEE Trans. Intell. Transp. Syst.* **2014**, *15*, 2063–2073. [CrossRef]
10. Schindler, A.; Maier, G.; Janda, F. Generation of high precision digital maps using circular arc splines. In Proceedings of the IEEE Intelligent Vehicles Symposium, Alcala de Henares, Spain, 3–7 June 2012; pp. 246–251.
11. Guo, C.; Meguro, J.I.; Kojima, Y.; Naito, T. Automatic lane-level map generation for advanced driver assistance systems using low-cost sensors. In Proceedings of the IEEE International Conference on Robotics and Automation, Hong Kong, China, 31 May–7 June 2014; pp. 3875–3982.
12. Bétalille, D.; Toledo-Moreo, R. Creating enhanced maps for lane-level vehicle navigation. *IEEE Trans. Intell. Transp. Syst.* **2010**, *11*, 786–798. [CrossRef]
13. Zhang, T.; Arrigoni, S.; Garozzo, M.; Yang, D.; Cheli, F. A lane-level road network model with global continuity. *Transp. Res. Part C Emerg. Technol.* **2016**, *71*, 32–50. [CrossRef]
14. Jo, K.; Sunwoo, M. Generation of a precise roadway map for autonomous cars. *IEEE Trans. Intell. Transp. Syst.* **2014**, *15*, 925–937. [CrossRef]
15. Gwon, G.-P.; Hur, W.-S.; Kim, S.-W.; Seo, S.-W. Generation of a precise and efficient lane-level road map for intelligent vehicle systems. *IEEE Trans. Veh. Technol.* **2016**, *66*, 4517–4533. [CrossRef]
16. DARPA. *Urban Challenge: Route Network Definition File (RNDF) and Mission Data File (MDF) Formats*; Defense Advanced Research Projects Agency: Arlington, VA, USA, 2007.
17. NDS Open Lane Model 1.0 Release. Available online: http://www.openlanemodel.org/ (accessed on 19 June 2019).
18. Dupuis, M.; Hekele, E.; Biehn, A. *OpenDRIVE® v1.4 Format Specification*; VIRES Simulationstechnologie GmbH: Bad Aibling, Germany, 2015.
19. Zhu, Z.; Li, L.; Wu, W.; Jiao, Y. Application of improved Dijkstra algorithm in intelligent ship path planning. In Proceedings of the Chinese Control and Decision Conference, Kunming, China, 22–24 May 2021; pp. 4926–4931.
20. Pandika, I.K.L.D.; Irawan, B.; Setianingsih, C. Apllication of optimization heavy traffic path with Floyd-Warshall algorithm. In Proceedings of the International Conference on Control Electronics Renewable Energy and Communications, Bandung, Indonesia, 5–7 December 2018; pp. 57–62.

21. Candra, A.; Budiman, M.A.; Hartanto, K. Dijkstra's and AStar in finding the shortest path: A tutorial. In Proceedings of theInternational Conference on Data Science, Artificial Intelligence and Business Analytics, Medan, Indonesia, 16–17 July 2020; pp. 28–32.
22. Jangra, R.; Kait, R. Analysis and comparison among ant system ant colony system and Max-Min ant system with different parameters setting. In Proceedings of the International Conference on Computational Intelligence & Communication Technology, Ghaziabad, India, 9–10 February 2017; pp. 1–4.
23. Xu, Z.; Liu, X.; Chen, Q. Application of improved Astar algorithm in global path planning of unmanned vehicles. In Proceedings of the Chinese Automation Congress, Hangzhou, China, 22–24 November 2019; pp. 2075–2080.
24. Lee, M.; Yu, K. Dynamic path planning based on an improved ant colony optimization with genetic algorithm. In Proceedings of the IEEE Asia-Pacific Conference on Antennas and Propagation, Auckland, New Zealand, 5–8 August 2018; pp. 1–2.
25. Jiang, C.; Fu, J.; Liu, W. Research on vehicle routing planning based on adaptive ant colony and particle swarm optimization algorithm. *Int. J. Intell. Transp. Syst. Res.* **2021**, *19*, 83–91. [CrossRef]
26. Lan, X.; Lv, X.; Liu, W.; He, Y.; Zhang, X. Research on robot global path planning based on improved A-star ant colony algorithm. In Proceedings of the 2021 IEEE 5th Advanced Information Technology, Electronic and Automation Control Conference (IAEAC), Chongqing, China, 12–14 March 2021; pp. 613–617.
27. Rossi, F.; Zhang, R.; Hindy, Y.; Pavone, M. Routing autonomous vehicles in congested transportation networks: Structural properties and coordination algorithms. *Auton. Robot.* **2018**, *42*, 1427–1442. [CrossRef]
28. Zheng, W.; Thangeda, P.; Savas, Y.; Ornik, M. Optimal routing in stochastic networks with reliability guarantees. In Proceedings of the IEEE International Intelligent Transportation Systems Conference (ITSC), Indianapolis, IN, USA, 19–22 September 2021; pp. 3521–3526.
29. Bailey, C.; Jones, B.; Clark, M.; Buck, R.; Harper, M. Electric Vehicle Autonomy: Realtime Dynamic Route Planning and Range Estimation Software. In Proceedings of the 2022 IEEE 25th International Conference on Intelligent Transportation Systems (ITSC), Macau, China, 8–12 October 2022; pp. 2696–2701.
30. Bucher, D.; David, J.; Raubal, M. A heuristic for multi-modal route planning. In *Progress in Location-Based Services 2016*; Springer: Cham, Switzerland, 2017; pp. 211–229.
31. Dibbelt, J.; Strasser, B.; Wagner, D. Customizable contraction hierarchies. *J. Exp. Algorithmics* **2016**, *21*, 1–49. [CrossRef]
32. Wu, Y.; Song, W.; Cao, Z.; Zhang, J.; Lim, A. Learning improvement heuristics for solving routing problems. *IEEE Trans. Neural Netw. Learn. Syst.* **2021**, *33*, 5057–5069. [CrossRef] [PubMed]
33. Kanakagiri, A. Development of a Virtual Simulation Environment for Autonomous Driving Using Digital Twins. Ph.D. Thesis, Technische Hochschule Ingolstadt, Ingolstadt, Germany, 2021.
34. That, T.N.; Casas, J. An integrated framework combining a traffic simulator and a driving simulator. *Procedia Soc. Behav. Sci.* **2011**, *20*, 648–655. [CrossRef]
35. Available online: https://github.com/ApolloAuto/apollo (accessed on 12 October 2022).

World Electric Vehicle Journal

MDPI

Article

Research on Reinforcement-Learning-Based Truck Platooning Control Strategies in Highway On-Ramp Regions

Jiajia Chen [1], Zheng Zhou [1], Yue Duan [1] and Biao Yu [2,*]

[1] School of Automotive and Transportation Engineering, Hefei University of Technology, Hefei 230009, China; chenjiajia@hfut.edu.cn (J.C.); 2021110991@mail.hfut.edu.cn (Z.Z.); 2021171021@mail.hfut.edu.cn (Y.D.)
[2] Hefei Institutes of Physical Science, Chinese Academy of Sciences, Hefei 230031, China
* Correspondence: byu@hfcas.ac.cn

Abstract: With the development of autonomous driving technology, truck platooning control has become a reality. Truck platooning can improve road capacity by maintaining a minor headway. Platooning systems can significantly reduce fuel consumption and emissions, especially for trucks. In this study, we designed a Platoon-MAPPO algorithm to implement truck platooning control based on multi-agent reinforcement learning for a platooning facing an on-ramp scenario on highway. A centralized training, decentralized execution algorithm was used in this paper. Each truck only computes its actions, avoiding the data computation delay problem caused by centralized computation. Each truck considers the truck status in front of and behind itself, maximizing the overall gain of the platooning and improving the global operational efficiency. In terms of performance evaluation, we used the traditional rule-based platooning following model as a benchmark. To ensure fairness, the model used the same network structure and traffic scenario as our proposed model. The simulation results show that the algorithm proposed in this paper has good performance and improves the overall efficiency of the platoon while guaranteeing traffic safety. The average energy consumption decreased by 14.8%, and the road occupancy rate decreased by 43.3%.

Keywords: truck platoon; reinforcement learning; Platoon-MAPPO algorithm; on-ramp region

Citation: Chen, J.; Zhou, Z.; Duan, Y.; Yu, B. Research on Reinforcement-Learning-Based Truck Platooning Control Strategies in Highway On-Ramp Regions. *World Electr. Veh. J.* **2023**, *14*, 273. https://doi.org/10.3390/wevj14100273

Academic Editor: Joeri Van Mierlo

Received: 28 August 2023
Revised: 14 September 2023
Accepted: 18 September 2023
Published: 1 October 2023

1. Introduction

Truck platooning refers to a driving state in which two or more vehicles drive on the highway, controlling the distance and driving state between the control vehicles to form a tight vehicle cluster. It uses sensors or communication technologies to obtain information between adjacent trucks and can autonomously maintain speed or distance. As early as 1996, the European Commission (EC) Information Society Technologies (IST) launched the CHAUFFEUR project to assess truck platooning [1], which studied either all trucks with automatic driving, or only the head truck with a human driving. Similar projects have also been carried out in other countries, such as California PATH, Energy-ITS, KONVOI, SARTRE, etc. [2–6]. The focus of these projects is mostly on truck control methods, platoon stability, and the external communication of trucks; all require automatic driving strategies to control all or some of the trucks in the platoon.

Platooning offers several benefits, one of which is enhancing traffic efficiency. By keeping smaller gaps between vehicles at the same speed, platooning can increase road capacity compared with normal traffic flow. Another benefit is improving driving safety. Trucks in the platoon communicate and coordinate with each other using autonomous driving control, resulting in smaller speed variations and faster reaction times than human drivers. A third benefit is reducing the environmental impact. The transportation sector is a major contributor to energy consumption and greenhouse gas emissions, with road transport accounting for 72.8% and 73.4%, respectively [7]. Platooning can significantly lower fuel use and emissions, especially for large vehicles such as trucks.

Early research projects used sensors to measure the distance from the preceding vehicle and a lower-level controller to maintain fixed spacing. For instance, refs. [2,4,5] achieved spacings of 6 m, 10 m, and 4.7 m, respectively. However, this method does not minimize the total energy consumption when the speed varies, as the platoon might need more energy to keep the preset spacing. Subsequent research formulated cost functions based on time and energy consumption and applied optimization and optimal control theory methods to find optimal solutions [8–11]. However, many methods were not scalable to large platoons, as the calculation cost increased with the number of vehicles.

In the past decade, there has been a lot of research based on deep learning (DL) to automate driving tasks as much as possible [12,13]. However, collecting decision-making and planning datasets for autonomous driving is costly and challenging, as it is hard to cover complex real-world traffic scenarios.

Reinforcement learning (RL) algorithms have emerged as a promising alternative for vehicle decision-making, planning, and control problems [14–16]. These algorithms do not require manual data collection and can adapt to various tasks. Some research has used RL to achieve autonomous driving for single vehicles [17], but few have applied it to multi-vehicle cooperative control [18]. This is because treating other vehicles as part of the environment violates the Markovian property and makes the learning unstable and difficult.

Multi-agent reinforcement learning (MARL) has been explored to control multi-vehicle behavior in recent studies. For example, ref. [19] only considered optimal strategies for stop-and-go under flat roads, while [20] applied DQN to vehicle grouping strategies in road networks. Ref. [21] used Q-learning to find optimal insert points for vehicles entering fleets.

This paper proposes a Platoon-MAPPO algorithm based on MARL for truck platoon control in a highway with an on-ramp region, which is the most complex traffic situation on highways. Existing platoon control methods, such as traditional control methods and DL-based methods, are not effective in this scenario. The Platoon-MAPPO algorithm has the following features:

(1) It uses a MAPPO-based algorithm with centralized training and decentralized execution to control the platoon in the on-ramp area. Each truck only computes its own action, avoiding the data computation delay caused by centralized calculation.
(2) It considers the driving status of the trucks in front and behind each truck, maximizing the overall platoon gain and improving the global operational efficiency.
(3) It does not require communication, and it is scalable to any number of vehicles and communication devices.

Compared with traditional control methods and existing methods based on DL and RL, Platoon-MAPPO can effectively reduce road occupancy and energy consumption. Moreover, Platoon-MAPPO is more flexible than existing RL-based methods because it does not require intra-platoon communication. Additionally, the use of distributed computing means that the computational cost of each vehicle is independent of the number of platooning members, making it more practical and avoiding the waste of computing resources. Traditional algorithms have strong interpretability and can be used as a safety guarantee to assist Platoon-MAPPO operation.

2. Related Work

Most traditional methods in platooning rely on optimal control theory, which aims to find the optimal control inputs for each vehicle to achieve certain objectives, such as fuel efficiency, safety, and comfort. For instance, [8] proposed a serial distributed model predictive control (MPC) approach that ensured local stability and multi-criteria string stability for connected automated vehicles. Local stability means that each vehicle can track its desired speed and position, while multi-criteria string stability means that the errors in speed and position do not propagate along the platoon. In Ref. [10], a multi-anticipative controller was devised that enabled an equipped vehicle to use information from its direct predecessor to predict the behavior of its pre-predecessor. In this way, the vehicle can anticipate future actions of the leader vehicle and adjust its own control accordingly. Subsequently, some methods based on DL and RL emerged, which can learn from data and experience without relying on explicit models or rules. Ref. [18] was one of the first papers to apply RL to platooning. The authors in [22] improved the reward function to make the training more reasonable, but they neglected the behavior of other vehicles and failed to optimize the overall efficiency of the platoon. In Ref. [23], a platoon control algorithm based on centralized RL was designed; however, its single-agent training strategy might still result in unstable training. In Ref. [19], a method based on multi-agent RL was developed, although the scenario was relatively simple and required communication between platoons.

For highway on-ramps, ref. [24] suggested an optimal trajectory optimization strategy for connected and automated vehicles to cooperatively perform mainline platooning and on-ramp merging, in which each platoon could obtain its optimal control input in terms of fuel consumption. Ref. [25] transformed the complex 2D multi-platoon cooperation problem into a 1D platoon following control problem and derived an analytical solution to the optimal control. Ref. [26] proposed a platoon-based hierarchical merging control algorithm for on-ramp vehicles to achieve automated merging control under a connected traffic environment. Some studies attempted to use machine learning to enhance the overall traffic efficiency of highway on-ramps. Ref. [27] presented an RL framework that adaptively adjusted the intra-platoon gap of an individual platoon member. However, both traditional and machine learning methods mostly assumed that vehicles were intelligent connected vehicles that could exchange information freely. However, it is unrealistic to expect all vehicles on the road to maintain good wireless communication at present. Communication problems could pose serious risks, such as collisions, delays, or breakdowns. Many studies have mentioned this issue, such as [28], which presents a summary of the key technologies and challenges in platooning. Ref. [29] discussed different communication technologies in the platoon, and further examined security issues related to communication, such as jamming, spoofing, or hacking.

3. Preliminaries and Methods

3.1. Preliminary Knowledge

RL is learning what to do—how to map situations to actions—so as to maximize a numerical reward signal [30]. Unlike supervised learning, RL does not require labeled datasets to train neural networks. Instead, it allows agents to explore their environments and choose actions based on feedback from the environments to maximize long-term gains. At each time step, t, the environment provides the current state, s_t, from the state space, S, and the agent selects the action a_t, $a_t \in A$, based on the observation value, o_t, of the state and its own policy, $\pi(a_t|o_t)$, where A is the action space. Then, the environment generates the reward $r(s_t, a_t)$ at this time, and gives s_{t+1} for the next time step. $p(s_{t+1}|s_t, a_t)$ is called the state transition probability. When the environment satisfies Markov properties, the state transition probability is stable, and this process can be described as a Markov decision process (MDP). After a period of time, the agent will collect a trajectory $(s_t, a_t, r_t, s_{t+1}, a_{t+1})$. The return, $g_t = r_t + \gamma g_{t+1}$, is defined as accumulative discount rewards, where γ is the discount factor and g_t is the return value at time step t. The ultimate goal of RL is to optimize policy, π, so that $E(g_t)$ is maximized as much as possible. This means that agents will evolve toward the goals expected by designers.

RL can be divided into value-based methods and policy-based methods. Q-learning is a classic example of value-based methods. It can use dynamic programming to solve the state values if the environment is a Markov decision process. It can also use Monte Carlo methods to approximate the state transition values. In actual applications, most states are continuous variables; therefore, a state-value function, $V(s_t)$, can be used to fit the predicted state values more accurately. DQN is a well-known method of this kind.

Policy-based methods were proposed to deal with continuous action space problems. They directly optimize policy, π, to maximize the expected state values at each state,

$\underset{\pi}{\text{argmax}}(E_S[V_\pi(S)])$. A3C, TRPO, PPO, and other methods follow this basic principle and incorporate advanced optimization techniques, especially PPO [31–33].

MARL has attracted more attention than single-agent reinforcement learning (SARL) in recent years. When the agents cooperate with each other, the optimization goal is to maximize the total return. Some early studies relied on optimizing independent agents to maximize the global reward function. Each agent uses a TRPO network structure and treats other agents as part of the environment. However, this type of algorithm suffers from non-stationarity problems, because each agent is also part of the environment; this means that the environmental state transition probabilities change continuously as the agent policies change, which violates the RL assumptions.

To address this problem, later studies mostly used centralized training decentralized execution (CTDE) algorithms, which significantly improved the MARL performance. Agents learn their own policy network under the guidance of a centralized feedback network, and then generate actions independently. Recent studies have applied CTDE algorithms to continuous action spaces, such as MAPPO [34] and MADDPG [35]. The two articles share the idea that each critic can obtain all action information. During training, a critic that can observe the global situation guides actor training, while during inference, each actor only uses its own local observation value to calculate actions. In MADDPG, the authors argue that in cooperative and competitive scenarios, CTDE can enable agents to discover and utilize various information in the environment to produce better strategies than other algorithms. Additionally, MAPPO achieves excellent performance in multiple test projects without making significant changes to PPO.

3.2. Methods

3.2.1. Truck Platoon Communication Topology

Figure 1 shows the overall framework of the Platoon-MAPPO. The vehicles that are not controlled by the truck platoon are called interactive vehicles (IVs), which may affect the driving status of the truck platoon by traveling in front of it. The trucks in the truck platoon are called platoon members (PM), with PM_0 being the first truck and PM_N being the last one. The other trucks are numbered sequentially from PM_1 to PM_{N-1}, and they can be referred to as PM_n collectively.

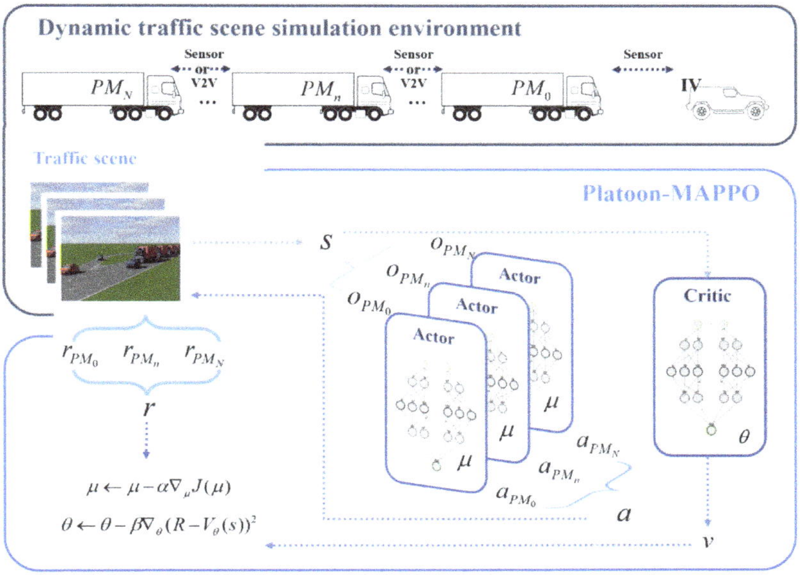

Figure 1. Framework of Platoon-MAPPO.

The platoon communication structure can be classified into three types: centralized communication topology, decentralized communication topology, and hybrid communication topology. In centralized communication topology, each vehicle only communicates with the lead vehicle, which is easy to control but may cause high latency problems. In decentralized communication topology, each vehicle only receives information from the front vehicle, which has low latency. This study considers using hybrid communication topology, as shown in the topology diagram. Each truck receives information from both front and rear vehicles, but not from the environmental vehicle detected by the lead vehicle. Compared with algorithms that only consider the front vehicle's driving status, this algorithm is more effective in optimizing the overall performance of the truck platoon.

This communication strategy enables the trucks in the platoon to switch positions, adjust the number of vehicles, and dynamically select the lead vehicle, enhancing the applicability and flexibility of the algorithm.

A truck platoon consists of three types of trucks: PM_0, PM_n, and PM_N. This design enables the platoon length to be scalable and the algorithms to be compatible with any number of trucks. The training process can simulate the performance of the whole platoon with any number of trucks. The scalability of the platoon size also enhances the sampling efficiency, as each interaction with the simulation environment yields more information about the action-value of the intelligent agents. The algorithmic computation grows linearly with the platoon size in both training and application phases. Therefore, using distributed computing, the computational load of each truck remains constant regardless of the platoon size, which significantly reduces the overall computational burden on the platoon and avoids computational waste due to changes in the platoon size.

3.2.2. Crucial Elements

Platoon-MAPPO is a stochastic gradient policy algorithm that is trained centrally and executed distributivity. Each agent observes and acts independently. The reinforce method is applied to estimate the unknown action value functions, using the actual return from a collected trajectory as a substitute. PPO introduces an advantage function to stabilize the training, which is usually defined as the difference between the action value function and the state value function. The state value function represents the expected return that an agent will receive in a certain state. The advantage function can intuitively show how much better the selected action is than the mean of all actions selected in this state. The actor, $\pi_\mu(a|o)$, is a neural network function and the critic, $V_\theta(s)$, is another neural network function. Under the multi-agent architecture, the critic observes the global information, i.e., the observations of all agents, and predicts the expected return values for all agents based on these observations.

(1) Observation and state: the observation value of each truck in the platoon is the part of the platoon's overall state that can be observed by the truck. The observation value of each truck comprises several elements: (a) the type of the truck, which can be PM_0, PM_n, or PM_N, indicating the position of the truck in the platoon; (b) the relative speed of the truck with respect to the vehicle in front of it; (c) the relative speed of the truck with respect to the vehicle behind it; (d) the relative distance between the truck and the vehicle in front of it; (e) the relative distance between the truck and the vehicle behind it; (f) the speed of the truck itself; and (g) if the truck is PM_0, it also obtains the relative distance and speed between itself and the IV that is detected within its detection range. RL algorithms are not suitable for applying batch normalization to the input values. Therefore, the input values are scaled directly to $[-1, 1]$ according to their value range. The position identifiers are 0, -1, and 1, representing PM_0, PM_n, and PM_N, respectively. When the platoon consists of more than three trucks, all the position identifiers of the middle trucks, PM_n, are set to 1. The training process can accommodate any number of trucks in the platoon.

State refers to various factors that influence driving in traffic environments. As a centralized training algorithm, PPO needs to predict the state value function of the entire environment. Therefore, state values include the observation values of all the trucks in the platoon.

(2) Action: the main application scenario of truck platoon control technology is on highways, where the traffic environment is relatively simple. To minimize the interference of truck platoons with other human drivers, truck platoons will travel on the rightmost lane. Therefore, this article focuses on the longitudinal control of truck platoons. After obtaining the observation values, the decision-making algorithms will determine the driving force for each truck at each time step. The driving force indicates the acceleration or deceleration of the truck. When the driving force is positive, it means that the truck accelerates forward and is bounded by the maximum driving force value. When the driving force is negative, it means that the truck decelerates.
(3) Reward: autonomous driving is typically a multi-objective optimization problem. These objectives may encompass various aspects, such as speed, travel time, collision, regulations, energy consumption, vehicle wear and tear, and passenger experience. For freight trucks, the main considerations are speed, energy consumption, and safety. The reward function should reflect the goals that autonomous driving vehicles aim to pursue, rather than the methods and techniques used to attain these goals. For instance, adding extra penalties for low speeds to motivate trucks to move forward is known as reward shaping. Reward shaping may facilitate learning in the early stages, but may also constrain the performance potential of the algorithm [36]. Therefore, our reward function only consists of three components: speed, energy consumption, and safety.

The reward function is as follows:

$$r = r_1 \max\left(\frac{v_{PM}}{v_{\mathrm{IV}} + r_\varepsilon}, 1\right) - r_2 D - r_3 \max(F_t, 0) \quad (1)$$

where r represents the reward at a certain moment, v_{PM} represents the instantaneous speed of a member of the platoon at a certain moment, v_{IV} represents the instantaneous speed of IV at a certain moment. r_ε is added to prevent the denominator from being zero and is set to 1×10^{-5}. D is the collision penalty term. F_t is the driving force. r_1, r_2, and r_3 are the coefficients used to adjust the weights of each item.

The reward function defines the objective of optimizing truck platoon driving strategies. The objective of truck platoon driving is efficiency and safety. Efficiency means enhancing travel speed while reducing energy consumption as much as possible. We aim to make the speed of truck platoon close to the traffic flow speed. If the traffic flow speed is fast, the platoon should be incentivized to increase their speed. If the traffic flow speed is slow, the platoon should not receive extra benefits from high speeds because high speeds are irrelevant at this time and may pose a risk instead. Therefore, we add a speed reward in reward function. When $v_{PM} < v_{\mathrm{IV}}$, the reward will be proportional to the speed of the platoon. When $v_{PM} > v_{\mathrm{IV}}$, no additional rewards will be given.

To ensure that policy optimization is carried out under safe driving conditions, a collision penalty term, D, is added to reward function. If a collision occurs, this value will be 1; otherwise, it will be 0.

For electric trucks, driving force size is almost proportional to energy consumption. Let instantaneous energy consumption be $c = c_\varepsilon F_t$ and c be the instantaneous fuel consumption. c_ε represents the linear coefficient mapping driving force to energy consumption. When the driving force equals zero, instantaneous energy consumption also equals zero, so there are no bias terms in reward function. An energy consumption penalty term was added to the reward function, and c_ε was absorbed into r_3, i.e.,

$$J(\mu) = \sum_{(s,a)} \min\left(\left(\frac{\pi_\mu(a|s)}{\pi_{\mu'}(a|s)}\right)A^\mu(s,a), clip\left(\frac{\pi_\mu(a|s)}{\pi_{\mu'}(a|s)}, 1-\varepsilon, 1+\varepsilon\right)A^\theta(s,a)\right) \tag{2}$$

Algorithm 1 shows the training process of the algorithm; we used the Monte Carlo method to estimate the action value function, and calculated the advantage function A from it. ε is a hyperparameter that controls the size of the clipping term.

Algorithm 1: Platoon-MAPPO

Initialize actor network, $\pi_\mu(a|o)$, and critic network, $V_\theta(s)$, with weights μ and θ
Initialize batch size B, iterative step α,β, and *done* = 0
for episode = 0, 1, 2, ... until convergence **do**
  while not *done* **do**
    initialize actor and critic states
    for all agents **i**, perform
      $a_{t,mean}{}^i, a_{t,std}{}^i = \pi(o_t^i; \mu)$
      Sampling $a_t{}^i$ from normal distribution $N(a_{t,mean}{}^i, a^2{}_{t,std}{}^i)$
      $v_t{}^i = V(s_t; \theta)$
    end for
    Send a_t to the simulation environment, obtain $[r_t, s_{t+1}, o_{t+1}]$
    Save $[s_t, o_t, a_t, r_t, s_{t+1}, o_{t+1}]$
  Compute the advantage function, A_t, and return, R_t
  end for
for for k = 0, ..., PPO epoch **perform**
  Compute $J(\mu)$
  Update network parameters using the gradient method
  $\mu^{k+1} \leftarrow \mu^k + \alpha \nabla_\mu J^{\mu^k}(\mu)$
  $\theta^{k+1} \leftarrow \theta^k - \beta \nabla_\theta (R_t - V_\theta(s_t))^2$
  end for

4. Simulation

4.1. Simulation Platform

Traffic simulation is the application of computer technology to mimic real traffic systems, which can virtually reproduce realistic traffic scenarios at low cost. RL algorithms depend on vehicles to learn the optimal driving strategies in traffic environments. This also implies that it is infeasible to train vehicle strategies in real traffic environments. Therefore, we established a road simulation environment in SUMO. SUMO is an open-source micro-continuous traffic simulation platform widely used in traffic research. SUMO enables users to construct various types of road networks and define basic parameters and behaviors of vehicles. It also provides interfaces for third-party programs, allowing algorithms to interact directly with the simulation environment.

To make the platoon control strategy more compatible with the demands of realistic situations, a dynamic and variable traffic environment was constructed that aligns with the actual platoon control algorithm.

Slopes affect the acceleration and deceleration capabilities of trucks, and have a significant effect on the fuel consumption and safety performance of platoons. Some studies have evaluated the energy consumption of truck platooning under different slope conditions; the results show that the road slope has a considerable impact on the energy consumption of truck platooning [37]. Therefore, when devising energy saving driving strategies for truck platooning, the impact of road slope must be taken into account. A normal highway slope is within 5%; thus, the road slope was set to $\alpha \in [-0.05, 0.05]$, α denotes the tangent value.

The interference that platooning encounters varies depending on the traffic flow state. The most influential vehicle for platooning is the individual vehicle (IV) in front of it. The speed of an IV may fluctuate when it cruises or accesses ramps in real-world traffic scenarios. Moreover, surrounding vehicles may alter their speed more frequently in unstable traffic flow. These situations require platooning to adjust accordingly; otherwise, it may increase energy consumption and pose safety risks.

We considered two scenarios of traffic flow in ramp areas. In the first case, there were vehicles accessing ramps in front of the IV, and the IV adopted the IDM-following strategy. In the second case, the IV accessed ramps in front of the platoon, and the platoon faced disturbance from vehicles that suddenly appeared. To simulate these scenarios, we assumed that the platoon could only observe ramp vehicles when they were 50 m away from the ramp entrance. Ramp vehicles entered the main road at random speeds before the platoon reached the ramp, and then accelerated to the speed limit after entering the main road. We set a 3 km long road with a truck queue starting at an interval of 30 m and a speed of 20 m/s at the beginning of the road. The maximum speeds of the PM and IV were 25 km/h and 27.6 km/h, respectively. IVs started at random positions within one hundred meters in front of the truck fleet with the same initial speed. To simulate different traffic conditions for the platoon, IVs first cruised at a stable speed and then randomly decelerated to a lower value to simulate interference from ramps in real traffic scenarios. Finally, they accelerated to the maximum speed limit to simulate free-flowing traffic.

4.2. Longitudinal Dynamics of Trucks

Trucks have more degrees of freedom, which makes it extremely complex to establish an accurate dynamic model. However, this article focuses on decision-making planning algorithms; dynamic modeling is not its main concern. Moreover, too complex models may hinder the subsequent simulations. This article studies scenarios for straight or approximately straight regions of highways, establishing a simplified longitudinal truck dynamics model. This model takes into account the main factors affecting vehicle operation, so that it does not deviate from reality. Compared with sophisticated models, simplified models are adequate to verify the effectiveness of the algorithm.

An electric truck experiences force while driving:

$$\begin{aligned}
F_t &= F_f + F_w + F_i + F_j \\
F_f &= mg\cos\omega(f_a + f_b v_c) \\
F_w &= 0.5a_D C_D U \rho {v_c}^2 \\
F_i &= mg\sin\omega \\
F_j &= ma_c \\
F_{bmax} &= \varphi_b mg\cos\omega + mg\sin\omega
\end{aligned} \tag{3}$$

where F_t is the traction force required by the truck, which consists of four parts: rolling resistance force F_f, air drag force F_w, gravity force F_i, and acceleration provided by the vehicle F_j; f_a and f_b are rolling resistance coefficients; and a_c is the acceleration of the truck.

When the truck platoon is moving, the air resistance experienced by trucks in the platoon will be significantly reduced. Therefore, it was necessary to modify the actual wind resistance coefficient, C_D, of the truck. Studies [38] have shown that the overall air resistance coefficient decreases as the distance between vehicles decreases, and a method for modifying the air coefficient, a_D, has been proposed.

$$a_D = \begin{cases} 0.7231(x_{PM_0} - x_{PM_1} - l)^{0.0919} \ for\ PM_0 \\ 0.2241(x_{PM_n} - x_{PM_{n-1}} - l)^{0.1369} + 0.5016\ for\ PM_n \end{cases} \tag{4}$$

When calculating a_D for PM_0, $(x_{PM_0} - x_{PM_1} - l)$ is the distance between PM_1 and the first PM_0, and when calculating a_D for PM_n, $(x_{PM_n} - x_{PM_{n-1}} - l)$ is the distance between PM_n and the previous PM_{N-1}. The values of parameters in formula (3) and (4) are shown in Table 1.

The neural network comprises fully connected network layers and activation function layers. The actor network outputs normal distributions $a_{t,mean}$ and $a_{t,std}$. We observed that using two independent networks to output them separately improves the performance. Hence, we set up two identical actor networks with fully connected network layers in the middle; each layer had 64, 128, 128, 64, 64 neurons. The dimensions of the first and last layers matched those of the observation and action.

Table 1. Truck longitudinal dynamics parameter settings.

Parameter	Value	Description
mg	200,000 N	Gravity of truck
l	17 m	Length of truck
f_a	0.0041	Rolling resistance coefficient
f_b	0.000025	Rolling resistance coefficient
φ_b	(0.3, 0.6)	Adhesion factor of truck
C_D	0.564	Air resistance coefficient of truck
U	5.8 m^2	Front projection area of truck
ρ	1.2258 kg/m^3	Density of air
ω	[−0.05, 0.05]	Road slope
v_c	[0 m/s, 25 m/s]	Velocity of truck

Furthermore, we employed the IDM [39] model and the CACC [40] control method as the baselines to control the platoon behavior, and compared them with the platoon-MAPPO algorithm. IDM is a well-established traffic flow simulation model; it can emulate the driving behavior of human drivers in traffic flow. CACC is a widely used control method for autonomous vehicle following, which enables cooperative driving by communicating with the preceding vehicle.

5. Results

We performed five simulation experiments with different initial random seeds. Figure 2 shows the training results.

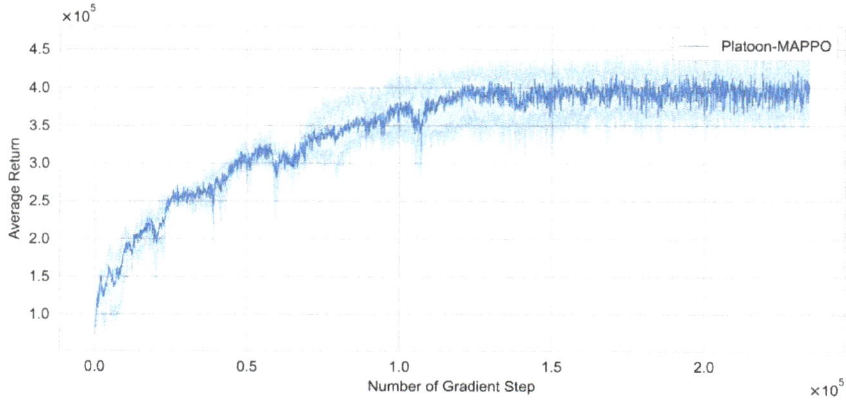

Figure 2. Return curve.

We set up a simple test scenario to intuitively demonstrate the effectiveness of the Platoon-MAPPO algorithm. The IV initially accelerates to 15 m/s. At 20 s, it decelerates to 5 m/s to simulate the interference from a ramp. After 15 s, it accelerates again to 25 m/s. All vehicles have an initial speed of 10 m/s and a maximum acceleration of 1 m/s^2. The PM_0 is 40 m away from the IV. The platoon consists of 10 trucks, and the initial headway distance within the convoy is 30 m. The test lasts for 1000 s.

We plotted a spatiotemporal map for each truck, as depicted in Figure 3, which shows the positional changes of the truck over time. We also plotted a heatmap of the speed value of each truck at each moment in Figure 4, to show the speed changes of all trucks in the fleet more intuitively. In the test, the PM_0 first observes the speed change of the IV. When the IV changes its speed, the PM_0 reacts first and then passes it on to each subsequent truck. To observe the reaction speed of each truck, we counted the reaction time difference according to the moment when the speed of each vehicle crosses 10 m/s. The reaction time differences during deceleration for IDM, CACC, and Platoon-MAPPO are 14 s, 9.2 s, and 4.1 s, respectively. The platoon reaction time of Platoon-MAPPO is significantly shorter than that of IDM and CACC.

Under IDM and CACC, the speed changes cause traffic oscillations that propagate through the entire fleet and have a lasting impact on the fleet. Especially for IDM, after experiencing an interference, the inter-vehicle distance keeps increasing. However, Platoon-MAPPO can effectively filter traffic oscillations. From the heatmap, we can see that the rear trucks in the fleet have smaller speed changes than the front trucks, which indicates that they can reduce energy consumption through smoother acceleration. Smoother speed changes reduce energy consumption in acceleration and smaller headway reduces overall air resistance of the fleet. In this test, based on IDM and CACC's energy consumption as a benchmark, Platoon-MAPPO reduces energy consumption by 14.8% and 32.7%.

More specifically, under the same simulation duration, the last truck of Platoon-MAPPO travels the longest distance, followed by CACC and then ordinary IDM. This indicates that our proposed model has the highest overall travel speed and does not suffer from significant speed reduction due to energy saving needs.

Table 2 shows average traction forces and their components for each control strategy in this test scenario. Traction force is an approximate linear function of energy consumption for electric trucks; therefore, we can compare energy consumption accordingly.

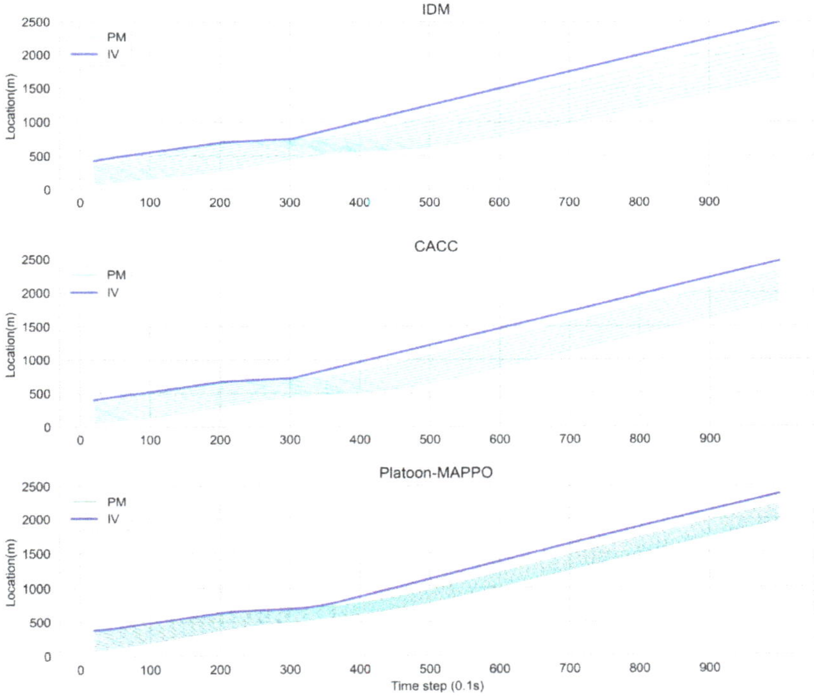

Figure 3. Spatiotemporal map for each truck.

Table 2. Traction force composition.

Control Strategy	F_t (N)	F_w (N)	F_j (N)	F_f (N)
IDM	7368	562	5690	1115
CACC	8334	712	6464	1157
Platoon-MAPPO	6278	643	4480	1155

For CACC, its control strategy may aim for higher instantaneous speed, but often, increasing instantaneous speed does not reduce travel time. Figure 4 shows that CACC did not eliminate traffic oscillations in this test scenario. The speed oscillations in the front of the platoon propagate to the rear, causing unnecessary acceleration and deceleration of the rear trucks. This may explain its higher energy consumption. At the same time, Platoon-MAPPO's F_w is also higher than IDM's because Platoon-MAPPO has a higher average speed. However, the platoon effectively filters traffic oscillations and has the least unnecessary acceleration and deceleration, resulting in the smallest F_j.

As mentioned above, the Platoon-MAPPO algorithm can be applied to truck platoons of any number of trucks. Another significant feature of platoons is that they reduce road occupancy and improve road capacity. To verify the effect of different numbers of vehicles on platoons, we simulated the fleet length as a variable based on the previous test. We set the minimum number of trucks in the fleet to 3 and the maximum to 20. We repeated the previous test and counted the length of the fleet occupying the road. Figure 5 shows the distance from the head of the PM_0 truck to the tail of the PM_N truck at each time step.

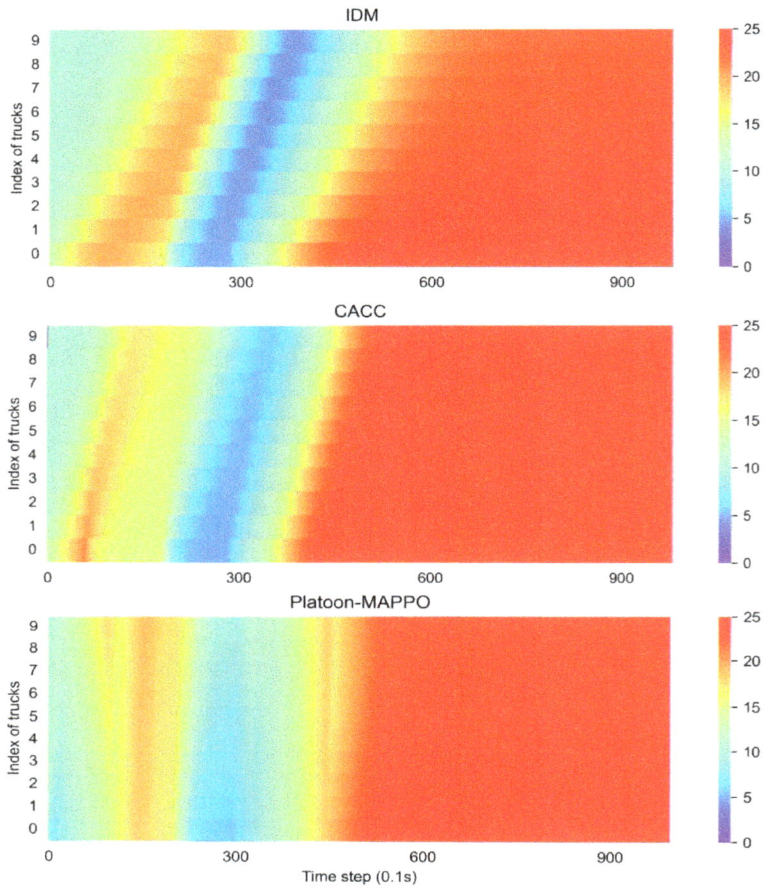

Figure 4. Heatmap of platoon speed.

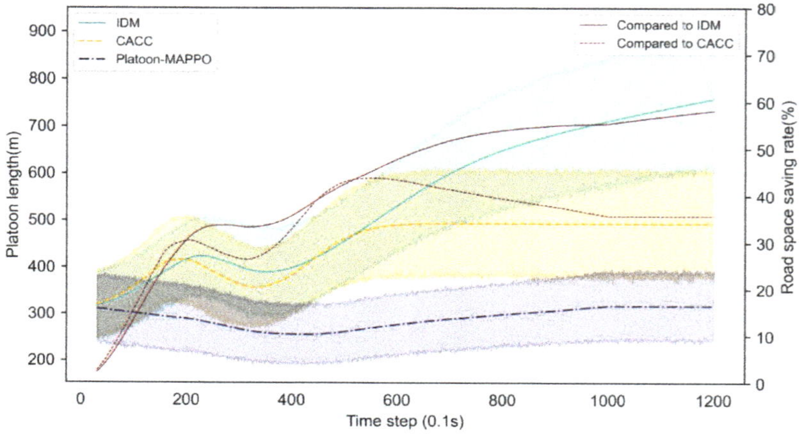

Figure 5. Platoon length.

Green, orange, and purple represent IDM, CACC, and Platoon-MAPPO control strategies, respectively. The shaded area represents the fleet length when there are 3 to 20 trucks. The dark line in the middle represents the average value. The two red lines represent the ratios of road space saved by Platoon-MAPPO compared with IDM and CACC at each time step.

The initial platoon lengths of IDM, CACC, and Platoon-MAPPO were identical. During the acceleration phase, the fleet lengths of IDM and CACC increased gradually due to the lag effect of truck acceleration. The trucks that accelerated earlier had higher speeds and covered longer distances in a given time, resulting in an increased gap between the trucks until the leading truck encountered an obstacle or a speed limit and stopped accelerating. In contrast, the platoon length of Platoon-MAPPO decreased during this phase because Platoon-MAPPO enabled each truck to have a very low reaction delay and a small safe distance under cooperative control, allowing the rear trucks to accelerate slightly more than the front trucks at low speeds until they reached an optimal gap, as shown in Figure 5. During the deceleration phase, the platoon lengths of IDM and CACC changed dramatically compared with Platoon-MAPPO's due to the lag effect. During the re-acceleration phase, the fleet lengths of IDM and CACC increased significantly as the speed increased, and this change was proportional to the number of vehicles in their platoons; more trucks led to a faster increase in platoon length (see Figure 5 for the upper and lower bounds of the shaded area). The platoon length of Platoon-MAPPO was relatively stable and increased slowly with the speed increase, but it was not significantly affected by the number of vehicles in their fleets. The higher the number of vehicles, the higher the instantaneous speed; thus, Platoon-MAPPO saved more road space. In this test scenario, Platoon-MAPPO reduced the occupied road space by averages of 43.3% and 34.2% compared with IDM and CACC, respectively, with maximum reductions of 58.2% and 35.7%, respectively.

We tested the Platoon-MAPPO algorithm in various dynamic traffic scenarios to further evaluate its adaptability. Based on the previous test scenario, we set the target speed value for IV deceleration as a variable with the same range of values as in the training process. For quantitative analysis, we selected five values within the range and used each value as a test scenario and measured the energy consumption. Figure 6 shows some cases extracted from the test scenarios with target speed values of 5 km/h and 10 km/h for IV deceleration, and road slopes of 0 and 0.03. Each control strategy can extract four curves, represented by different linear gray curves. To facilitate observation and analysis, we also plotted the mean energy consumption ratio values of IDM, CACC, and Platoon-MAPPO with green, yellow, and blue curves, respectively. The horizontal axis in the figure is the number of trucks in the platoon, and the vertical axis is the energy consumption ratio. The energy consumption ratio was calculated by taking the maximum total energy consumption value in the test scenario as a benchmark and comparing other energy consumption values with this maximum value. This eliminated the linear relationship coefficient between traction force and energy consumption.

Platoon-MAPPO consistently exhibited the lowest energy consumption in various scenarios. The energy consumption of all three control strategies decreased with the increase in the number of vehicles in their platoon. This was especially noticeable in IDM, which might be because IDM's control strategy could effectively reduce traffic oscillations, while CACC's excessive pursuit of instantaneous speed made its acceleration change too sensitive, leading to high energy consumption. In the dynamic traffic scenarios tested, Platoon-MAPPO's energy consumption was reduced by an average of 3.3% compared with IDM, with a maximum reduction of 4.0%.

Acceleration and climbing are extremely energy-consuming for trucks. For instance, when the slope was 0.02, the truck speed was 10m/s, and the acceleration was 0.5 m/s^2, the rolling resistance force, the air drag force, the gravity force, and the force providing vehicle acceleration accounted for 5.8%, 1.2%, 34.9%, and 58.1% of the total force, respectively. Therefore, the air resistance has a very small effect during acceleration or climbing. When the slope and the acceleration were both zero and the speed was 20 m/s with an acceleration of 0.5 m/s^2, the rolling resistance force and the air drag force accounted for 59.5% and

40.5% of the total force, respectively. In high-speed uniform motion on flat roads, the air resistance has a very high share but its absolute value is not large. Therefore, in dynamic traffic scenarios, Platoon-MAPPO's advantage over IDM may not seem obvious, but this is actually because large numerical values that are hard to reduce through control strategies mask the efforts to reduce additional consumption.

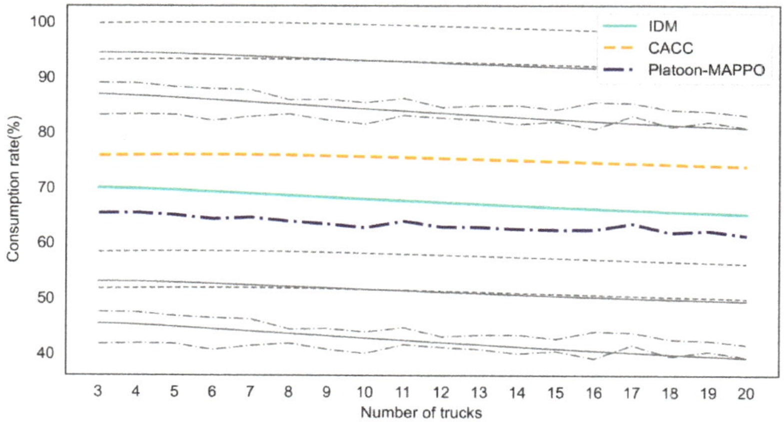

Figure 6. Energy consumption of the platoon.

6. Conclusions

Platoon-MAPPO is a truck platooning control algorithm based on MARL that can achieve efficient and safe platooning in highway ramp regions. Platoon-MAPPO adopts a centralized training and decentralized execution framework, allowing each truck to act independently without relying on a central controller, making the allocation of computing resources more concise. For the driving status of each truck in the platoon, each truck only refers to the driving status of the front and rear trucks, without requiring internal communication within the platoon or with the road network. This makes it easier for the truck's driving strategy to maximize the overall interests of the platoon; this architecture can greatly improve the flexibility of the platoon formation compared with architectures that rely on communication and centralized computing.

According to the experimental results, Platoon-MAPPO reduced average energy consumption and road occupancy by 14.8% and 32.7%, respectively, compared with uncontrolled IDM models and CACC control methods. From the experimental results, it can be seen that this method significantly filters traffic oscillations, greatly reduces meaningless acceleration and deceleration, and maintains driving speed while reducing energy consumption. Compared with IDM and CACC, Platoon-MAPPO has a faster overall speed at lower energy consumption. At the same time, Platoon-MAPPO can maintain a smaller headway distance, which not only reduces energy consumption, but also significantly reduces road occupancy. Compared with IDM and CACC, Platoon-MAPPO's road occupancy is reduced by 43.3% and 34.2% on average, and up to maximums of 58.2% and 35.7%, respectively. Therefore, in conclusion, Platoon-MAPPO can significantly improve the energy efficiency of truck platoons and improve road capacity.

Platoon-MAPPO benefits from a good basic model and reward function design, achieving good performance in experiments. However, similarly to general DL algorithms, RL cannot guarantee reasonable output results for all inputs, which is one of the reasons why learning-based algorithms have not yet been widely used in vehicle control fields. To address this issue, future research can start from two aspects: one is to use traditional methods to assist in the real-time monitoring of unsafe behavior; the other is to improve the interpretability of RL models and train them in more comprehensive and diverse traffic scenarios to improve robustness.

Author Contributions: Conceptualization, J.C.; methodology, J.C. and Z.Z.; software, Z.Z. and Y.D.; validation, Y.D.; formal analysis, Z.Z.; in-vestigation, J.C.; resources, J.C.; data curation, Y.D.; writing—original draft preparation, Y.D.; writing—review and editing, Z.Z.; visualization, Z.Z.; supervision, B.Y.; project administration, B.Y.; funding acquisition, J.C. All authors have read and agreed to the published version of the manuscript.

Funding: This research was funded by the National Nature Science Foundation of China, Grant No.51805133 and the Innovation Project of New Energy Vehicle and Intelligent Connected Vehicle of Anhui Province.

Data Availability Statement: The data used to support the findings of this study are available from the corresponding author upon request.

Conflicts of Interest: The funders had no role in the design of the study; in the collection, analyses, or interpretation of data; in the writing of the manuscript; or in the decision to publish the results.

References

1. Harker, B.J. PROMOTE-CHAUFFEUR II & 5.8 GHz vehicle to vehicle communications system. In Proceedings of the 2001 ADAS. International Conference on Advanced Driver Assistance Systems, (IEE Conf. Publ. No. 483), Birmingham, UK, 17–18 September 2001. Available online: https://digital-library.theiet.org/content/conferences/10.1049/cp_20010504 (accessed on 15 August 2023).
2. Shladover, S.E. AHS research at the California PATH program and future AHS research needs. In Proceedings of the 2008 IEEE International Conference on Vehicular Electronics and Safety, Columbus, OH, USA, 22–24 September 2008; pp. 4–5.
3. Shladover, S.E. PATH at 20—History and major milestones. *IEEE Trans. Intell. Transp. Syst.* **2007**, *8*, 584–592. [CrossRef]
4. Kunze, R.; Tummel, C.; Henning, K. Determination of the order of electronically coupled trucks on German motorways. In Proceedings of the 2009 2nd International Conference on Power Electronics and Intelligent Transportation System (PEITS), Shenzhen, China, 19–20 December 2009; pp. 41–46.
5. Tsugawa, S.; Kato, S.; Aoki, K. An automated truck platoon for energy saving. In Proceedings of the 2011 IEEE/RSJ International Conference on Intelligent Robots and Systems, San Francisco, CA, USA, 25–30 September 2011; pp. 4109–4114.
6. Robinson, T.; Chan, E.; Coelingh, E. Operating platoons on public motorways: An introduction to the sartre platooning programme. In Proceedings of the 17th World Congress on Intelligent Transport Systems, Busan, Republic of Korea, 25–29 October 2010; p. 12.
7. Zhao, W.; Ngoduy, D.; Shepherd, S.; Liu, R.; Papageorgiou, M. A platoon based cooperative eco-driving model for mixed automated and human-driven vehicles at a signalised intersection. *Transp. Res. Part C Emerg. Technol.* **2018**, *95*, 802–821. [CrossRef]
8. Zhou, Y.; Wang, M.; Ahn, S. Distributed model predictive control approach for cooperative car-following with guaranteed local and string stability. *Transp. Res. Part B Methodol.* **2019**, *128*, 69–86. [CrossRef]
9. Zhou, Y.; Ahn, S.; Wang, M.; Hoogendoorn, S. Stabilizing mixed vehicular platoons with connected automated vehicles: An H-infinity approach. *Transp. Res. Part B Methodol.* **2020**, *132*, 152–170. [CrossRef]
10. Wang, M.; Daamen, W.; Hoogendoorn, S.P.; van Arem, B. Rolling horizon control framework for driver assistance systems. Part II: Cooperative sensing and cooperative control. *Transp. Res. Part C Emerg. Technol.* **2014**, *40*, 290–311. [CrossRef]
11. He, X.; Liu, H.X.; Liu, X. Optimal vehicle speed trajectory on a signalized arterial with consideration of queue. *Transp. Res. Part C Emerg. Technol.* **2015**, *61*, 106–120. [CrossRef]
12. Grigorescu, S.; Trasnea, B.; Cocias, T.; Macesanu, G. A survey of deep learning techniques for autonomous driving. *J. Field Robot.* **2020**, *37*, 362–386. [CrossRef]
13. Hu, Y.; Yang, J.; Chen, L.; Li, K.; Sima, C.; Zhu, X.; Chai, S.; Du, S.; Lin, T.; Wang, W.; et al. Planning-oriented Autonomous Driving. In Proceedings of the IEEE/CVF Conference on Computer Vision and Pattern Recognition, Online, 1 September 2022.
14. Ye, F.; Zhang, S.; Wang, P.; Chan, C.Y. A Survey of Deep Reinforcement Learning Algorithms for Motion Planning and Control of Autonomous Vehicles. In Proceedings of the 2021 IEEE Intelligent Vehicles Symposium (IV), Nagoya, Japan, 11–17 July 2021; pp. 1073–1080.
15. Kiran, B.R.; Sobh, I.; Talpaert, V.; Mannion, P.; Al Sallab, A.A.; Yogamani, S.; Pérez, P. Deep reinforcement learning for autonomous driving: A survey. *IEEE Trans. Intell. Transp. Syst.* **2021**, *23*, 4909–4926. [CrossRef]
16. Talpaert, V.; Sobh, I.; Kiran, B.R.; Mannion, P.; Yogamani, S.; El-Sallab, A.; Perez, P. Exploring applications of deep reinforcement learning for real-world autonomous driving systems. *arXiv* **2019**, arXiv:1901.01536.
17. Hoel, C.-J.; Wolff, K.; Laine, L. Automated speed and lane change decision making using deep reinforcement learning. In Proceedings of the 2018 21st International Conference on Intelligent Transportation Systems (ITSC), Maui, HI, USA, 4–7 November 2018; pp. 2148–2155.
18. Desjardins, C.; Chaib-draa, B. Cooperative Adaptive Cruise Control: A Reinforcement Learning Approach. *IEEE Trans. Intell. Transp. Syst.* **2011**, *12*, 1248–1260. [CrossRef]

19. Li, M.; Cao, Z.; Li, Z. A Reinforcement Learning-Based Vehicle Platoon Control Strategy for Reducing Energy Consumption in Traffic Oscillations. *IEEE Trans. Neural Netw. Learn. Syst.* **2021**, *32*, 5309–5322. [CrossRef] [PubMed]
20. Chen, C.; Jiang, J.; Lv, N.; Li, S. An Intelligent Path Planning Scheme of Autonomous Vehicles Platoon Using Deep Reinforcement Learning on Network Edge. *IEEE Access* **2020**, *8*, 99059–99069. [CrossRef]
21. Chen, C.; Zhang, Y.; Khosravi, M.R.; Pei, Q.; Wan, S. An Intelligent Platooning Algorithm for Sustainable Transportation Systems in Smart Cities. *IEEE Sens. J.* **2021**, *21*, 15437–15447. [CrossRef]
22. Zhu, M.; Wang, Y.; Pu, Z.; Hu, J.; Wang, X.; Ke, R. Safe, efficient, and comfortable velocity control based on reinforcement learning for autonomous driving. *Transp. Res. Part C Emerg. Technol.* **2020**, *117*, 102662. [CrossRef]
23. Chu, T.; Kalabić, U. Model-based deep reinforcement learning for CACC in mixed-autonomy vehicle platoon. In Proceedings of the 2019 IEEE 58th Conference on Decision and Control (CDC), Nice, France, 11–13 December 2019; pp. 4079–4084.
24. Gao, Z.; Wu, Z.; Hao, W.; Long, K.; Byon, Y.J.; Long, K. Optimal Trajectory Planning of Connected and Automated Vehicles at On-Ramp Merging Area. *IEEE Trans. Intell. Transp. Syst.* **2022**, *23*, 12675–12687. [CrossRef]
25. Kumaravel, S.D.; Malikopoulos, A.A.; Ayyagari, R. Decentralized Cooperative Merging of Platoons of Connected and Automated Vehicles at Highway On-Ramps. In Proceedings of the 2021 American Control Conference (ACC), New Orleans, LA, USA, 25–28 May 2021; pp. 2055–2060.
26. Xue, Y.; Ding, C.; Yu, B.; Wang, W. A Platoon-Based Hierarchical Merging Control for On-Ramp Vehicles Under Connected Environment. *IEEE Trans. Intell. Transp. Syst.* **2022**, *23*, 21821–21832. [CrossRef]
27. Yadavalli, S.R.; Das, L.C.; Won, M. RLPG: Reinforcement Learning Approach for Dynamic Intra-Platoon Gap Adaptation for Highway On-Ramp Merging. *arXiv* **2022**, arXiv:2212.03497.
28. Jia, D.; Lu, K.; Wang, J.; Zhang, X.; Shen, X. A Survey on Platoon-Based Vehicular Cyber-Physical Systems. *IEEE Commun. Surv. Tutor.* **2016**, *18*, 263–284. [CrossRef]
29. Willke, T.L.; Tientrakool, P.; Maxemchuk, N.F. A survey of inter-vehicle communication protocols and their applications. *IEEE Commun. Surv. Tutor.* **2009**, *11*, 3–20. [CrossRef]
30. Sutton, R.S.; Barto, A.G. *Reinforcement Learning: An Introduction*, 2nd ed.; Adaptive Computation and Machine Learning; The MIT Press: Cambridge, MA, USA; Cambridge, London, 2018.
31. Mnih, V.; Badia, A.P.; Mirza, M.; Graves, A.; Lillicrap, T.; Harley, T.; Silver, D.; Kavukcuoglu, K. Asynchronous methods for deep reinforcement learning. In Proceedings of the 33rd International Conference on Machine Learning, Online, 16 June 2016; pp. 1928–1937.
32. Schulman, J.; Levine, S.; Abbeel, P.; Jordan, M.; Moritz, P. Trust region policy optimization. In Proceedings of the 32nd International Conference on Machine Learning, Online, 6–11 July 2015; pp. 1889–1897.
33. Schulman, J.; Wolski, F.; Dhariwal, P.; Radford, A.; Klimov, O. Proximal policy optimization algorithms. *arXiv* **2017**, arXiv:1707.06347.
34. Yu, C.; Velu, A.; Vinitsky, E.; Gao, J.; Wang, Y.; Bayen, A.; Wu, Y. The surprising effectiveness of ppo in cooperative multi-agent games. *Adv. Neural Inf. Process. Syst.* **2022**, *35*, 24611–24624.
35. Lowe, R.; Wu, Y.I.; Tamar, A.; Harb, J.; Abbeel, O.P.; Mordatch, I. Multi-agent actor-critic for mixed cooperative-competitive environments. In Proceedings of the 31st Annual Conference on Neural Information Processing Systems (NIPS 2017), Long Beach, CA, USA, 4–9 December 2017; Volume 30.
36. Knox, W.B.; Allievi, A.; Banzhaf, H.; Schmitt, F.; Stone, P. Reward (mis) design for autonomous driving. *Artif. Intell.* **2023**, *316*, 103829. [CrossRef]
37. Alam, A.; Besselink, B.; Turri, V.; Mårtensson, J.; Johansson, K.H. Heavy-duty vehicle platooning for sustainable freight transportation: A cooperative method to enhance safety and efficiency. *IEEE Control Syst. Mag.* **2015**, *35*, 34–56.
38. Hussein, A.A.; Rakha, H.A. Vehicle Platooning Impact on Drag Coefficients and Energy/Fuel Saving Implications. *IEEE Trans. Veh. Technol.* **2022**, *71*, 1199–1208. [CrossRef]
39. Treiber, M.; Hennecke, A.; Helbing, D. Congested traffic states in empirical observations and microscopic simulations. *Phys. Rev. E* **2000**, *62*, 1805–1824. [CrossRef] [PubMed]
40. Milanés, V.; Shladover, S.E. Modeling cooperative and autonomous adaptive cruise control dynamic responses using experimental data. *Transp. Res. Part C Emerg. Technol.* **2014**, *48*, 285–300. [CrossRef]

World Electric Vehicle Journal

MDPI

Article

Coordinated Control of Unmanned Electric Formula Car

Hua Tao * and Baocheng Yang

Department of Mechanical and Electrical Engineering, Sichuan Engineering Technical College, Deyang 618000, China
* Correspondence: thua@scetc.edu.cn

Abstract: The coordinated control method of Unmanned Electric Formula Racing (UEFC) was studied to improve the handling stability of UEFC. The UEFC's mechanical structure, which is based on the driving system and transmission system, was designed. In accordance with mechanical structure of the designed racing car, a seven-degree of freedom mathematical model of the UEFC was established. In accordance with the built mathematical model of racing car, the lateral controller of racing car was designed by using a fuzzy neural network method. The longitudinal controller of the racing car was designed by using the method of incremental PID control, and the coordination controller of the racing car was designed by combining the lateral controller and the longitudinal controller so as to realize the lateral and longitudinal coordination control of the UEFC. The experimental results showed that the output parameters such as yaw rate, vehicle speed and heading angle were slightly different from the expected output. It was confirmed that the research method can enhance the handling stability of the UEFC.

Keywords: unmanned; electric equation; racing cars; coordination control; PID control

1. Introduction

Unmanned platforms include unmanned aerial vehicles, unmanned boats, unmanned submersibles and ground unmanned vehicles. Unmanned ground vehicles are also called autonomous ground vehicles and autonomous ground mobile robots. With the rapid development of advanced smart sensors, fast response actuators, high-performance ECUs, advanced control strategies, computer-based network technology, radar technology, mobile communication technology and other advanced technologies, intelligent driving vehicles have stepped from conceptualization to production [1]. With the fast growth of social productivity and automobile technology, the number of cars is increasing progressively year after year, and the traffic problem of traditional drivers driving cars is also increasing. Unmanned technology and unmanned vehicles are the developmental result of the human automobile industry [2], which represents the highest technical level in the field of automobile research [3], and it is also the future development trend of automobiles. In addition, with the advancement of conventional unmanned vehicle technology, the unmanned trend of formula car is increasingly obvious. The longitudinal control system, which is one of the key technologies of unmanned racing, is the basis for the realization of unmanned formula racing [4]. Thus, the design of control methods with high stability and reliability is of great significance.

Vehicle motion control technology is the most key technology for unmanned intelligent vehicle at this stage, and it is the basis to realize stable, reliable and autonomous driving of vehicles. Motion control involves unmanned vehicle dynamics theory and is an important content of unmanned vehicle technology [5]. In accordance with environmental information obtained by the vehicle perception layer, the vehicle acceleration, braking, steering and other systems are under automatic control. So, the vehicle is controlled to drive in accordance with desired path. Unmanned vehicle motion control can be either longitudinal or lateral control. Additionally, these two types are different and correlated.

Citation: Tao, H.; Yang, B. Coordinated Control of Unmanned Electric Formula Car. *World Electr. Veh. J.* **2023**, *14*, 58. https://doi.org/10.3390/wevj14030058

Academic Editors: Biao Yu, Linglong Lin and Jiajia Chen

Received: 29 December 2022
Revised: 8 February 2023
Accepted: 16 February 2023
Published: 24 February 2023

Lateral control, in essence, refers to the control of the unmanned vehicle's steering angle, mainly considering the safety and stability of controlled vehicle [6]. Longitudinal control mainly refers to the control of unmanned vehicle's speed and distance. The purpose is to make the controlled vehicles drive at the expected speed or maintain the expected distance. Unmanned vehicle is a highly integrated nonlinear system [7]. There are many interactive and restrictive factors between lateral and longitudinal control systems. Therefore, the design of lateral or longitudinal control system cannot satisfy an unmanned vehicle's actual driving requirements alone. During actual driving of the vehicle, various driving conditions need to take into account the lateral and longitudinal movements [8]. For example, the direction and speed of the vehicle need to be controlled at the same time in sharp corners, overtaking, parking alongside and other situations. Most unmanned vehicles only combine lateral and longitudinal control without coupling them, and do not take into account the interaction between lateral and longitudinal dynamics and kinematics. Therefore, the control effect is not ideal. During driving of an unmanned vehicle, there is a complex and strong coupling relationship between longitudinal direction and lateral direction [9]. The whole vehicle is a highly nonlinear motion constraint system, and its model and environment have uncertainty and measurement inaccuracy, which makes the lateral and longitudinal control more complex. Designing the lateral control or longitudinal control system of intelligent vehicle cannot satisfy the intelligent vehicle's driving needs alone. Simple working conditions should be designed to move laterally and longitudinally at the same time, so that actions such as overtaking, pulling over and so on can be carried out. Most vehicle control systems are synthesized based on the simultaneous loading of lateral and longitudinal independent controllers [10]. However, since the interaction of lateral and longitudinal dynamics is not considered in the dynamic modeling, and such a vehicle control effect is not necessarily the same as that in the independent control.

Numerous scholars have carried out intensive research and achieved certain results. Research shows that most control methods need high-precision vehicle models or complex optimization algorithms to optimize the controller. Ge L and his team used MPC to directly solve the longitudinal and lateral coupling control problems, and used convex optimization methods to solve the reference value problem of predictive control. Additionally, they proposed a systematic method to prevent the reference value from exceeding the feasible region. The research results showed that the algorithm could solve the lateral and longitudinal disturbance problems. Because longitudinal and lateral coupling constraints are considered, the algorithm has better tracking accuracy and high-speed stability [11]. Wang, Y and their team proposed an integrated algorithm for longitudinal and lateral trajectory planning and tracking control based on vehicle to vehicle communication. The algorithm includes two levels: trajectory planning and path tracking control. A three-step nonlinear method of multi input and multi output is proposed to design longitudinal and lateral path tracking controllers. Finally, the stability of the closed-loop system is strictly proved based on Lyapunov function, and the effectiveness of the algorithm is verified through a high fidelity full vehicle model on the veDYNA (Vehicle dynamics) platform [12]. To sum up, the research on the control algorithm of driverless vehicles has made certain achievements. The control system of driverless vehicles is not just composed of a simple longitudinal and lateral independent control system. To have sufficient practical value in practical projects, it is necessary to consider the interaction between the longitudinal and lateral directions of vehicles. The lateral and longitudinal control can improve the path tracking ability of the driverless car. The previous lateral and longitudinal control only considered the parameters that affect each other between the lateral control and longitudinal control directions. Additionally, it did not take into account the actual working conditions.

Therefore, based on the study of lateral and longitudinal control, this paper proposes a lateral and longitudinal control method that takes into account various actual working conditions of a driverless car. It aims to improve the path tracking control effect of driverless cars. The research contribution is as follows: (1) the proposal of driverless electric formula racing is to promote the development of driverless control technology. (2) It is proposed that the driverless control method considering horizontal and vertical directions can achieve intelligent control of driverless racing through coordinated control, which is of great significance to the development of driverless technology. (3) In addition, the performance of driverless control technology is verified with the help of driverless electric formula racing, which provides ideas for the intelligent development of transportation. Additionally, the innovation points in this study are (1) a lateral controller based on a fuzzy neural network; (2) a racing longitudinal controller based on an incremental PID; (3) the coordination controller of the transverse and longitudinal direction of the racing car is proposed; (4) the effectiveness of the controller is verified by simulation experiments.

The main contents of this study are shown in Figure 1. Firstly, the mechanical structure of UEFC is briefly introduced, and then the mathematical model of UEFC is analyzed. At the same time, the lateral controller based on the fuzzy neural network and the longitudinal controller of the racing car based on incremental PID are proposed. Then, the lateral and longitudinal controllers of UEFC are proposed according to the abovementioned research. Finally, the proposed research methods are verified and analyzed by experiments.

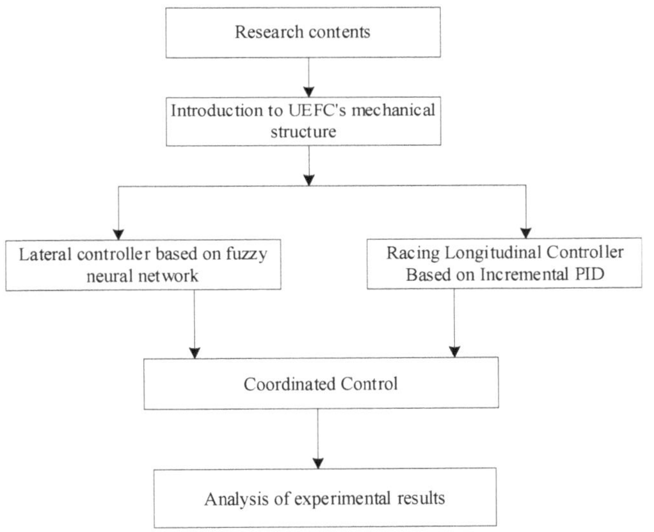

Figure 1. Main contents of this study.

2. Materials and Methods

2.1. Introduction to UEFC's Mechanical Structure

On the basis of the electric formula racing car, the driverless electric formula racing car is composed of environment sensing system and wire control mechanism. The whole car conforms to the vehicle system dynamics. The environment awareness system consists of a series of sensors and processors such as cameras, laser radars, GPS/INS, etc. These components are responsible for sensing the environment around the racing track and the real-time status of the racing car. At the same time, these components are responsible for planning the corresponding reference path through algorithms. The control by wire system is composed of steering by wire, drive by wire and brake by wire. The vehicle controller is used to coordinate and control the executive components of the wire control system. The vehicle controller receives the data from the perception system and processes it to get the control command. According to the command, each part of the wire control system makes corresponding actions to complete the tracking control of the car. In order to improve the path tracking accuracy of the car, this chapter introduces a hardware structure and software structure of the car. At the same time, this chapter analyzes the role of each part as well as the design and calculation process for the research of driverless car's control strategy. The hardware structure of the unmanned system is composed of sensing sensor, wire control mechanism, processor, controller, etc. Perceptual data processing, mapping and planning, and control constitute the software structure of the unmanned system. At the same time, in order to ensure the accuracy and stability of the driverless car in the lateral and longitudinal control, the vehicle dynamics model, tire model, motor model, etc. are built. Figure 2 shows the mechanical structure of UEFC [13].

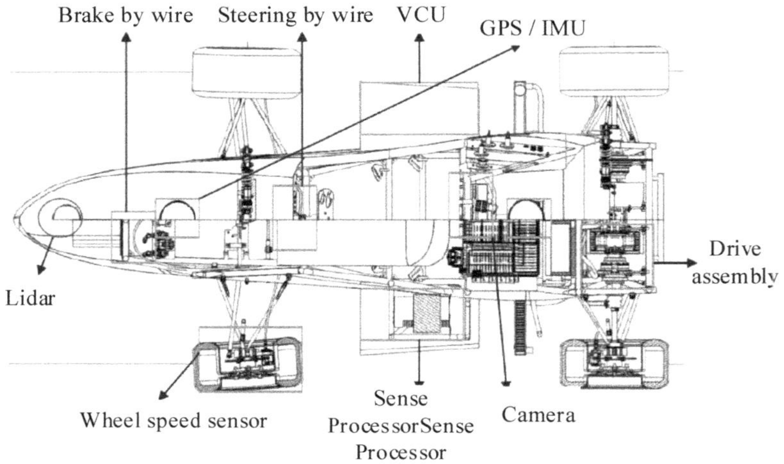

Figure 2. Mechanical structure of UEFC.

By obtaining and storing the racing point cloud information through the cone barrel, the perceptual data processes the point cloud information from the LiDAR acquisition cone barrel, and uses the point cloud filtering algorithm to filter and identify the target cone barrel point cloud. The camera collects the cone barrel color information, and uses the depth learning mode to accurately identify the target cone barrel color. The joint calibration algorithm of LiDAR and camera is designed independently. The 3D LiDAR point cloud data is mapped to the image through the internal and external parameter transformation matrix through timestamp matching, and finally the data fusion between the two sensors is realized.

The drive-by-wire drive system of driverless racing car is partially driven by two independent motors, which can improve the controllability of the racing car. The torque output of each motor can be controlled separately according to the actual working conditions to improve the dynamic performance of the racing car. The vehicle controller sends the torque command calculated by the algorithm to the motor controller through CAN communication to realize the drive-by-wire control of the driverless racing car. For steering-by-wire, on the basis of ensuring the original mechanical steering, a drive motor is added in the middle of the steering shaft to realize steering-by-wire. In order to facilitate the disassembly of the brake drive system and integrate the assembly, the brake drive motor mounting frame is designed. Two connecting rods are used between the motor and the brake pedal to transmit the torque of the motor so that the brake pedal can rotate and brake. There is an empty stroke structure between the motor and the drive connecting rod. When the driver steps on the brake, the drive connecting rod rotates relative to the empty stroke of the motor, ensuring that it is not interfered by the motor. The motor and connecting rod are connected with flat keys, and the two connecting rods are connected with bolts. When the motor brakes, the connecting rod is connected with the motor rotates, which drives the brake pedal to rotate for braking. After braking, the brake pedal is reset under the action of the brake master cylinder.

This research is the basis of the longitudinal and transverse controller's design. In order to have a better understanding of the driverless car, its structural design is given. Considering the control accuracy and complex working conditions of the driverless car, three degree of freedom dynamics model of the whole car is used, and the tire model of the car is established.

2.2. Mathematical Model of UEFC

As the track is smooth and dry asphalt pavement, and no driver is allowed to sit in the car when the car is driverless. No in-depth study is made on the suspension system that affects the ride comfort of the car [14]. According to the above analysis, when modeling the vehicle dynamics, the plane motion of the car is mainly considered in two directions, namely, lateral and yaw motion. Therefore, a three-degree-of-freedom vehicle monorail dynamics model is established.

(1) Lateral motion

$$m(\dot{v}_y + v_x\omega) = F_{yfr}\cos\delta + \left(F_{xfr} + F_{xf1}\right)\sin\delta + F_{yf1} + F_{yr1} + F_{yrr} \tag{1}$$

(2) Longitudinal motion

$$m(\dot{v}_x + v_y\omega) = F_{xfr}\cos\delta + F_{xr1} - m\left(F_{xfr} + F_{xf1}\right)\sin\delta + F_{xrr} \tag{2}$$

(3) Yaw motion

$$\begin{aligned} I\dot{\omega} = \tfrac{d}{2}\left(F_{yf1} + F_{yfr}\right)\sin\delta - bF_{yf1} - F_{yrr} + a\left(F_{yf1} + F_{yfr}\right) \\ +\tfrac{d}{2}\left(F_{xfr} + F_{xf1}\right)(F_{xrr} + F_{xr1})\cos\delta + a\left(F_{xf1} + F_{xfr}\right)\sin\delta \end{aligned} \tag{3}$$

(4) Rotational movement of four wheels

$$J_\varepsilon\omega_\varepsilon = T_{dij} - R_\varepsilon T_{bij} - F_{zij}R_\varepsilon \tag{4}$$

In the formulas, m refers to the vehicle mass; δ refers to the front wheel angle; I refers to the inertia moment of the vehicle; a and b refer to the distance from the front axis and rear axis of the vehicle to the mass center respectively; v_x refer to the longitudinal speed; v_y refer to the lateral velocity; ω refers to the yaw rate. F_{xf1}, F_{xfr}, F_{xr1}, F_{xrr}, F_{yf1}, F_{yfr}, F_{yr1} and F_{yrr} respectively refer to the front right, front left, rear right and rear left tire force components along the lateral direction; d refers to the track width of left and right wheels (given that the front and rear wheels' track widths are equal); J_ε refers to the rolling inertia moment of the wheel; ω_ε refers to the angular velocity of the wheel; T_{bij} refers to the braking torque of the wheel ($i = f, r$ refers to the front wheel and rear wheel, $j = l, r$ refers to the left wheel and right wheel, the same below); T_{dij} is the rear wheel drive torque; F_{zij} is the wheel's longitudinal force; R_ε and is the wheel's rolling radius.

In the actual process, the tire structure is more complex, and the dynamic characteristics are nonlinear. The steering characteristics and driving stability of the racing car are closely related to the nonlinear characteristics of the tire. Therefore, appropriate tire model plays a key role in establishing vehicle model and conducting dynamic simulation, i.e., real vehicle control.

The magic tyre formula has the characteristics of unity, convenient fitting and high fitting accuracy. Besides, its parameters can be easily determined. Therefore, the formula is widely accepted by researchers and engineers in the automotive industry. The empirical tire model based on the magic formula is adopted in this paper, as shown in Formula (5) [15].

$$Y(x) = D\sin\{C\arctan[Bx - E(Bx - \arctan(Bx))]\} \tag{5}$$

In Formula (4): Y is the output variable, which can be represented by the longitudinal force F_l or the lateral force F_c or the return moment M_z; x is the input variable, which can be determined by the tire side slip angle α or longitudinal slip ratio s; B. C, D and E are stiffness factor, shape factor, peak factor and curvature factor, respectively.

2.3. Lateral Controller Based on Fuzzy Neural Network

2.3.1. Torque Control Layer

The stability of UEFC can be evaluated with the deviation of yaw rate and the deviation of the centroid side deflection angle. The smaller deviation will lead to stable control better. The yaw moment control layer adopts fuzzy neural network algorithm. A five layer fuzzy neural network is adopted [16]. The input layer includes the deviation $e(\gamma)$ of yaw rate and the deviation $e(\beta)$ of centroid side deflection angle. The output layer is the yaw moment M_{zf}, and the membership functions of the input and output layers are Gaussian functions. By comparing the yaw moment M_{zf} with the maximum yaw moment M_{zmax} provided by the ground, the final yaw moment M_z is obtained for controlling the UEFC. Fuzzy neural network can quickly and effectively calculate the optimal parameters of membership function by learning the relationship between input variables $e(\gamma)$, $e(\beta)$ and output variable M_{zf}. Additionally, 30 fuzzy rule statements can be generated to form the fuzzy rules of the controller. During the driving of the UEFC, the fuzzy neural network controller will apply a yaw moment to the UEFC to make the vehicle drive stably [17]. If the torque exceeds the torque limit of the ground acting on the wheel, the wheel will slip excessively, so the yaw moment needs to be limited. The maximum yaw moment that the ground can provide is shown in Formula (6).

$$M_{zmax} = B\left(F_{zfr} + F_{zrr} - F_{zf1} - F_{zr1}\right) \tag{6}$$

In Formula (6), B is the track width; $F_{zfr}, F_{zrr}, F_{zf1}, F_{zr1}$ are the longitudinal force of the right front, rear wheels, the left front and rear wheels respectively. In accordance with comparison between the yaw moment output by the fuzzy neural network and the maximum yaw moment that can be provided by the ground, the smaller moment is selected as the final yaw moment M_z, and the expression is as follows:

$$M_z = \min\left(M_{zf}, M_{zmax}\right) \tag{7}$$

2.3.2. Distribution of Torque and Control Layer of Slip Rate

To distribute the torque to the driving wheel, the distribution of torque and the control layer of slip rate receive the expected torque and the yaw torque output by the yaw control layer. Additionally, it should take into account the factors of motor failure and driving wheel slip ratio. In case of motor failure, the driving wheels on both sides will output different torques, which will lead to difficult tracking or even out of control [18]. If the slip rate of the vehicle is too large, it is easy to cause the vehicle out of control and accidents. Therefore, the factors of motor failure and slip rate should be considered in the torque distribution strategy to ensure the vehicle's stable driving.

(1) Torque distribution and restraint

To ensure the vehicle's stability, a target yaw moment M_z is used to control the driving wheel torque. In the process of turning, the tire on the same side as the steering wheel is called the inside wheel, which turns differently and is interchangeable inside and outside. As for tires, the side facing the outside of the vehicle body is the outside of the tyres. It is controlled by increasing ΔT for the outer wheel and decreasing ΔT for the inner wheel. ΔT is calculated as follows:

$$\Delta T = M_z \times R/B \tag{8}$$

In Formula (8), R refers to the racing car's tire radius.

Considering the limitation of the maximum torque of the motor, the expected torque of the driving wheels on both sides is shown in Formula (9).

$$T_1 = \begin{cases} T_{req}/2 - \Delta T & (\text{if } T_{req}/2 - \Delta T \leq T_{1\_max}) \\ T_{1\_max} & (\text{if } T_{req}/2 - \Delta T > T_{1\_max}) \end{cases} \quad (9)$$

$$T_2 = \begin{cases} T_{req}/2 - \Delta T & (\text{if } T_{req}/2 - \Delta T \leq T_{2\_max}) \\ T_{2\_max} & (\text{if } T_{req}/2 - \Delta T > T_{2\_max}) \end{cases} \quad (10)$$

$$T_{i\_max} = \begin{cases} T_{max} & (n \leq n_r; i = 1,2) \\ P_{max}/n & (n > n_r; i = 1,2) \end{cases} \quad (11)$$

In the formula, T_1 and T_2 represent the actual output torque of internal and external motors respectively; T_{1_max} and T_{1_max} respectively indicate the maximum torque that the inner and outer sides of the motor can output at this time; n_r refers to the motor's rated speed; n refers to the motor's actual speed.

(2) Motor failure

In case of failure, the motor will send a fault code to the motor controller. According to different fault levels, the motor fault factor ξ needs to be designed to restrict the torque of the driving wheel and prevent damage to the motor and driver. The range of fault factor ξ is [0, 1]; 0 indicates that the motor has the highest fault level: the motor fails and needs to be stopped immediately; 1 indicates that the motor has no fault and operates normally. The fault level of the motor is judged by the fault code sent by the motor encoder to the motor controller. Under the constraint of motor fault factor, the output torque of inner and outer drive wheels is

$$T_i \leq \xi_i T_{i\_max} \quad (12)$$

(3) Slip rate control

In accordance with actual conditions of road, set the target slip rate to [0.06, 0.22]. The slip ratio s of the tire on the ground can be expressed by Formula (13).

$$s = \begin{cases} \frac{Rv\omega_t}{v} - 1 & (v > R\omega_t, v \neq 0) \\ 1 - \frac{v}{R\omega_t} & (v \leq R\omega_t, v \neq 0) \end{cases} \quad (13)$$

In Formula (13), R is the wheel radius, ω is the wheel rotation angle speed, v is the tire speed. When the slip rate of the motor exceeds the slip rate set by the logic, the torque PID control of the motor will be started, and the difference of slip rate is taken as the input to adjust the motor torque output.

$$S_{goal} = \frac{S_{min} + S_{max}}{2} \quad (14)$$

In the formula, S_{min} and S_{max} respectively are the minimum value and the maximum value of the target slip rate range; S_{goal} is the target slip rate representing the actual slip rate. In the algorithm of PID control, the motor torque after adjustment is

$$\Delta T_i = k_p\left(S_i - S_{goal}\right) + k_i \int \left(S_i - S_{goal}\right)dt + k_d \frac{d\left(S_i - S_{goal}\right)}{dt} \quad (15)$$

In the formula, k_p, k_i and k_d respectively are proportional coefficient, integral coefficient and differential coefficient. The UEFC's final torque output is as follows:

$$\begin{cases} T_{in} = T_1 + \Delta T_1 \\ T_{out} = T_2 + \Delta T_2 \end{cases} \quad (16)$$

In Formula (16), T_{in} and T_{out} are the torques of the inner and outer wheels of the UEFC, respectively.

The basic structure of the fuzzy controller is shown in Figure 3.

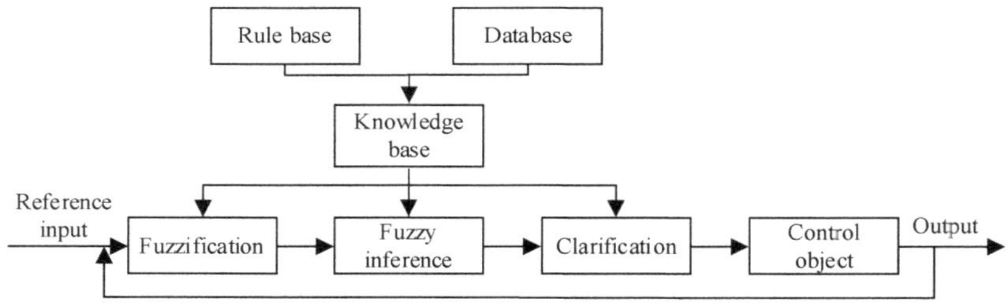

Figure 3. Structure of the fuzzy controller.

2.4. Longitudinal Controller of Racing Car Based on Incremental PID

In essence, the UEFC's longitudinal control is a speed control system, which aims to track the expected speed through the control of acceleration and deceleration. In accordance with deviation between the vehicle's expected speed and its actual speed, the main control principle of longitudinal control is to obtain the ideal acceleration and deceleration control quantity. Then, it must be input into the corresponding controller for execution. The longitudinal control of racing car includes the switching control of acceleration control system, braking control system and acceleration and deceleration system. Therefore, this research introduces the incremental PID control method, which is a basic form of digital PID control algorithm and a control algorithm for PID control. This can be achieved through the increment of control quantity (the difference between the current control quantity and the last control quantity). The longitudinal control structure of racing car on the basis of incremental PID is presented in Figure 4 [19].

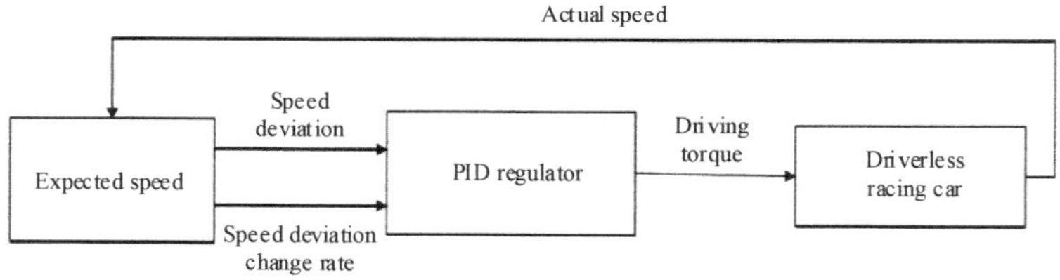

Figure 4. Incremental PID longitudinal control of the racing car.

The longitudinal control of the driverless car is mainly the control of car's longitudinal axis direction. It generally includes the control of the car's displacement, longitudinal speed and acceleration. When the car is running on a straight track, it needs to increase its speed and decelerate before entering the curve. Therefore, in order to make the car complete the race quickly and stably, it needs to adjust the longitudinal speed in real time according to the track conditions. As a nonlinear system, driverless car brings many difficulties to the design of longitudinal controller, and it is difficult to ensure its control accuracy and robustness. The longitudinal control principle of the driverless car is to control and adjust the torque of the driving system in real time. At the same time, it should indirectly control the speed and acceleration of the car, and realize the tracking and control of the expected speed of the driverless car.

PID control, which is a classical control technology in modern control, has advantages including simple structure, reliable operation, good stability, convenient parameter adjustment and so on. In accordance with deviation of the controlled system, PID control combines the integral coefficient I, proportional coefficient Pas well as differential coefficient D linearly to control the controlled object. $r(t)$ and $c(t)$ are used to represent the expected output and actual output, respectively. The calculation formula of PID control algorithm is shown as follows:

$$u(t) = K_p\left[r(t) - c(t) + \frac{1}{K_i}\int_0^t [r(t) - c(t)]dt + K_D d[r(t) - c(t)]/dt\right] \quad (17)$$

In the formula, K_p refers to the proportional adjustment coefficient; K_i refers to the integral adjustment coefficient; K_D refers to the differential time constant.

The proportional regulation coefficient K_p in the PID control algorithm mainly regulates the system's response speed, eliminates the error at the fastest speed, and makes the controlled quantity reach the expected value of the system [20,21]. The integral adjustment coefficient K_i is mainly to remove the steady-state error. The value of K_i corresponds to the speed of removing the system's steady-state error. The differential adjustment coefficient K_d is mainly used to improve the system's dynamic characteristics and keep the system deviation changing to the set point at all times.

The incremental PID controller has advantages including simple structure and small influence range on the system in case of failure. Additionally, the incremental PID control is selected as the control algorithm for the UEFC's longitudinal control. The calculation formula of incremental PID controller is as follows:

$$\Delta u = K_p[e(k) - e(k-1)] + K_i e(k) + K_d[e(k) - 2e(k-1)] \quad (18)$$

In Formula (18), Δu refers to the opening control increment of the electronic accelerator pedal; $e(k)$ refers to the velocity deviation at the k-th sampling; e refers to the deviation between the expected speed and the actual speed. The incremental PID control takes the speed deviation input at the k-th and $k-1$-th sampling times as the independent variable. The incremental PID control compares the size, change direction and change frequency of the speed deviation during the last three sampling times. Additionally, the incremental PID control grasps the change trend of the UEFC's speed at the next time. Through parameters K_p, K_i and K_D, the system error is rapidly reduced so that the UEFC's actual speed approaches the expected one, and finally reaches the target value stably.

Fuzzy PID adaptive controller is an intelligent algorithm that combines fuzzy control and PID control, and can adjust PID parameters online through empirical fuzzy rules. The input of the controller is error e and error change rate e_c. The fuzzy controller adjusts the changes of the three parameters of PID online according to the experience rules accumulated in the past [22]. Then, the adjusted three changes K_p, K_i and K_D are added to the previously set initial value of the PID controller to obtain the final output, as shown in Equation (19).

$$\begin{cases} K_p = K_{p0} + \Delta K_p \\ K_i = K_{i0} + \Delta K_i \\ K_D = K_{D0} + \Delta K_D \end{cases} \quad (19)$$

Among them, K_p, K_i and K_D are the initial parameters set by the PID controller, and ΔK_p, ΔK_i and ΔK_D are the transformation values set according to the fuzzy control rules. The characteristics of fuzzy controller can make up for the defects of traditional PID. It does not need to establish an accurate model of the controlled object, but only need to convert the previously accumulated experience into corresponding control rules for control. The fuzzy adaptive PID controller is more flexible, and usually has more obvious advantages for the object with large nonlinearity and time variation.

According to the difference between the expected and the actual speed as well as the difference change rate, fuzzy PI control is to correct the values of K_p and K_i so as to realize the purpose of online real-time adjustment of the proportional value P and integral value I. Therefore, the rule selection basis is as follows: when the speed error $|e|$ is large, select a larger K_p value to obtain a larger driving torque value. When $|e|$ and $|ec|$ values are moderate, select smaller K_i and appropriate K_p to obtain moderate driving torque. When

$|e|$ is very small. In order to prevent over control of the speed, the K_p value output is zero, that is, the fuzzy controller does not intervene.

2.5. Lateral and Longitudinal Controller of UEFC

When either the UEFC's lateral control system or its longitudinal control system acts alone, UEFC cannot operate perfectly and the needs of various race tracks cannot be met. Due to the strong coupling relationship between the lateral and longitudinal control of the UEFC, a coordination controller is built to improve the lateral control effect. The coordination controller is built for combining the lateral control and longitudinal control to achieve the coordination control of the UEFC. The input parameter of the lateral control is the longitudinal speed, and that of the longitudinal control is the lateral position deviation. The lateral and longitudinal parameters are constrained by each other. In accordance with the analysis UEFC's working conditions, the corresponding coordination law is established. In addition, the parameters are adjusted to achieve the lateral and longitudinal coordination control. The overall structure of lateral and longitudinal control of UEFC is shown in Figure 5.

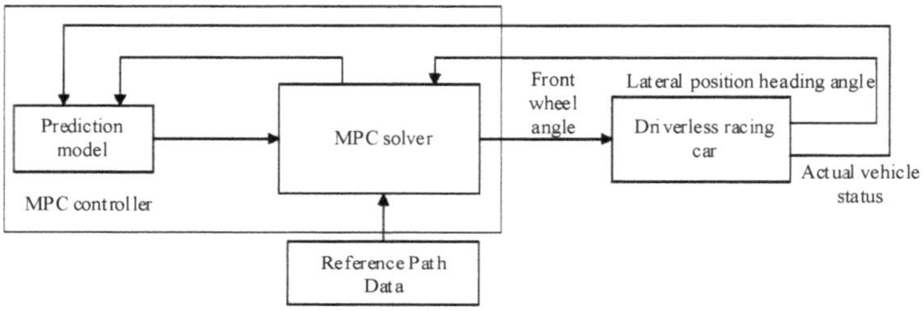

Figure 5. Coordinated control of lateral control and longitudinal control.

The coordination control of UEFC calculates the additional driving torque in accordance with actual position and lateral deviation of the racing car, and judges whether the racing car needs to accelerate or decelerate so as to realize the lateral and longitudinal coordination control of UEFC.

The longitudinal controller based on fuzzy PI is coupled. Through analyzing various working conditions of the car in the actual race, the lateral and longitudinal control laws are summarized, and the lateral and longitudinal control method of the driverless car is constructed. Compared with reference [23], after the longitudinal control being added, the tracking effect of the car on the path was significantly improved. The car performs well in different curves, and the frequency and amplitude of the swing were also reduced a lot. This shows that the fuzzy PI adaptive controller is suitable for the longitudinal control of an unmanned car.

In case of a lateral position deviation, if the reference path is too far from the position, it should be properly decelerated to make the UEFC track the reference path under the action of lateral control. In case of a lateral position deviation between the reference path and the car, if the car approaches the reference path, it should be properly accelerated to make the car quickly track the reference path [22–24]. The coordination control process of UEFC is as follows:

(1) It should judge whether the UEFC has a lateral deviation. Where the lateral deviation ΔY is more than a certain value γ, judge the φ angle. It is the included angle between the vehicle speed direction v_1 and the prediction range edge with the radius r at the intersection of the reference path. v_1 is taken as the fixed axis. If the angle between vehicle speed v and v_1 is formed by φ clockwise, the positioning is positive, while the counterclockwise is negative.

(2) It should determine which side the car is on. When the car is on the left side with regular and tends to be far away from the reference path, the car should reduce its speed,

that is, reduce its driving torque. When the car is on the right side with regular φ and tends to approach the reference path, it should speed up, that is, increase its driving torque. When it is on the left side with negative φ, if the car tends to approach the reference path, it should speed up, that is, increase its driving torque. When it is on the right side with negative φ, if the car tends to approach the reference path, it should slow down, that is, reduce its driving torque.

γ represents the threshold for judging whether there is overlarge lateral deviation. In the case that the lateral deviation $\Delta Y < \gamma$, it is deemed that the car has carried out tracking of the reference path and the control is not involved. In the case that the lateral deviation $\Delta Y < \gamma$, it is deemed that the car lateral deviation to certain extent, and the controller should intervene for reducing the lateral deviation. The lateral deviation ΔY calculates an additional driving torque T_1 through the PID controller, and then judges whether the target driving torque T needs to add or subtract the additional driving torque in accordance with coordination controller so as to finally realize the lateral control and longitudinal control.

3. Results

To verify the designed mechanical structure of the UEFC and the effectiveness of the coordination control method, the experimental analysis was conducted by the MATLAB/SIMULINK software. The coordinated control of the controller was carried out in Simulink. The parameters of the control model were coupled through the coordinated control module, so that the parameters in the joint controller were mutually constrained. The simulation model is the UEFC lateral and longitudinal coordination control model proposed in this study, and the basic parameters are as follows: the vehicle mass is 1620 kg; the distance from the mass center to the front axle is 1.35 m; the distance from the mass center to the rear axle is 1.43 m; the inertia moment is 2480 kg/m^2; the wheel radius of the racing car is 0.355 m and the height of the mass center is 0.565 m.

The corner input of UEFC is shown in Table 1.

Table 1. Racing steering wheel angle input.

Time/s	Steering Wheel Angle/Deg
0	0
1	20
2	40
3	20
4	0
5	−20
6	−40
7	−20
8	0

In accordance with steering wheel input of UEFC in Table 1, the angle for UEFC was input to check the coordination control performance of the proposed method.

This method is used to coordinate and control the longitudinal and lateral control changes of UEFC when the vehicle is running. The vehicle coordinated control results of the proposed method were compared with those of UEFC under lateral control and longitudinal control. Among them, "Expected output" is the reference standard path. The lateral and longitudinal control tracking comparison results of the car are shown in Figure 6.

In Figure 6, the results of UEFC path tracking under different control modes are compared. Because the linear approximate expression of the tire is used in the process of simplifying the dynamic model, the side slip angle of the tire is also required to be limited so as to ensure that the force on the tire is in the linear region during driving.

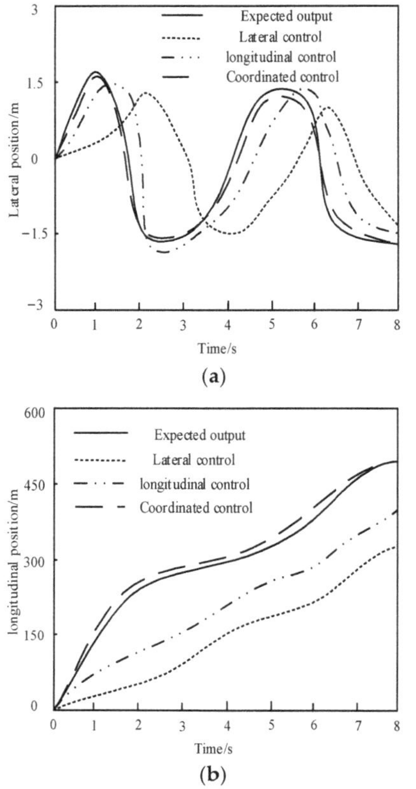

Figure 6. Comparison results of racing position tracking. (**a**) Lateral position. (**b**) Longitudinal position.

In the lateral control, the output of the front wheel angle is controlled, and the steering motor drives the steering mechanism to achieve the change of the front wheel angle. Therefore, the limit position of the steering mechanism and the responsiveness of the steering motor need to be considered at the same time. The angle of front wheels turning from the middle position to the leftmost limit position and the rightmost limit position, as well as the turning angle of the steering motor, were obtained from the real vehicle test. It is defined that turning left is counterclockwise and the angle is negative, turning right is clockwise and the angle is positive and turning the steering motor 110° corresponds to a front wheel angle of 30°. The continuous working speed of the steering motor is 50 r/min, so when the front wheel rotates 30°, the steering motor works for 0.4 s.

According to the comparison results, the deviation between UEFC's actual driving path and its reference path is large and swings unsteadily when under the lateral control only. Only when the UEFC's speed is controlled longitudinally is the deviation reduced, but there is still a significant deviation. When the proposed method is used to coordinate and control the UEFC laterally and longitudinal control, the car can track the reference path well. In Figure 5, further analysis of the experimental results shows that when only lateral control is adopted do the lateral position and longitudinal position of UEFC deviate greatly. During longitudinal control, the position deviation of the UEFC is maintained between the two control methods. When the proposed method is used for lateral control and longitudinal control of UEFC, the position deviation of UEFC is significantly reduced. The abovementioned experimental results indicate that the coordination control by this method can greatly cut down the path tracking deviation of UEFC.

Three methods are used for the control of the UEFC's heading angle. The results are shown in Figure 7.

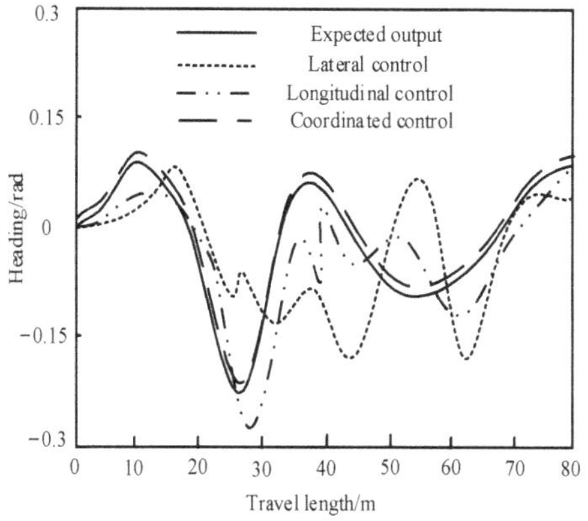

Figure 7. Results of heading angle control.

Figure 7 is a comparison of heading angle control results of UEFC under three control modes. When there is only lateral control, the course of the UEFC changes more frequently. It can be seen that the UEFC swings seriously when tracking the reference path. When the longitudinal control is adopted for the UEFC, the variation range of heading angle of the UEFC is significantly reduced. When the method in this paper is used for lateral control and longitudinal control of UEFC, the heading angle of UEFC can track the reference heading angle well, and UEFC can track the expected path of the experiment well.

Under the condition of constant speed, driverless racing cars can basically track the reference path through the control of the lateral controller, but there will be instability in the starting and turning places. The racing cars swing back and forth on both sides of the reference path. The swing amplitude and frequency vary at different speeds. When the speed is higher, the swing amplitude is greater. When the swing frequency is lower, the car's actual heading angle can basically track the ideal heading angle. However, when the speed increases, the range of the racing car's heading angle will increase. At the same time, the difference between the ideal heading angle and the racing car's heading angle will become larger. By comparison, when the speed increases, the change frequency of the front wheel angle of the car increases. The lateral position can basically be maintained within the constraint range. There are some deficiencies in the lateral control of a given speed. It is necessary to control the longitudinal direction of the car accordingly to improve the control effect.

After longitudinal control is added and the lateral and longitudinal directions are decoupled, the car can track the reference path well. Additionally, the unstable situation is much less than that in Section 3, where there is only lateral control. The course tracking error is reduced a lot; under the condition of longitudinal control, the lateral position deviation is maintained between −0.4 m and 0.3 m, which can be maintained within the comfortable distance defined in Section 3. The deviation is smaller than that of only lateral control. With the participation of longitudinal control, the speed is adjusted accordingly under different path curvature. In order to achieve a better tracking effect, it is necessary to couple the lateral control with the longitudinal control.

Three methods are used to control the front wheel angle of UEFC. The results are shown in Figure 8.

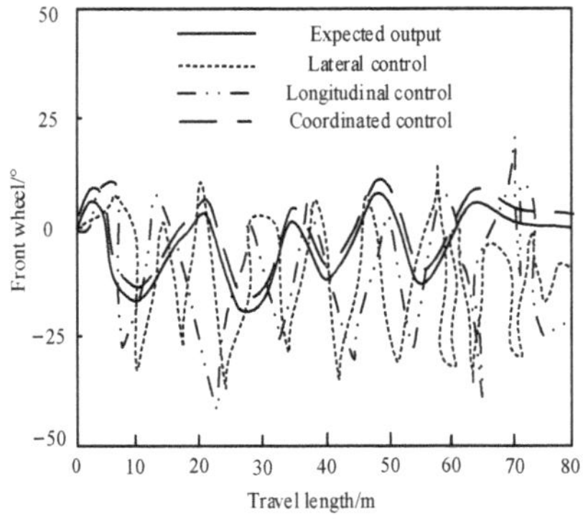

Figure 8. Control results of front wheel angle.

The experimental results in Figure 8 show that compared with the other two methods, the difference between the front wheel angle of the UEFC controlled by the proposed method and the expected output front wheel angle is small. In addition, the coordination control of UEFC can be realized using this method so as to keep the UEFC in an ideal running state. If the lateral control and longitudinal control are used alone, the change of the front wheel angle of the output is quite different from the expected output. The coordination control performance of the proposed method is significantly better than that under the lateral control and longitudinal control alone. The UEFC's front wheel angle controlled by this method can control car steering in a favorable way, thus ensuring the car stability in path tracking.

Three methods are used for the control of the UEFC's speed. The results are shown in Figure 9.

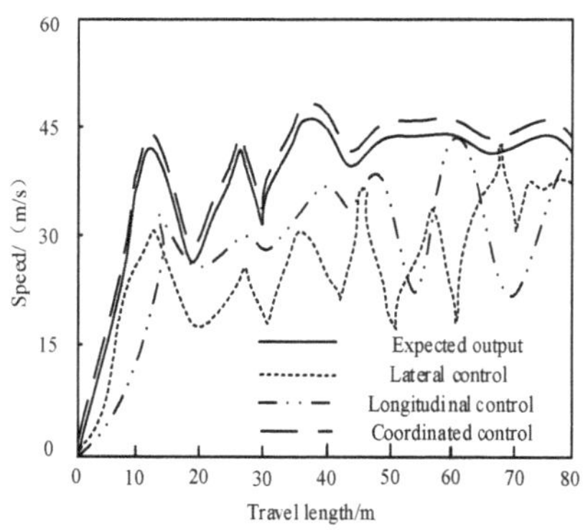

Figure 9. The results of vehicle speed control.

Figure 9 shows the speed tracking comparison diagram of UEFC under three control modes. When the lateral control or longitudinal control method is used alone to control the UEFC, the car speed is quite different from the expected output speed. When using the lateral or the longitudinal control of the proposed method alone to control the UEFC, the actual car speed is consistent with the reference car speed on the whole.

Three methods are used for the control of the UEFC's yaw rate, and the control results are shown in Figure 10.

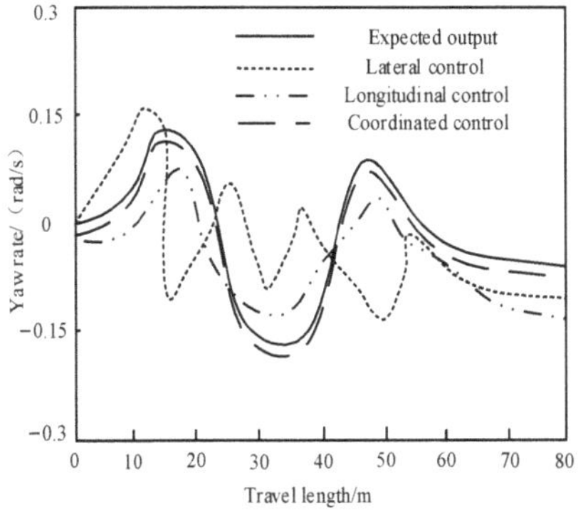

Figure 10. Yaw rate control results.

The comparison of yaw rate control results of UEFC in Figure 10 shows that the yaw rate of UEFC is very close using this method for coordination control of UEFC. It is verified that this method has coordination control performance and can ensure a yaw rate running within the expected range. To sum up, the coordination control method adopted in the proposed method has superior control performance for UEFC. This method can enhance the UEFC's path tracking effect, which is feasible.

4. Discussion

This paper studied the coordination control method of UEFC, and realized the coordination control of the mechanical structure of UEFC through combining lateral control and longitudinal control of UEFC. Unmanned vehicle technology is the achievement of the highly intelligent development of modern vehicles. It brings together the technologies of multiple disciplines, such as network and communication, sensing and detection and electronics and control.

The simulation results showed that the car can basically track the reference path at different constant vehicle speeds. Although the lateral deviation increased slightly with the increase of vehicle speed, the car always followed the reference path under the action of the lateral controller. Additionally, it remained within the constraint range of trajectory tracking error, indicating the feasibility of model predictive control for lateral control. This study assumes conditions similar to the real road conditions. To test in this and verify the tracking performance of the lateral controller under variable speed conditions, the corresponding reference speed was set according to the curvature of the reference path. In order to ensure the real-time adjustment of the speed, a longitudinal controller based on fuzzy PI was designed. The simulation results showed that the car can adjust the speed and track the reference speed in real time according to the track conditions. At the same time, the feasibility of the lateral controller was demonstrated. When designing the lateral and longitudinal integrated control system, the lateral controller needs the vehicle speed as the input of racing state information. Additionally, the longitudinal controller needs the lateral deviation as the reference value, so the longitudinal speed and lateral deviation were taken as the coupling points of the lateral and longitudinal control system. At the same time, this study analyzed the actual multiple working conditions of the car and formulating the coordinated control law. The longitudinal controller based on fuzzy PI control strategy was coupled with the lateral controller based on model predictive control strategy. Through analyzing the various working conditions of the car in the actual competition process, the lateral control and longitudinal control laws were summarized, and the lateral and longitudinal control of the driverless car was constructed. By using Carsim and MATLAB/Simulink software, a joint simulation model was built. Additionally, the simulation test of longitudinal and lateral coordinated control was completed. The simulation results showed that under the role of the coordinated controller, the lateral tracking effect was significantly improved. There was no unstable swing state, the reference path could be tracked smoothly, and the lateral deviation remained within the set constraint range. The longitudinal speed control can track the reference speed and adjust it in real time according to the track and car status. Therefore, the lateral and longitudinal control proposed in this paper can eliminate the interaction between lateral and longitudinal control. This model can meet the requirements of lateral and longitudinal control, and conduct comprehensive control safely, stably and quickly so as to ensure the independent driving of the car.

5. Conclusions

This paper investigated the mechanical structure coordination control method of UEFC, designed a mechanical structure of UEFC, and studied the coordination control method for the designed UEFC. The coordination control method of UEFC proposed in this paper enhances the handling stability of UEFC when turning at a high speed. Experiments showed that this method can realize the coordination control performance of UEFC. When the vehicle turns and changes lanes at high speed, the coordination control strategy can effectively control the lateral stability of UEFC, which improves the state response and handling stability of car of UEFC.

Author Contributions: Methodology, H.T.; Formal analysis, B.Y.; Data curation, B.Y.; Writing—original draft, H.T.; Writing—review & editing, H.T. All authors have read and agreed to the published version of the manuscript.

Funding: This research received no external funding.

Institutional Review Board Statement: Not applicable.

Informed Consent Statement: Not applicable.

Data Availability Statement: All data generated or analysed during this study are included in this published article.

Conflicts of Interest: The authors declare no conflict of interest.

References

1. Dolara, A.; Leva, S.; Moretti, G.; Mussetta, M.; De Novaes, Y.R. Design of a Resonant Converter for a Regenerative Braking System Based on Ultracap Storage for Application in a Formula SAE Single-Seater Electric Racing Car. *Electronics* **2021**, *10*, 161. [CrossRef]
2. Li, Y.; Hong, H.; D'Apolito, L. Reliability Control of Electric Racing Car's Accelerator and Brake Pedals. *World Electr. Veh. J.* **2020**, *12*, 1. [CrossRef]
3. Xie, Y.; Wang, C.; Hu, X.; Lin, X.; Li, W. An MPC-based control strategy for low-temperature electric vehicle battery cooling considering energy saving and battery lifespan. *IEEE Trans. Veh. Technol.* **2020**, *69*, 14657–14673. [CrossRef]
4. Henao-Muñoz, A.C.; Pereirinha, P.; Bouscayrol, A. Regenerative Braking Strategy of a Formula SAE Electric Race Car Using Energetic Macroscopic Representation. *World Electr. Veh. J.* **2020**, *11*, 45. [CrossRef]
5. Karmakar, G.; Chowdhury, A.; Das, R.; Kamruzzaman, J.; Islam, S. Assessing Trust Level of a Driverless Car Using Deep Learning. *IEEE Trans. Intell. Transp. Syst.* **2021**, *22*, 4457–4466. [CrossRef]
6. Zhang, H.; Yang, P.; Zhang, G.W.; Sun, H. Research on stability of electric vehicle based on terminal sliding mode control. *Comput. Simul.* **2020**, *37*, 134–138.
7. Li, W.; Xie, Z.; Wong, P.K.; Mei, X.; Zhao, J. Adaptive-Event-Trigger-Based Fuzzy Nonlinear Lateral Dynamic Control for Autonomous Electric Vehicles under Insecure Communication Networks. *IEEE Trans. Ind. Electron.* **2020**, *68*, 2447–2459. [CrossRef]
8. Yang, J.; Zhang, T.; Hong, J.; Zhang, H.; Zhao, Q.; Meng, Z. Research on driving control strategy and Fuzzy logic optimization of a novel mechatronics-electro-hydraulic power coupling electric vehicle. *Energy* **2021**, *233*, 121221. [CrossRef]
9. Wu, J.; Ji, Y.; Sun, X.; Xu, Y. Price regulation mechanism of travelers' travel mode choice in the unmanned transportation network. *J. Adv. Transp.* **2020**, *2020*, 9191834.
10. Subroto, R.K.; Wang, C.Z.; Lian, K.L. Four-Wheel Independent Drive Electric Vehicle Stability Control Using Novel Adaptive Sliding Mode Control. *IEEE Trans. Ind. Appl.* **2020**, *56*, 5995–6006. [CrossRef]
11. Sungyoul, P.; Kwangseok, O.; Yonghwan, J. Model predictive control-based fault detection and reconstruction algorithm for longitudinal control of autonomous driving vehicle using multi-sliding mode observer. *Microsyst. Technol.* **2020**, *26*, 239–264.
12. Ge, L.; Zhao, Y.; Ma, F.; Guo, K. Towards longitudinal and lateral coupling control of autonomous vehicles using offset free MPC. *Control Eng. Pract.* **2022**, *121*, 105074. [CrossRef]

13. Wang, Y.; Shao, Q.; Zhou, J.; Zheng, H.; Chen, H. Longitudinal and lateral control of autonomous vehicles in multi-vehicle driving environments. *IET Intell. Transp. Syst.* **2020**, *14*, 924–935. [CrossRef]
14. Vicente, B.A.H.; James, S.S.; Anderson, S.R. Linear System Identification Versus Physical Modeling of Lateral–Longitudinal Vehicle Dynamics. *IEEE Trans. Control Syst. Technol.* **2020**, *29*, 1380–1387. [CrossRef]
15. Guo, J.H.; Li, K.Q.; Luo, Y.G. Coordinated Control of Autonomous Four Wheel Drive Electric Vehicles for Platooning and Trajectory Tracking Using a Hierarchical Architecture. *J. Dyn. Syst. Meas. Control* **2015**, *137*, 101001. [CrossRef]
16. Zhao, H.; Sun, D.; Zhao, M.; Pu, Q.; Tang, C. Combined Longitudinal and Lateral Control for Heterogeneous Nodes in Mixed Vehicle Platoon Under V2I Communication. *IEEE Trans. Intell. Transp. Syst.* **2021**, *23*, 6751–6765. [CrossRef]
17. Han, H.; Liu, H.; Liu, Z.; Qiao, J. Interactive Transfer Learning-Assisted Fuzzy Neural Network. *IEEE Trans. Fuzzy Syst.* **2021**, *30*, 1900–1913. [CrossRef]
18. Sohn, C.; Andert, J.; Manfouo, R.N.N. A Driveability Study on Automated Longitudinal Vehicle Control. *IEEE Trans. Intell. Transp. Syst.* **2020**, *21*, 3273–3280. [CrossRef]
19. Huang, Y.; Chen, Y. Vehicle Lateral Stability Control Based on Shiftable Stability Regions and Dynamic Margins. *IEEE Trans. Veh. Technol.* **2020**, *69*, 14727–14738. [CrossRef]
20. Sun, X.; Wang, G.; Fan, Y. Adaptive trajectory tracking control of vector propulsion unmanned surface vehicle with disturbances and input saturation. *NonlinearDyn.* **2021**, *106*, 2277–2291. [CrossRef]
21. Verma, B.; Padhy, P.K. Robust Fine Tuning of Optimal PID Controller with Guaranteed Robustness. *IEEE Trans. Ind. Electron.* **2020**, *67*, 4911–4920. [CrossRef]
22. Zhao, P.; Chen, J.J.; Song, Y.; Tao, X.; Xu, T.; Mei, T. Design of a Control System for an Autonomous Vehicle Based on Adaptive-PID. *Int. J. Adv. Robot. Syst.* **2012**, *9*, 44–46. [CrossRef]
23. Victor, S.; Receveur, J.-B.; Melchior, P.; Lanusse, P. Optimal Trajectory Planning and Robust Tracking Using Vehicle Model Inversion. *IEEE Trans. Intell. Transp. Syst.* **2021**, *23*, 4556–4569. [CrossRef]
24. Attia, R.; Orjuela, R.; Basset, M. Combined longitudinal and lateral control for automated vehicle guidance. *Veh. Syst. Dyn.* **2014**, *52*, 261–279. [CrossRef]

World Electric Vehicle Journal

MDPI

Article

Interpolation-Based Framework for Generation of Ground Truth Data for Testing Lane Detection Algorithm for Automated Vehicle

Swapnil Waykole, Nirajan Shiwakoti * and Peter Stasinopoulos

School of Engineering, RMIT University, Melbourne, VIC 3000, Australia
* Correspondence: nirajan.shiwakoti@rmit.edu.au

Abstract: Automated vehicles, predicted to be fully electric in future, are expected to reduce road fatalities and road traffic emissions. The lane departure warning system, an important feature of automated vehicles, utilize lane detection and tracking algorithms. Researchers are constrained to test their lane detection algorithms because of the small publicly available datasets. Additionally, those datasets may not represent differences in road geometries, lane marking and other details unique to a particular geographic location. Existing methods to develop the ground truth datasets are time intensive. To address this gap, this study proposed a framework for an interpolation approach for quickly generating reliable ground truth data. The proposed method leverages the advantage of the existing manual and time-slice approaches. A detailed framework for the interpolation approach is presented and the performance of the approach is compared with the existing methods. Video datasets for performance evaluation were collected in Melbourne, Australia. The results show that the proposed approach outperformed four existing approaches with a reduction in time for generating ground truth data in the range from 4.8% to 87.4%. A reliable and quick method for generating ground truth data, as proposed in this study, will be valuable to researchers as they can use it to test and evaluate their lane detection and tracking algorithms.

Keywords: lane detection algorithm; electric vehicle; custom database; image processing; advanced driver assistance systems

Citation: Waykole, S.; Shiwakoti, N.; Stasinopoulos, P. Interpolation-Based Framework for Generation of Ground Truth Data for Testing Lane Detection Algorithm for Automated Vehicle. *World Electr. Veh. J.* **2023**, *14*, 48. https://doi.org/10.3390/wevj14020048

Academic Editors: Biao Yu, Linglong Lin and Jiajia Chen

Received: 14 December 2022
Revised: 2 February 2023
Accepted: 7 February 2023
Published: 9 February 2023

1. Introduction

Traffic crashes are increasing and becoming one of the critical issues worldwide with the rapid development of expressways and the growth of motor vehicle numbers. In recent years, automated ground vehicles have emerged as an essential component of intelligent transportation systems (ITS). Further, automated vehicles, predicted to be fully electric in future, will reduce road traffic emissions. A fully automated vehicle will drive people to their destination without any shared control with the driver, including controlling safety-critical tasks. It is necessary to integrate computers, controls, communications, and different automation technologies in ITS in order to improve transportation safety, throughput, and efficiency, while lowering energy consumption and environmental impact [1]. The car communicates with the driver, the surroundings, and the infrastructure. These interactions are enhanced in intelligent cars by sensing, sharing information, and actuating different primary or secondary driving activities. The Advanced Driving Assistant Technology (ADAS) system is the foundation of ITS. The primary goal of ADAS systems is to enhance road safety, reduce traffic congestion, and increase driving comfort by increasing various ADAS features [2].

Nowadays, most automobile advances are driven by embedded technologies and software solutions that identify potentially unsafe driving scenarios. It is argued that ADAS may reduce human driving mistakes [3]. One key breakthrough in the automotive industry is the advent of electric cars, which presents various potentials for ADAS development, but also many obstacles. Much research is being conducted in automated vehicles to develop enhanced ADAS to provide safety to drivers. Lane detection and tracking have been critical features of ADAS for safe driving and avoiding accidents.

Automated ground vehicle control systems can be divided into three components: environment prediction and planning; decision-making; and vehicle motion control. Vehicle motion control is separated into two categories: longitudinal motion control and lateral motion control. Longitudinal motion control is also known as longitudinal velocity tracking control, and it is a critical component of an intelligent driving control system [4]. Several control algorithms have been applied to longitudinal velocity tracking control. The longitudinal controller controls the vehicle speed, while the lateral controller controls the lateral offset and relative heading between the planned trajectories and the vehicle. The controller design challenge is addressed using a mathematical model that describes both lateral and longitudinal motions. The controller would ensure that the vehicle follows a speed profile while automatically regulating the vehicle speed to maintain a safe distance from the preceding vehicle [5]. A mixture of lane maintaining and overtaking in the left and right scenarios is utilized to test the controllers, which provide excellent tracking and comfort preservation [6,7].

Ground truth data in the context of computer vision involves a series of images, a set of labels on the images, and establishing a model for object recognition, including the number, location, and relationships of important characteristics, among other things (environmental factors). Depending on the complexity of the challenge, the labels are placed manually or automatically using image analysis.

The lane departure warning system (LDWS) utilizes lane detection and tracking algorithms and is now an important feature of ADAS or automated vehicles. The availability of high-quality datasets is important for developing new computer vision algorithms. In the literature, lane detection algorithms are usually classified under three categories: the model-based approach; a features-based approach; and a learning-based approach [8]. Learning-based approaches, on the one hand, require a large amount of ground truth data during their training process, whereas high-quality data are required for a detailed and equal assessment and comparison of other methods. Datasets and benchmark evaluations, which allow us to quantify progress in lane detection and tracking systems, have stimulated much innovation in the computer vision and machine learning fields. Some data types, such as camera images or depth images in indoor scenarios, are relatively simple to obtain with high precision. Low-cost sensors that can quickly produce a large amount of data are available for such image modalities. Manual user annotation can be used to obtain other ground truth data, such as object annotations or multiclass pixel-wise image annotations [8].

There are many large-scale datasets available for the performance evaluation of an algorithm of optical flow and object detection [9]. There are very few extensive dataset available for lane detection and tracking algorithms with a high-quality dataset of outdoor scenes, especially in the case of degraded lane markings. High-precision dense videos and huge annotated images are required to test an algorithm in ADAS. One way to obtain a large high-precision dataset is the ground truth dataset, which can be collected by mounting a camera and other equipment at the front of the vehicle. This is a time-consuming and costly procedure unless the ground truth data can be generated in a reasonable time. Other types of ground truth data, such as outdoor depth images and optical flow, are also rare to pull out. Stereo cameras and laser scanners may provide depth information, but they are either inaccurate or provide very sparse data.

Objective and Scope of the Study

This study aims to develop high-precision ground truth data based on an interpolation approach, which takes advantage of the existing manual and time-slice approaches. Researchers are constrained in testing their lane detection algorithms because of the low number of publicly available datasets. Additionally, those datasets may not represent differences in road geometries, lane marking, and other details unique to a particular geographic location. Researchers have developed a few procedures for collecting ground truth (custom) datasets, but it is time-consuming to generate high-precision data to test the algorithm. Therefore, to address this knowledge gap, this study develops an interpolation approach to quickly generate high-precision data for testing and evaluating the performance of lane detection and tracking algorithm.

This study provides methodological advancement in terms of quickly generating ground truth data (i.e., data with annotations added to them), which researchers can then use for testing lane detection algorithms. Without ground truth annotation, lane detection algorithms cannot be trained and tested. At the present time, researchers are constrained in testing their lane detection algorithms because of the low number of publicly available datasets. Further, the generation of ground truth annotation is mainly limited to a time-consuming and less reliable manual approach. In the manual approach, annotation depends on the user's skill or judgement; thus, the performance of lane annotation varies from person to person. For example, if two people with varied experience perform the manual approach, the annotation result may differ. This is one of the limitations we overcome with our proposed approach, besides the reduction in time for annotation.

Since ground truth annotation is a crucial step before testing lane detection algorithms, any method that can overcome the disadvantages of the time-consuming task of manual annotation is a methodological advancement. With our proposed semiautomatic interpolation-based framework for generating ground truth annotation, researchers can quickly make a reliable dataset to be ready for training and testing lane detection algorithms. Interpolation is a method by which related known values are used to estimate an unknown value or set of values. We have defined finite number rows and identified the unknown value through an interpolation approach to connect a smooth line or curve on the images by connecting these points. Our proposed approach allows researchers to quickly generate a variety of reliable ground truth datasets that they can deploy to test their lane detection algorithms for different road geometries, lane marking, and other details unique to a particular geographic location. Video datasets for performance evaluation of our proposed approach are collected in Melbourne, Australia. The performance of the proposed approach is tested by comparing results with other existing approaches (including manual annotation as a benchmark) to demonstrate the superiority of our proposed approach.

The structure of the paper is organized as follows. Section 2 presents a literature review on related works for the generation of ground truth data, a comparison of existing datasets, and approaches to develop ground truth data. It is followed by Section 3, which explains the data collection procedure adopted for this study and the proposed interpolation approach framework. Section 4 discusses the proposed approach's advantage over other existing approaches. Finally, conclusions and recommendations for future works are presented.

2. Literature Review

The following subsections discuss related work on generating ground truth or custom data, and the comprehensive analysis of available datasets. Data used for lane detection and tracking and the evaluation of algorithms are explained, along with existing approaches to develop ground truth data.

2.1. Related Work to a Generation of Ground Truth Data

Veit et al. [10] used the ROMA dataset to test a number of feature extractors for road markings. With camera details and ground truth for the lane markers, the ROMA dataset offer high-quality color images. The total length of its image sequence is below 20 s. In addition, it appears that the dataset is compiled from a set of random images that would make it inappropriate to test lane marking that combines algorithms with lane tracking. Nevertheless, datasets are also available that contain image sequences for testing. For instance, while driving on local city roads, Leibe et al. [11], Wanf et al. [12], and Brostow et al. [13] generated image sequences with different scenarios. Although the recorded image content of the dataset had a fluctuation in illumination, different lane marking and surfaces, and the neighboring vehicles, the overall length of the image content is less than 10 min. In addition, ground truth files are not provided, which is this dataset's main drawback. While the dataset generated by Aly et al. [14] contains a ground truth file, scenes are not included in sequences of the image on the recorded videos on the highway. The PESTS2001 dataset [15] included image sequences of a highway, whereas the dataset from Sivaraman et al. [16] generated image sequences of a street and highway. Unfortunately, both datasets have compromised image quality by applying lossy compressions such as JPEG and MPEG-4. Finally, the datasets by Santos et al. [17], Lim et al. [18], and EISATS [19], provide a comprehensive collection of images that would be ideal for lane marking testing. The images in [17–19] display variations in the types of illumination, traffic flow conditions, road texture, and lane marking, that reflect the conditions. In addition, in these datasets, the total length of the sequence is more than 10 min, and even the camera parameters that are required for generating camera models are also given. These datasets are applicable for different applications, except for lane detection. Consequently, files containing ground truth data are unavailable, so objective analysis may not be obtained. In addition, testing algorithms on the dataset listed in [16–18] appear to measure their finding based on visual inspection, and may be biased. A dataset containing a large amount of image data along with ground truth is required to make a systematic, objective evaluation.

2.2. *Comprehensive Analysis of the Available Dataset*

The development of lane detection and tracking for ADAS is of great interest to the automotive society. However, access to resources is extremely restricted because of the significant commercial involvement in this area of research. Data that are used for assessment and research is one of the most valuable tools. Compared to areas of study such as face detection or the identification of optical characters (OCR), where labeled and structured datasets are often available for training and testing, there appear to be no such tools for lane detection and tracking. The issue caused by the absence of common datasets is that it is difficult to currently compare available lane detection algorithms easily. Additionally, the lack of common testing data make verifying other implementation results difficult. In this scenario, one of the viable solutions is to enforce each algorithm and its performance verification. However, the sheer quantity of algorithms that have already been published in journals and conferences makes this challenge difficult to achieve. Table 1 shows a comparison of different dataset features available in the existing literature for lane detection, including CU Lane [20], Caltech [21], NEXET [22], DIML [23], KITTI [24], TuSimple [25], UAH [26], and BDD100K [27].

Table 1. Comparison among different lane datasets.

Sr. No	CU Lane [20]	Caltech [21]	NEXET [22]	DIML [23]	KITTI [24]	TuSimple [25]	UAH [26]	BDD100K [27]
Sequences	More than 55 hrs videos	4 clips	Not available	Not available	22 sequences	7000 sequences	500 min videos	100,000 videos
Images	133,235	1225	50,000	470	14,999	140,000	N/A	120,000,000
Multiple cities	-	-	√	√	-	√	√	-
Multiple weathers	-	√	√	√	-	-	√	√
Multiple times of day	√	√	√	√	-	-	√	√
Multiple scene types	√	√	√	√	√	√	√	√
Multiple cameras	√	-	√	√	√	-	-	√
Multiple street	-	-	√	√	√	-	√	√
Labelled streets	-	√	√	√	-	√	√	√

Note: √: Item under "Sr. No" column is included, and -: Item under "Sr. No" is excluded.

Researchers have used improvised datasets in terms of visual content to explore multitask learning for a self-driving car. In earlier decades, researchers were constrained because of a small set of datasets available for research. However, recently, Berkeley Deep Drive, University of California, Berkeley, developed the BDD100K dataset, consisting of 100K driving videos with diverse scene types for researchers. A dataset for structure has been covered in BDD100K according to different types of lanes, such as lane category, lane direction, lane continuity, and drivable area. In the lane category, the following types of lane markings have been covered with annotation: road curb, double white lanes, double yellow lanes, double other, single white lanes, single yellow lines, and other single and crosswalks. Additionally, parallel lanes and vertical lanes are annotated in the lane direction. Annotated images with full and dashed lanes are covered in lane continuity, and derivable and alternative types of lanes come under the drivable area.

Based on the review of studies on lane detection and tracking [8,9], and the summary of datasets presented in Table 1, it can be observed that there are limited datasets in the literature that researchers have used to test lane detection and tracking algorithms. It was also observed that there are a lack of specific datasets from Australian roads.

In future, more datasets may be available for researchers as this field continues to grow, especially with the development of fully automated electric vehicles. The verification of the performance of the algorithms for lane detection and tracking system is carried out based on the ground truth dataset. Ground truth data require a collection of images and a set of labels on the image in the context of computer vision and models for object recognition. The number, position, and relationships of key characteristics need to be included. Labels are added to the image either automatically or manually, depending on the complexity of the problem. The label set includes interesting points, the corners of lane boundaries and descriptors of features. Using a range of machine learning methods, a model can be trained, and the detected characteristics can be fed into a classifier at run time to calculate the correspondence between detected characteristics and modeled features.

2.3. *Approaches for Developing Ground Truth Data*

There are four popular approaches to developing ground truth data, as briefly described below:

(1) Manual approach

This method is transparent, straightforward, and often used to generate accurate ground truth. It is conducted as follows:

- The user can manually annotate the lane markings of the left lane's boundary at various points in an image.
- Same steps are repeated for the right lane boundary.
- All the steps are repeated for the video clip.
- Finally, the ground truth is generated and saved in the file.

This approach, however, has two issues: it is a slow procedure, and it requires proper details to annotate a single curve in an image, which can take up a lot of time. Gaps between two dashed lines are not annotated, because the lane boundary is difficult to estimate in these areas [28].

(2) Time-slice approach

By stacking an empty image with a row of pixels from frames in the video collection, a time-sliced (TS) image is produced. To further explain, each frame of a video set with F frames can be considered as an image with M × N dimensions [29].

(3) E-NET approach

In a study by [30], 168 images with a resolution of 1024 × 1280 pixels and normal RGB channels are included in the complete dataset. Around five to seven mast cells make up each image. About 40% of all cells have connected boundaries and appear to be near to one another. Images were divided into 240 × 560 tile segments, in accordance with the size of the neural networks. An expert manually segmented the dataset using the ground truth.

The complete dataset was divided into test and validation parts at a ratio of 74:51, with the samples being randomly shuffled [30].

(4) Automated test approach

The automatic data-labeling method for creating crack ground truths (GTs) within concrete images is presented in this approach [31]. The primary technique entails producing first-round GTs, pretraining a model based on deep learning, and producing second-round GTs. A learning-based crack detection model can be trained in a self-supervised way using the generated second-round GTs of the training data. After being retrained using the second set of GTs, the deep learning-based model that was previously trained is successful at detecting cracks. This method's primary purpose is the suggestion of an automated GT generating approach for pixel-level fracture detection model training [31].

3. Proposed Framework

In this work, we have proposed a framework for generating a ground truth dataset that can be used for testing lane detection algorithm. Our proposed approach leverages the advantages of manual and time-slice approaches through the interpolation approach. While the philosophy of the interpolation approach is applicable to all types of roads, we have used linear interpolation for straight roads and cubic spline interpolation for curved roads. The linear interpolation is the straight line connecting the two known positions, shown in Figure 1a. The equation of slopes gives the value of y along the straight line for x in the interval:

$$\frac{y^* - y_1^*}{y_2^* - y_1^*} = \frac{x^* - x_1^*}{x_2^* - x_1^*} \tag{1}$$

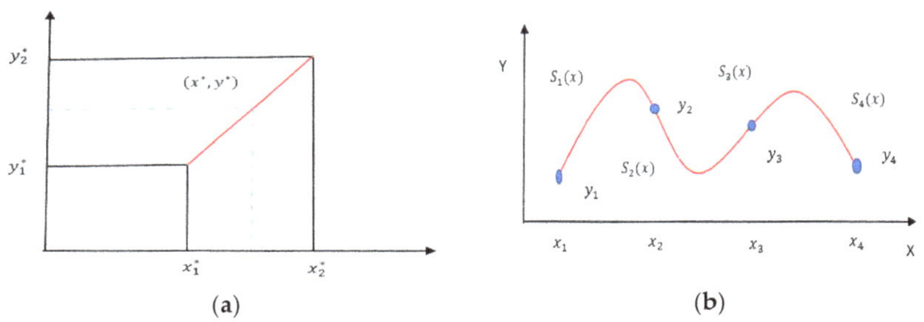

Figure 1. The concept of linear interpolation (**a**) and cubic spline interpolation (**b**).

y^* = linear interpolation values.
x^* = independent variable.
x_1^* & x_2^* = values of the function at one point.
y_1^* & y_2^* = values of the function at another point.

Cubic spline interpolation is a mathematical approach for creating new points within the boundary of a given collection of points, shown in Figure 1b. These new points are function values of an interpolation function composed of numerous cubic piecewise polynomials. The cubic spline is the function $S(x)$ with given properties $S(x)[x_i,\ x_i + 1] = S_i(x)$.

On the basis of the interpolation principle, the ground truth dataset was developed based on the videos obtained using monocular camera image sequences of traffic scenes in Melbourne, Australia. We have used the Driving Scenario Designer app (available in MATLAB R2021a) and MATLAB tools to annotate images.

It is difficult to estimate lane boundaries in the gaps, and to address this challenge, we have selected user-annotated points to specify a maximum number of rows close to each other. Then, we created cubic interpolation in these annotated points to get a smooth curve because lane markings are not straight everywhere (for straight roads, we used linear interpolation), so the controller can predict the lane boundaries based on the optimization. In this way, we can generate a dataset for lane detection and tracking algorithms in arbitrary scenarios with different traffic flow conditions. Finally, a comparison of the performance of the proposed approach with other existing methods is conducted.

3.1. The Requirement of Ground Truth Dataset

The image data and the metadata are required to formulate a full dataset. The image data must consist of a series of image sequences captured at a speed of over 20 km/h while driving. Additionally, the time between the end of one sequence and the start or second recording is suggested to be a minimum of 2 min. This prevents the image data from being a collection of random images in the dataset. The criteria for the dataset from captured images are the following:

1. Duration of sequences: each recorded image sequence must have a length of at least 20 s. This approach will help test algorithms to use a frame-to-frame detection consistency.
2. Color: recorded image must be in 24-bit RGB format.
3. Uncompressed storage: recorded image sequence must store in a lossless format. There are two reasons for putting this provision in place. Some researchers have used real-time systems for research, in this case, real-time systems analyze files directly from the camera. So, the dataset's images must represent the camera's performance. Lossy compression creates artefacts of the compression and damage image quality.
4. Image format: to provide sufficient detail of the road surface, image must be at least (480 * 640) resolution.
5. Camera parameters: information about the intrinsic and extrinsic camera's parameters should be provided for camera models to be recreated. Focal length, the field of view, and image format, are among the intrinsic parameters. The extrinsic parameters include the ground plane's yaw, pitch, roll, and the height of the optical center of the sensor.
6. Lane marking: lanes are on straight and curving paths, with solid and dashed lane markers. They are the most common lane markers on roads in Australia. They must be present in the image details. The image details must also include images where lane markers are missing, to test for false positives.
7. Type of road: captured image data must include scenes from both streets and motorways.
8. Illumination effect: scenes in images that have various lighting changes, such as a vehicle driving via tunnels, under bridges, under shadows, etc., during the day time, night-time, and cloudy conditions. Data should be recorded to reflect variation in the levels of ambient light.
9. Weather condition: recorded image must include the weather condition, such as dry, rain, or snow.
10. Traffic flow condition: image must give traffic flow situations, such as heavy traffic, modern traffic, and light traffic, etc.
11. The ground truth describes the lane boundaries of the road in curve form, and these requirements are implemented by the shape of the roads and the position of the curve identified on the road. Curves identified on the lane boundaries should be smooth, and they must exhibit no visible kinks in their lane boundaries.
12. Generated curve must be at the center of the lane makers: if there are double lane markers, it should be between two lines. The curve must start at the top of the Region of Interest (ROI) and extend to the bottom of the image, along the lane boundary, after specifying a ROI.
13. In the presence of either solid or dashed markers, the lane boundary curve must be given.

3.2. Data Collection

To generate the ground truth datasets, we collected video clips covering more than 120 km of distance, including day and night, and under different weather conditions in four suburbs (Bundoora, Heidelberg, Kew, and Brighton) of Melbourne, Australia. Additionally, it includes both unstructured (no clear lane markings) and structured roads (clear lane markings). Some videos were taken during the day time in Bundoora, Heidelberg, and Kew, with structured and unstructured roads, while some videos were collected in Brighton during the night. The parameters of the lens of the monocular camera used for data collection are shown in Table 2. We used this camera system because the ultracompact set provides a range of CMOS image sensors. The camera allows some image processing features, such as gamma correction and color interpolation, to plot a smooth curve with connecting points. The parameters of the lens are shown in Table 2.

Table 2. Parameters of the lens used in the camera.

Attribute of Lens	Parameters of Lens
Name of the company	Optics
Model number	60-255
Focal length	5-55(Mm)
Aperture	F1.3-1.6 C
Sensor	Not available
Working distance	800-1000(Mm)
Lens design	Back surface

For creating a ground truth dataset, it includes the following steps:

Model design: the model explains the structure of the objects, such as intensity, count, and location, and the relationship between a group of scale-invariant feature transform (SIFT) features. The model should be properly adjusted to the problem and image knowledge in order to yield significant outcomes.

Training dataset: in order to work with the model, this set was collected and labelled, and includes both positive and negative images and different features. Negatives include images and attributes that are intended to generate false matches on the lanes [32].

Test set: several images are collected for testing against the training set to predict the accurate match for the model.

Classifier design: this is designed to meet the speed and accuracy of the application objectives, including data organization and model search optimizations.

Training and testing: this dataset was collected to verify a group of images against ground truth [33].

Table 3 illustrates various road traffic environment features captured in the video data, collected in Melbourne, while Figure 2 summarizes the overall data collection process to postprocessing and evaluation.

Table 3. Road traffic environment features captured in the video data collection.

Features	Clip 1	Clip 2	Clip 3	Clip 4	Clip 5
Weather condition	Cloudy	Heavy rain in the night-time	Sunny	Cloudy	Heavy rain in the day time
Road surface	Rough surface	Rough surface	Smooth	Smooth	Rough
Color of lane marking	White	White	White	White	White
Traffic situation	Modern	Modern	Light	Light	Light
Speed	60 km/h	68 km/h	70 km/h	64 km/h	66 km/h
Number of frames	14	17	21	25	11
Type of lane marking	Dash	Solid	Dash and solid	Dash and solid	Dash and solid
Timing	Day time	Night-time	Day time	Evening	Day
Location	Heidelberg	Brighton	Bundoora	Bundoora	Kew
Structured and unstructured roads	Yes	Yes	Yes	Yes	Yes
Clothoid roads	Yes	Yes	Yes	Yes	Yes

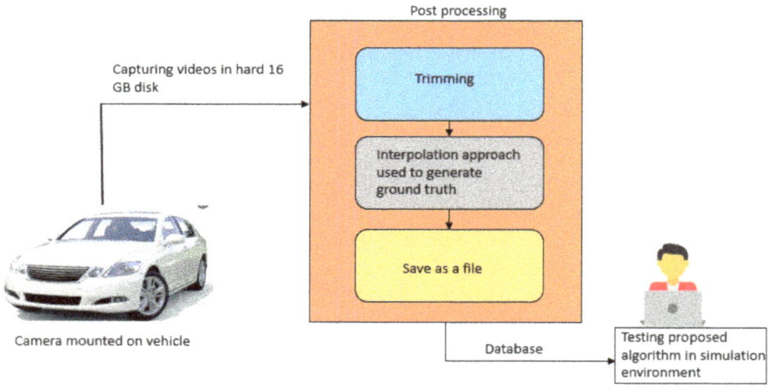

Figure 2. Summary of the procedure adopted to develop ground truth data.

3.3. Proposed Interpolation Approach

The proposed interpolation approach is a quick and effective technique for processing reliable ground truth data. One of the benefits of this method over the manual approach is that interpolation values give the spatiotemporal representation of the video clip; as a result, annotations can be carried out rapidly and with better accuracy.

We have divided this interpolation approach into three major steps.

Step 1:

In stage 1, we define a finite number of rows in which the boundary of the lane is annotated. These rows are shown in Figure 3. It is important to define at least three rows or more for producing a smooth curve of the lane boundaries for a good result. First, we started by defining only two rows, but noticed it was not of sufficient accuracy. So, we tested with three, or more than three rows (mostly three to four rows) using linear interpolation (for straight roads) or cubic spline interpolation (for curved roads).

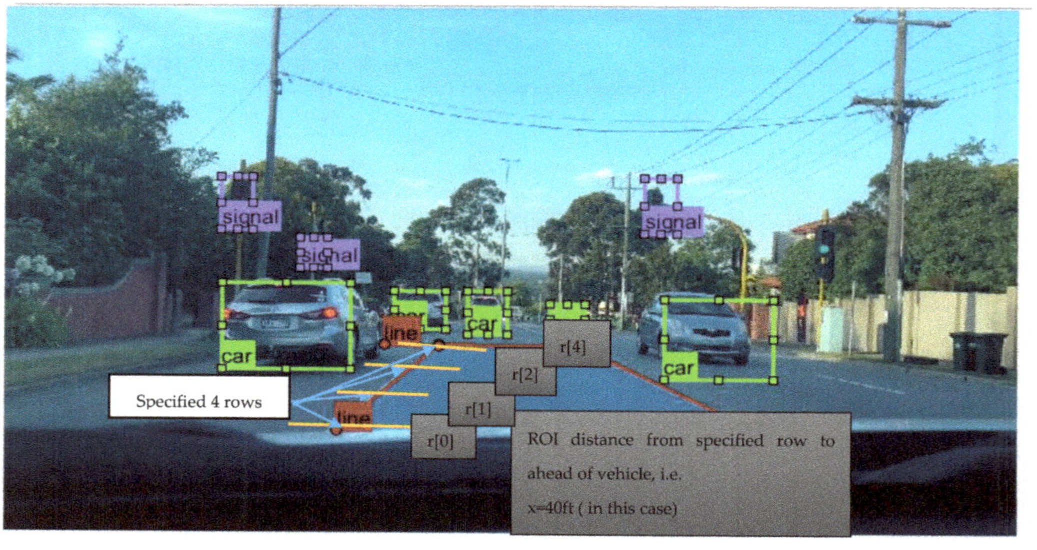

Figure 3. Defining a finite number of rows (four rows in this case) in the image.

It is to be noted that, in Figure 3, other markups, such as cars, signals, etc., are also shown. This is because we want to demonstrate (as a sample) how ground truth annotation for lane findings can be applied to a supervised lane detection and tracking algorithm, which also need to detect surrounding obstacles and objects to avoid a collision.

Step 2:

This stage is made up of two substages put within a frame. The specification of each substage is provided below.

1. Annotations: an interpolation image is generated by filling an empty image (without annotation) in the video clip with rows of image pixels. The sliced image, which is initially an empty image, is then generated with dimensions F × N, where F is the video clip's frame count, and each image has dimension M × N in the video clip. The sliced image has dimensions of F × N as F copies of N pixels wide, unlike the images in the video clip. We have used 240 × 640 image size in this study. In the empty image, rows will be specified. The row corresponds to the sliced image row to simplify the nomenclature in this explanation, while the row is a row in the image in the video clip. Next, in the image, we specify the row from which pixels will be copied. Then, a single row of pixels from each image of the video clip is copied to a specific row in the empty image by using a loop. To elaborate, from the first image in the video clip (Figure 3), row = 300, 400, and 500. The first row of the sliced image is copied. Then, from the row = 300, 400, and 500 in time-sliced, the second image in the video clip is copied to the second-row image. Likewise, row = 300 from the F, the image will be copied to the F row in the time-sliced. The rows of pixels appear to be stacked in the time-sliced image since they are sequentially copied and positioned. The generated time-slice image is shown in Figure 4, where rows are copied from the image on the left to the sliced image on the right.

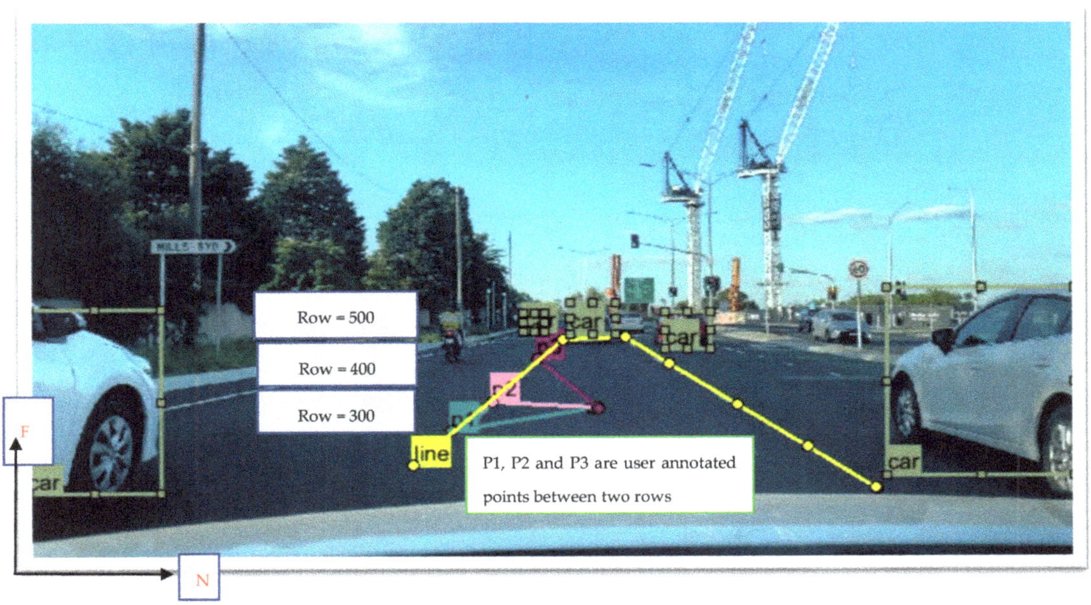

Figure 4. Connecting specified rows via interpolation to generate a smooth curve.

2. Interpolation: in this step, we first annotate a few points in the time-sliced image at the center of the left lane marker. Then, we perform curve fitting to link the points by interpolation (Figure 4). Instead of the nearest neighbor or linear interpolation, cubic spline interpolation is used here. It produces a smooth curve between the points, as shown in

Figure 4. In the interpolation image, curve fitting estimates lane marker positions between the annotated points (p1, p2, p3, along the yellow line), as shown in Figure 4.

Step 3:

At the end of stage 2, in each image of the video clip, we have lane boundary points in four rows. However, it will not create a smooth curve by simply connecting the four positions with straight lines. Therefore, again we use interpolation. However, this time, in each image, interpolation is performed across the rows, as shown in Figure 5.

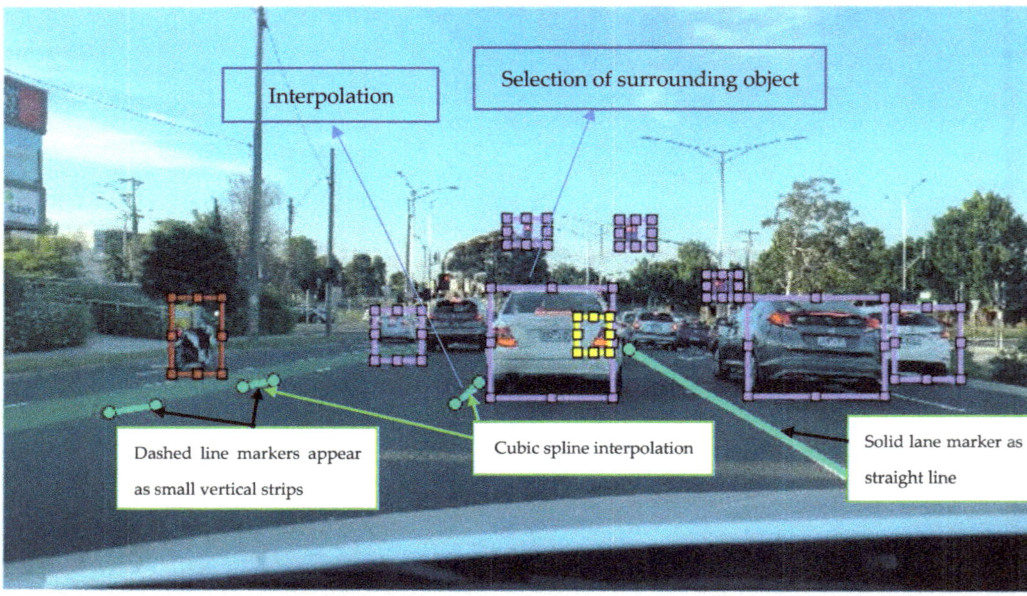

Figure 5. Generating ground truth annotation.

To elaborate, we have selected three points on the left lane boundary. The interpolation of the cubic spline binds the four rows and defines the position of the lane boundary on rows 300–500 at the same time, as shown in Figure 4. The cubic spline results in a smooth curve from r [0] to r [4] for the left lane boundary in the image.

Similarly, this procedure is carried out for the right lane boundary and then repeated for all the images in the video clip. Figure 4 displays the position on rows 300–500 for image 1 of the left and right lane boundaries. Furthermore, a sample interpolation image is shown in Figure 5 above this image, and two observations can be made:

1. Solid lane markers observed as a straight line;
2. Dashed lane markers observed small vertical strips.

Figure 6 summarizes the three steps involved in the interpolation approach. The XML file generated from the proposed interpolation approach is presented in the Appendix A (Figure A1).

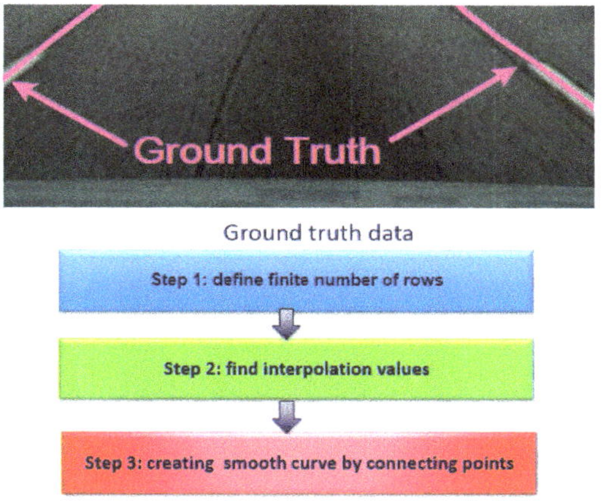

Figure 6. Summary of steps involved in the interpolation approach.

4. Discussion

The proposed approach focuses on interpolation to automate user-annotated points and, therefore, speed up the process of generating ground truth data. A reliable and quick method for generating ground truth data will be of immense importance to researchers as they can use it to test and evaluate their lane detection and tracking algorithms. We compared the performance (time taken to process and generate the ground truth data) of our proposed method with other existing methods: manual [28], time-slice [29], E-Net [30], and automated test [31], and the results are shown in Table 4. For comparison, we used different video clips with various duration, as shown in Table 4. All clips have solid and dashed lane markings. On average, from Table 4, it can be seen that our proposed interpolation approach took less time (5.78 min) as compared to other existing methods, such as the manual (46 min), time-slice (6.07 min), E-NET (6.35 min) and automated tests (6.97 min). This is a reduction of 87.4%, 4.8%, 8.9%, and 17.1%, respectively, in the processing time. When implementing lane marker classification in different weather conditions, the challenge is with ROI selection since the necessary image characteristics could be influenced by weather conditions, such as rain. While the proposed interpolation approach outperformed previous approaches in all environments, some extra time was needed for rainy conditions, compared to other normal days. For example, as compared to normal weather (sunny and dry: clip 3), it took around 16 s longer to extract the lane marker features and obtain interpolation values between defined rows in heavy rain at night-time (clip 2), and around 9 s longer in heavy rain during the day time (clip 5).

As interpolated values are obtained in our proposed approach, it has the capability to define the exact location of the lane boundaries. Further, the interpolation approach is not restricted to any specific lane marking, and road structure and weather conditions do not influence the ground truth dataset. The selection of an appropriate number of rows for interpolation may be a limitation, but can be overcome with the experience. At this stage of understanding, we believe that our proposed interpolation approach will provide a better result for testing the lane detection and tracking algorithm. This may explain why our approach is suitable for all types of lane marking. Table 4 shows the comparison of the interpolation approach with other approaches.

Table 4. Comparison of the proposed interpolation approach with other approaches.

Clip and Duration	Duration for Processing (mins)				
	Manual Approach	Interpolation Approach	E-NET	Automated Test Approach	Time-Slice Approach
Clip 1 [1 min]	47	5.00	6.00	8.00	6.01
Clip 2 [1.12 min]	41	4.56	5.00	7.32	5.00
Clip 3 [1.19 min]	53	7.00	7.30	6.53	7.09
Clip 4 [1.43 min]	44	6.37	6.70	7.00	8.00
Clip 5 [1.36 min]	45	6.00	6.78	6.04	7.04
Average	46	5.78	6.35	6.97	6.07
Standard Deviation	4.48	0.99	0.88	0.74	1.15

5. Conclusions

This study developed an interpolation approach for quickly generating reliable ground truth data. The proposed method takes advantage of the existing manual and time-slice approaches. The interpolation approach has been developed in three different steps: define a finite number of rows; find interpolation values; and create a smooth curve by connecting interpolation values. The system can be used to quickly produce large numbers of high-quality camera images, depth and optical flow videos, and textual annotations at the pixel level.

Furthermore, the proposed framework enables the creation of ground truth data for complex driving scenarios, which could be useful in developing lane detection and tracking systems for advanced driver assistance systems, or for testing algorithms for automated vehicles. Other applications of this proposed framework could be the determination of visual odometry, and the detection of motion models and scenes. Further, due to the semiautomatic nature of the proposed approach, the user can annotate each image by clicking a mouse button. Future work may explore combining data acquired from the front and rear cameras to determine the vehicle's position on the road, and identify lane detection and tracking.

Author Contributions: Conceptualization, investigation, data collection, methodology, writing—original draft preparation, S.W.; supervision, writing—review and editing, N.S.; supervision, P.S. All authors have read and agreed to the published version of the manuscript.

Funding: This research received no external funding.

Data Availability Statement: Not applicable.

Acknowledgments: The first author would like to acknowledge the Government of India, Ministry of Social Justice and Empowerment, for providing a full scholarship to pursue Ph.D. study at RMIT University.

Conflicts of Interest: The authors declare no conflict of interest.

Appendix A


```
<Ground Truth>
    <Author>Swapnil Waykole </Author>
    <Institute>RMIT</Institute>
    <Date>20/01/2021 6:16:21 AM</Date>
    <ID>S1C1</ID>
    <CamNum>0</CamNum>
    <FrameCount>1376</FrameCount>
    <Annotation>
        <Fr ID="1">
            <Left>
                <X>236.609 235.220 234.831 ... 65.873 65.000</X>
                <Y>290.001 292.000 293.000 ... 419.000 420.000</Y>
            </Left>
            <Right>
                <X> ... </X>
                <Y> ... </Y>
            </Right>
        </Fr>
        <Fr ID="2"> ...
```


Figure A1. XML file generated from the proposed interpolation approach.

References

1. Eskandarian, A. *Handbook of Intelligent Vehicles*; Springer: London, UK, 2012.
2. Gagliardi, G.; Lupia, M.; Cario, G.; Casavola, A. Optimal $H\infty$ Control for Lateral Dynamics of Autonomous Vehicles. *Sensors* **2021**, *21*, 4072. [CrossRef] [PubMed]
3. Galvani, M. History and future of driver assistance. *IEEE Instrum. Meas. Mag.* **2019**, *22*, 11–16. [CrossRef]
4. Hang, P.; Chen, X.; Zhang, B.; Tang, T. Longitudinal Velocity Tracking Control of a 4WID Electric Vehicle. *IFAC-PapersOnLine* **2018**, *51*, 790–795. [CrossRef]
5. Gagliardi, G.; Casavola, A.; Toscano, S. Linear Parameter Varying Control Strategies for Combined Longitudinal and Lateral Dynamics of Autonomous Vehicles. In Proceedings of the 2022 European Control Conference (ECC), London, UK, 12–15 July 2022 ; pp. 181–186. [CrossRef]
6. Hima, S.; Lusseti, B.; Vanholme, B.; Glaser, S.; Mammar, S. Trajectory Tracking for Highly Automated Passenger Vehicles. *IFAC Proc. Vol.* **2011**, *44*, 12958–12963. [CrossRef]
7. Hima, S.; Glaser, S.; Chaibet, A.; Vanholme, B. Controller design for trajectory tracking of autonomous passenger vehicles. In Proceedings of the 2011 14th International IEEE Conference on Intelligent Transportation Systems (ITSC), Washington, DC, USA, 5–7 October 2011; pp. 1459–1464. [CrossRef]
8. Waykole, S.; Shiwakoti, N.; Stasinopoulos, P. Review on Lane Detection and Tracking Algorithms of Advanced Driver Assistance System. *Sustainability* **2021**, *13*, 11417. [CrossRef]
9. Waykole, S.; Shiwakoti, N.; Stasinopoulos, P. Performance Evaluation of Lane Detection and Tracking Algorithm Based on Learning-Based Approach for Autonomous Vehicle. *Sustainability* **2022**, *14*, 12100. [CrossRef]
10. Veit, T.; Tarel, J.; Nicolle, P.; Charbonnier, P. Evaluation of Road Marking Feature Extraction. In Proceedings of the IEEE Conference on Intelligent Transportation Systems, Beijing, China, 12–15 October 2008; pp. 174–181.
11. Leibe, B.; Cornelis, N.; Cornelis, K.; Van Gool, L. Dynamic 3D Scene Analysis from a Moving Vehicle. In Proceedings of the IEEE Conference on Computer Vision and Pattern Recognition, Minneapolis, MN, USA, 17–22 June 2007; pp. 1–8.
12. Wang, C.C. CMU/VASC Image Database. 2003. Available online: http://vasc.ri.cmu.edu/idb/html/road/index.html (accessed on 15 December 2022).

13. Brostow, G.J.; Shotton, J.; Fauqueur, J.; Cipolla, R. Segmentation and Recognition using Structure from Motion Point Clouds. In Proceedings of the European Conference on Computer Vision, Marseille, France, 12–18 October 2008; pp. 1–14.
14. Aly, M. Real time detection of lane markers in urban streets. In Proceedings of the IEEE Intelligent Vehicles Symposium, Eindhoven, The Netherlands, 4–6 June 2008; pp. 7–12.
15. Makris, D. PETS2001 Dataset. 2001. Available online: http://www.cvg.cs.rdg.ac.uk/PETS2001/pets2001-dataset.html (accessed on 15 December 2022).
16. Sivaraman, S.; Trivedi, M.M. A General Active-Learning Framework for On-Road Vehicle Recognition and Tracking. *IEEE Trans. Intell. Transp. Syst.* **2010**, *11*, 267–276.
17. Santos, V.; Almeida, J.; Gameiro, D.; Oliveira, M.; Pascoal, R.; Sabino, R.; Stein, P. ATLASCAR Technologies for a Computer Assisted Driving System on board a Common Automobile. In Proceedings of the IEEE Conference on Intelligent Transportation Systems, Funchal, Portugal, 19–22 September 2010; pp. 1421–1427.
18. Lim, K.H.; Cat, A.; Ngo, L.E.; Seng, K.P.; Ang, L.-M. UNMC-VIER Auto Vision Database. In Proceedings of the 2010 International Conference on Computer Applications and Industrial Electronics, Kuala Lumpur, Malaysia, 5–8 December 2010; pp. 650–654.
19. Multimedia Imaging Technology, Image Sequence Analysis Test Site (EISATS). Available online: http://www.mi.auckland.ac.nz/EISATS/ (accessed on 18 November 2022).
20. Cu Lane Dataset. Available online: https://xingangpan.github.io/projects/CULane.html (accessed on 13 April 2020).
21. Gregory, G.; Holub, A.; Perona, P. Caltech-256 Object Category Dataset. California Institute of Technology. 2007. Available online: https://resolver.caltech.edu/CaltechAUTHORS:CNS-TR-2007-001 (accessed on 15 October 2022).
22. Klein, I. NEXET—The Largest and Most Diverse Road Dataset in the World. 2007. Available online: https://data.getnexar.com/blog/nexet-the-largest-and-most-diverse-road-dataset-in-the-world/ (accessed on 21 October 2022).
23. Lee, E. Digital Image Media Lab. Diml.yonsei.ac.kr. 2020. Available online: http://diml.yonsei.ac.kr/dataset/ (accessed on 13 April 2020).
24. Cvlibs.net. The KITTI Vision Benchmark Suite. Available online: http://www.cvlibs.net/datasets/kitti/ (accessed on 27 April 2020).
25. Tusimple/Tusimple-Benchmark. Available online: https://github.com/TuSimple/tusimple-benchmark/tree/master/doc/velocity_estimation (accessed on 15 April 2020).
26. Romera, E.; Luis, M.; Arroyo, L. Need Data for Driver Behavior Analysis? Presenting the Public UAH-Drive Set. In Proceedings of the 2016 IEEE 19th International Conference on Intelligent Transportation Systems, Rio de Janeiro, Brazil, 1–4 November 2016.
27. BDD100K Dataset. Available online: https://mc.ai/bdd100k-dataset/ (accessed on 2 April 2020).
28. Coudray, N.; Karathandu, A.; Chamborn, S. Multi resolution approach fir fine structure extraction—Application and validation on road images. In Proceedings of the Fifth International Conference on Computer Vision Theory and Applications, Angers, France, 17–21 May 2010; Volume 2.
29. Al-Sarraf, A.; Shin, B.S.; Xu, Z.; Klette, R. Ground Truth and Performance Evaluation of Lane Border Detection. In *Computer Vision and Graphics*; ICCVG 2014. Lecture Notes in Computer Science, Volume 8671; Chmielewski, L.J., Kozera, R., Shin, B.S., Wojciechowski, K., Eds.; Springer: Cham, Switzerland, 2014. [CrossRef]
30. Karimov, A.; Razumov, A.; Manbatchurina, R.; Simonova, K.; Donets, I.; Vlasova, A.; Khramtsova, Y.; Ushenin, K. Comparison of UNet, ENet, and BoxENet for Segmentation of Mast Cells in Scans of Histological Slices. In Proceedings of the 2019 International Multi-Conference on Engineering, Computer and Information Sciences (SIBIRCON), Novosibirsk, Russia, 21–27 October 2019; pp. 0544–0547. [CrossRef]
31. Chen, H.-C.; Li, Z.-T. Automated Ground Truth Generation for Learning-Based Crack Detection on Concrete Surfaces. *Appl. Sci.* **2021**, *11*, 10966. [CrossRef]
32. He, X.; Zemel, R.; Carreira-Perpin, M. Multiscale conditional random fields for image labeling. In Proceedings of the 2004 IEEE Computer Society Conference on Computer Vision and Pattern Recognition, Washington, DC, USA, 27 June–2 July 2004.
33. Shotton, J.; Winn, J.; Rother, C.; Criminisi, A. Textonboost: Joint appearance, shape and context modeling for multi-class object recognition and segmentation. In Proceedings of the European Conference on Computer Vision (ECCV), Graz, Austria, 7–13 May 2006.

World Electric Vehicle Journal

MDPI

Review

Exploring Computing Paradigms for Electric Vehicles: From Cloud to Edge Intelligence, Challenges and Future Directions

Sachin B. Chougule [1,2], Bharat S. Chaudhari [1,*], Sheetal N. Ghorpade [2] and Marco Zennaro [3]

[1] Department of Electrical and Electronics Engineering, Dr. Vishwanath Karad MIT World Peace University, Pune 411038, India; sachin.chougule@mitwpu.edu.in
[2] Rubiscape Private Limited, Pune 411045, India; sheetal.ghorpade@rubiscape.com
[3] Science, Technology and Innovation Unit, Abdus Salam International Centre for Theoretical Physics, 34151 Trieste, Italy; mzennaro@ictp.it
* Correspondence: bsc@ieee.org or bharat.chaudhari@mitwpu.edu.in

Abstract: Electric vehicles are widely adopted globally as a sustainable mode of transportation. With the increased availability of onboard computation and communication capabilities, vehicles are moving towards automated driving and intelligent transportation systems. The adaption of technologies such as IoT, edge intelligence, 5G, and blockchain in vehicle architecture has increased possibilities towards efficient and sustainable transportation systems. In this article, we present a comprehensive study and analysis of the edge computing paradigm, explaining elements of edge AI. Furthermore, we discussed the edge intelligence approach for deploying AI algorithms and models on edge devices, which are typically resource-constrained devices located at the edge of the network. It mentions the advantages of edge intelligence and its use cases in smart electric vehicles. It also discusses challenges and opportunities and provides in-depth analysis for optimizing computation for edge intelligence. Finally, it sheds some light on the research roadmap on AI for edge and AI on edge by dividing efforts into topology, content, service segments, model adaptation, framework design, and processor acceleration, all of which stand to gain advantages from AI technologies. Investigating the incorporation of important technologies, issues, opportunities, and Roadmap in this study will be a valuable resource for the community engaged in research on edge intelligence in electric vehicles.

Keywords: electric vehicles; artificial intelligence; edge intelligence; cloud computing; edge computing; internet of things; deep neural networks; energy efficiency; autonomous vehicles

Citation: Chougule, S.B.; Chaudhari, B.S.; Ghorpade, S.N.; Zennaro, M. Exploring Computing Paradigms for Electric Vehicles: From Cloud to Edge Intelligence, Challenges and Future Directions. *World Electr. Veh. J.* **2024**, *15*, 39. https://doi.org/10.3390/wevj15020039

Academic Editors: Biao Yu, Linglong Lin and Jiajia Chen

Received: 31 December 2023
Revised: 17 January 2024
Accepted: 22 January 2024
Published: 26 January 2024

1. Introduction

Four prominent technology trends are playing a pivotal role in driving innovation within the automotive industry: autonomous driving, connected vehicles, electric vehicles, and shared mobility. Electric Vehicles (EVs) are rapidly gaining ground in intelligent transportation systems (ITS) owing to their low driving costs and minimal carbon emissions [1–3]. Artificial Intelligence (AI) stands out as a critical component in enhancing the sophistication of EVs. ITS encompasses a range of technologies, including automation, computers, controls, and communication, all geared towards improving the safety, efficiency, energy efficiency, and environmental friendliness of transportation. The rise of Autonomous Vehicles (AV) introduces challenges in the realm of intelligent decision-making, often perceived as incomprehensible to humans. Such a lack of transparency impedes the widespread acceptance of AV technology within society. In the case of self-driving cars, AI systems face the dual challenge of making real-time and secure decisions while also providing explanations for those decisions, a necessity to comply with legal requirements in various jurisdictions.

A significant portion of AI solutions relies on cloud computing for data storage and algorithmic processing. Hence, the cloud-based Internet of Things (IoT) platform is essential for autonomous vehicles, and cloud computing encounters several challenges, as highlighted by existing research [4]. The surge in interconnected devices necessitates efficient data processing and robust decision-making within strict latency constraints. Despite the efficiency and speed of our networks, transporting the massive volume of information spawned by these devices to the cloud for investigation and storage is impractical. The transfer of such vast data over cloud networks introduces overheads that diminish throughput, escalate energy consumption, increase network traffic, and incur additional costs. The heterogeneous nature of data produced by a large number of IoT sensors and devices in AVs further complicates cloud processing [5].

The cloud's complexity is exacerbated by the diverse and real-time data streams from numerous AVs, significantly increasing the workload on the cloud infrastructure. In response to these challenges, edge computing emerges as a solution to a distributed computation model deployed in nearby proximity to the data source. By deploying an Edge Intelligence (EI) model, the inference computing of AVs can experience substantial improvements in accuracy and latency. This shift toward edge computing addresses the limitations posed by centralized cloud processing and aligns with the demands of processing diverse, real-time data from interconnected devices in AVs. The increasing demands of autonomous driving have brought together machine learning, explicitly AI and Mobile Edge Computing (MEC), giving rise to edge intelligence (EI) or edge AI. This convergence aims to enhance various routine activities [6–8] significantly. The core aim of edge intelligence is to orchestrate the collaboration among numerous edge devices and servers to handle data spawned in close vicinity. Simultaneously, AI seeks to replicate intelligent human behavior in devices and machines by learning from data. The fusion of AI and edge intelligence is a logical progression due to the evident overlap between these two technologies, collectively referred to as edge intelligence.

With AV's perspective, edge intelligence plays a crucial role by enabling the AV to recognize its backgrounds precisely. This is achieved by offloading the data to additional powerful edge server situated at the base station. The substantial volume of information spawned and offloaded to the edge necessitates robust AI algorithms for precise processing, thereby giving rise to the integration of edge intelligence. Consequently, the inference processing capabilities of AVs can be significantly enhanced by installing an edge intelligence model to enhance precision and reduce latency. Nevertheless, the research on edge intelligence is still in its early stages equally in academia and industries. The exploration of this field encounters notable challenges related to transmission, computation within restricted bandwidth, data safety, confidentiality concerns, and energy utilization [9,10]. These hurdles underline the complexity and evolving nature of edge intelligence, indicating the need for further exploration and innovative solutions in the integration of AI and edge computing for autonomous systems. The research journey for this study started with the in depth understanding of six levels of edge intelligence. The analysis focused on four pivotal components essential for enhancing the efficiency of edge intelligence: edge caching, edge training, edge interpretation, and edge offloading. Subsequently, the devised techniques for optimizing these components were examined. The document provides an outline of the applications of edge intelligence in electric smart vehicles. Finally, it discusses the hurdles and prospects related to the adoption of edge intelligence in electric vehicles, enhancements in performance, and emerging directions for future research.

The rest of the paper is organized as follows: In Section 2, we have discussed edge intelligence paradigms, which describe strategies for training and inference on cloud or edge. The advantages and applications of edge intelligence are presented in Section 3. Edge intelligence presents more intriguing opportunities but encounters numerous challenges during its implementation. Section 4 elaborates on challenges and opportunities in edge intelligence. Section 5 presents in-depth analysis of artificial intelligence-based solutions for optimizing the computation of edge intelligence. Architectural layers in the Roadmap for edge intelligence are discussed in detail in Section 6. Lastly, the paper is concluded in Section 7. The organization of the paper is illustrated in Figure 1.

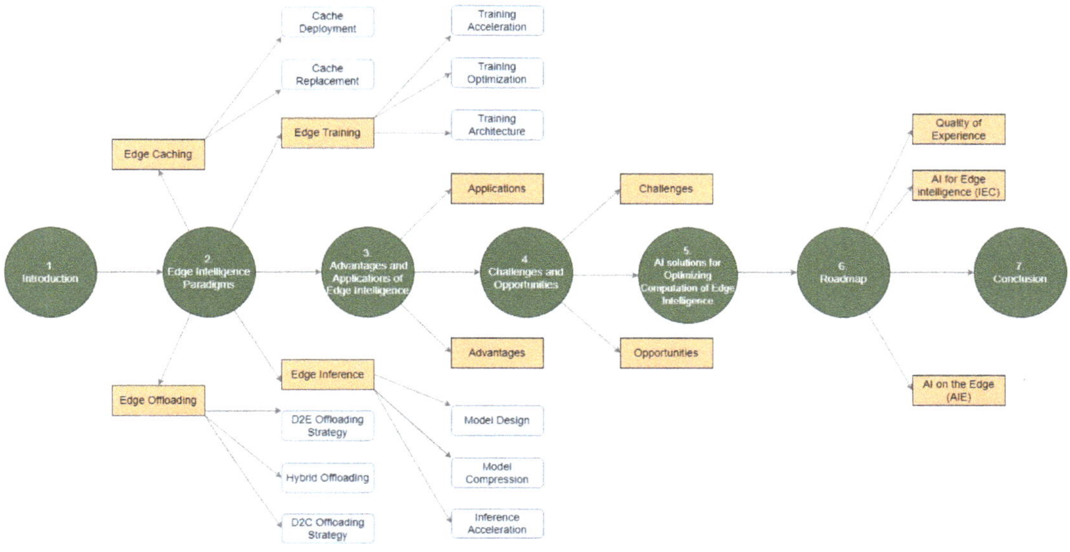

Figure 1. Organization of the paper.

2. Edge Intelligence Paradigms

Edge Intelligence (EI) is the execution of AI algorithms on edge devices using data generated on those devices and on sensor nodes [11,12]. This approach often involves high-performance AI chips but has limitations. It increases energy consumption as well as cost and is inappropriate for aged devices having restricted computation capabilities. However, it is essential to recognize that this narrow definition of EI does not fully leverage the potential of the technology. Recent studies have shown that for Deep Neural Network (DNN) models, an amalgamation of edge and cloud computing can reduce latency and energy consumption compared to local implementations [13–15]. EI should encompass a broader concept, utilizing available data and resources across various levels, as shown in Figure 2, from end nodes and edge devices to cloud data centers, for optimizing the training and inference of DNN models. These levels include:

- **L0_Cloud Intelligence:** Complete DNN model training and interpretation in the cloud.
- **L1_Cloud-Edge Cooperation and Cloud Training:** Train the deep neural network-based model (DNNM) in the cloud, then perform inference in collaboration with the edge, partly offloading new additional data to the cloud.
- **L2_In-Edge Cooperation and Cloud Training:** Train the DNNM in the cloud but perform interpretation at the edge, potentially offloading data on edge devices or else on the adjacent devices.
- **L3_On-Device Interpretation and Cloud Training:** Train the DNNM in the cloud but perform on end nodes for interpretation with partly data offloading from cloud to end nodes.
- **L4_Cloud–Edge Co-training and Interpretation:** Both training and interpretation of the DNNM model occur in cooperation between the cloud and edge.
- **L5_All In-Edge:** Both training and interpretation of the DNNM take place at the edge environment.
- **L6_All On-Device:** Both training and interpretation of the DNNM occur exclusively on the end node.

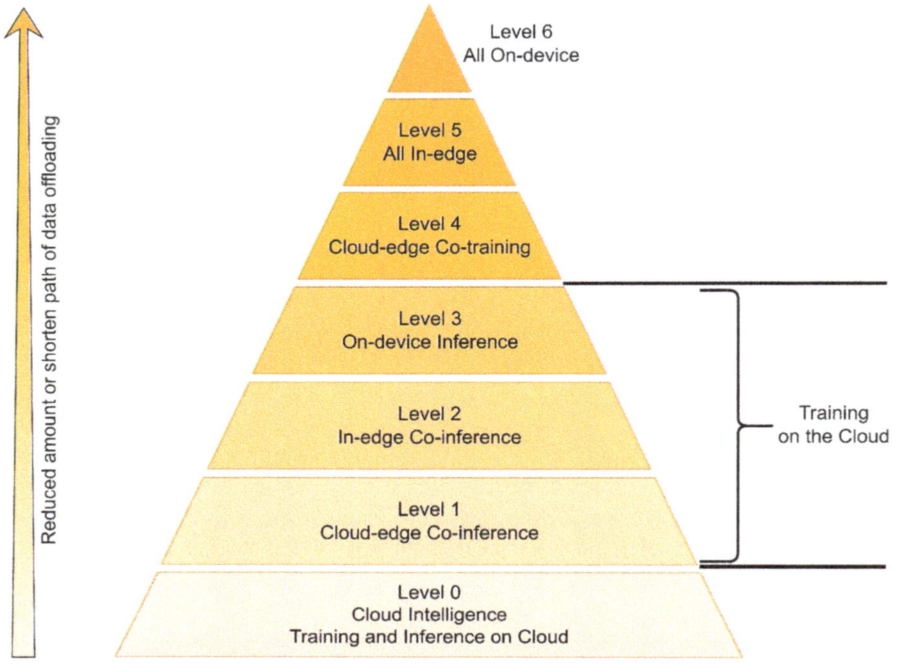

Figure 2. Six leveled ratings for edge intelligence.

The location for training and inferencing at each level is presented in Table 1.

Table 1. Training and Inferencing Location.

Level	Location	Training	Inferencing
Level 0	Coud Intelligence	Cloud	Cloud
Level 1	Cloud-edge Co-inference	Cloud	Cloud-edge
Level 2	In-Edge Co-inference	Cloud	Edge & adjacent devices
Level 3	On-device Inference	Cloud	Edge
Level 4	Cloud-edge Co-training	Cloud-edge	Cloud-edge
Level 5	All In-edge	Edge & adjacent devices	Edge & adjacent devices
Level 6	All On-device	Edge Device	Edge Device

The choice of EI level depends on various factors, including latency, energy efficiency, privacy, and WAN bandwidth cost, making it application-dependent. Four crucial elements of edge intelligence are edge caching, edge training, edge interpretation, and edge offloading, and their subclasses are shown in Figure 3. These elements are elaborated further in this section.

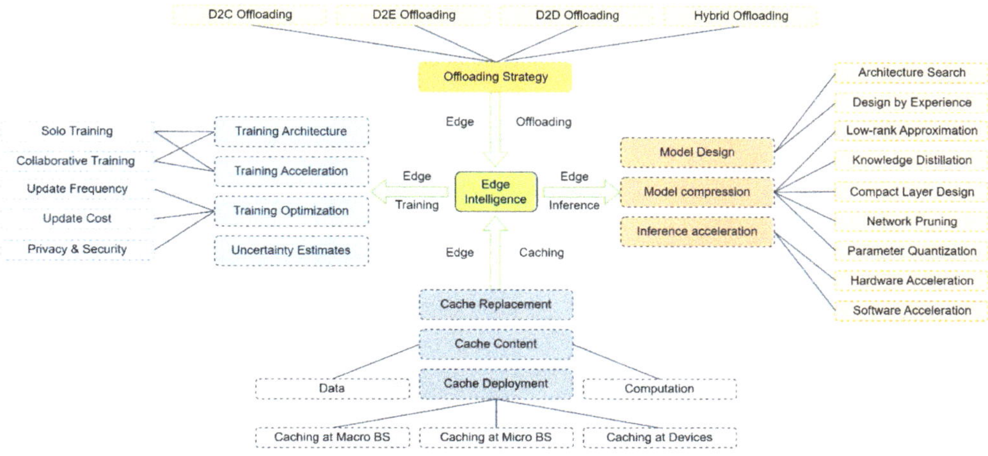

Figure 3. Elements of Edge Intelligence.

2.1. *Edge Caching*

Edge caching involves storing information generated by edge devices, sensors, and IoT devices closer to users to enhance performance and lower latency. This technique can reduce computational complexity and interpretation time, storing raw sensor data or previous computation results for reuse. Various caching methods have shown significant latency improvements [16–19].

2.1.1. Cache Deployment

Cache elements are deployed at edge or end entities such as macro base stations, micro base stations, and end nodes.

- *Caching at Macro Base Stations*

Broader coverage and substantial cache size are the characteristics of macro base stations. A macro base station typically covers a radius of approximately 500 m [20].

- *Caching at Micro Base Stations*

Micro base stations refer to a group of low-energy access points with a coverage span ranging from 20 to 200 m, including microcells, picocells, and femtocells [21]. By deploying small base stations or hot spots in strategic locations enhances the overall experience. This improvement is attributed to advantages like efficient spatial spectrum reuse, resulting in benefits such as higher end rates [22,23].

- *Caching at Devices*

Utilizing device-level caching takes advantage of the storage capacity within end devices, which can optimize local transmission and computing redundancy. Additionally, they have the capability to retrieve the desired contents or computing outcomes through nearby devices via device-to-device (D2D) transmission. [24,25]. However, this approach could be better for IoT as most of the end nodes in IoT are resource-constrained nodes having very low onboard memory, limited computational capability, and, most importantly battery, battery-operated.

2.1.2. Cache Replacement

In practical situations, the allocation of requests for cache access changes over time, and new content is continuously generated. Therefore, it is crucial to update caches periodically. This updating process is referred to as cache replacement. Various conventional strategies for cache replacement have been suggested, including first-in-first-out (FIFO), least frequently used (LFU), least recently used (LRU), and its modifications. [26]. The comparison of edge caching placement is shown in Table 2.

Table 2. Cache placement locations comparison.

Cache Locations	MBSs	SBSs	Devices
Coverage Range	500 m	20~200 m	10 m
Cache Size	Larger	Intermediate	Smaller
Number of Users	Massive	Smaller	Fewer
Topology Structure	Stable	Alters Vaguely	Alters Significantly
Redundancy Capability	Higher	Medium	Lower
Computing Power	Higher	Medium	Lower

2.2. *Edge Training*

Edge training allows devices to learn patterns from cached edge data. It can occur on an edge server or device and includes independent training and collaborative training strategies. Collaborative training involves multiple devices and requires communication updates, posing challenges to data privacy and security. Various factors related to edge training are discussed in this section.

2.2.1. Training Architecture

The training framework relies on the computational capabilities of both edge devices and servers. If a singular edge device or server possesses adequate power, it can employ the identical training structure as a centralized server, performing the training on a single device. Conversely, when the device or server lacks such capabilities, cooperation with other devices becomes imperative. This results in the emergence of two types of training frameworks: individual training, which entails executing training tasks on a lone edge device or server, and cooperative training, where limited devices and servers work together to execute training tasks.

A prevalent example of a cooperative training framework is the master-slave model, as illustrated by federated learning [27]. In federated learning, a server involves numerous devices and delegates training tasks individually. Another form of cooperative training architecture is peer-to-peer, where participants are regarded as equals in the training process.

2.2.2. Training Acceleration

The emphasis is on expediting training at the edge, with some initiatives [28,29] exploring transfer learning to enhance training speed. Transfer learning involves utilizing features learned from previous models, resulting in a significant reduction in learning time. In a cooperative training approach, edge devices have the capability to acquire from one another, thereby enhancing overall knowledge proficiency. A framework known as Recycle ML employs cross-modal transmission to expedite the training of neural networks along mobile platforms throughout the diverse sensory system. Federated learning can also be employed to hasten model training on distributed edge devices, particularly in scenarios where labeled data is insufficient [30].

2.2.3. Training Optimization

Training optimization involves streamlining the training procedure to attain specific goals, viz., minimizing energy consumption, enhancing precision, preserving secrecy, maintaining security, and more. The critical factors involved in optimization are communication frequency, communication cost, privacy, and security issues.

- *Communication Frequency*

Communication Frequency is a crucial aspect of federated learning, where the exchange of information among edge devices and the cloud server plays a vital role. This operation involves update uploading through edge devices to the cloud server and downloading the combined updates from the distributed model to local models. Given the potential for erratic network conditions in edge devices, it is essential to minimize update cycles.

- *Communication Cost*

Besides the frequency of communication, another factor influencing the efficiency of communication among edge devices and the central server is the cost of communication. Minimizing communication costs has the potential to save bandwidth substantially and enhance overall communication efficiency.

- *Privacy and Security Issues*

Upon catching updates from edge devices, the central server is tasked with aggregating these updates to construct unified updates for the distributed universal model. The concern arises that malicious hackers may scrutinize the updates, posing a threat to the privacy of participating edge users. To address this, an aggregation process is implemented to combine updates from all edge devices, rendering individual updates indiscernible by the central server [31]. More precisely, every edge device transmits encrypted updates to the server. The server later combines these encrypted updates. The counteraction of masks occurs when enough edge devices are involved. Consequently, the server gains the ability to unveil the aggregated update by unmasking it. Throughout this aggregation process, exclusive updates remain unscrutinized, and the server can solely approach the combined unmasked updates, thereby efficiently safeguarding the secrecy of participants.

The conventional deep learning approaches for classification and regression lack the ability to account for model uncertainty.

2.3. Edge Interpretation

Edge interpretation occurs during the use of the trained model for computing output on edge devices and servers. However, many deep learning algorithms are designed for high-performance hardware and are not suitable for edge environments. Challenges include designing models for edge deployment and accelerating edge inference for real-time responses. These issues can be addressed through new model designs or model compression techniques.

2.3.1. Model Design

The primary emphasis in model design revolves around creating neural network architectures that are lightweight and appropriate for execution on edge devices with lower hardware demands. This process involves either automated generation of the optimal architecture by machines or manual design by humans.

- *Architecture Search*

Exploration in architecture search is a thriving research field with broad applications in the future. A recent notable advancement in this area is differentiable architecture search (DARTS) [32], which offers the potential to substantially decrease reliance on hardware. DARTS relies on the constant easing of architecture description and employs gradient descent for the process of hunting for architecture.

- *Design by Experience*

The experience-driven design employs two distinct strategies. The initial approach involves the use of comprehensive detachable convolutions, which are utilized to construct a streamlined DNN known as MobileNets. This design specifically caters to the needs of mobile and embedded devices [33]. Another method employed is group convolution, which serves as an alternative means to diminish computation costs during the process of designing architecture.

2.3.2. Model Compression

Model compression seeks to reduce the dimensions of a model, enhance energy efficiency, and accelerate inference on edge devices with a restricted number of resources, all without compromising precision. The five important methods to model compression are lower Rank approximation, knowledge distillation, compact layer design, parameter quantization, and network pruning [34].

- *Lower Rank Approximation*

The fundamental concept behind low-rank approximation involves the replacement of high-dimensional kernels by the low-rank convolutional kernel multiplication.

- *Knowledge Refinement*

It relies on transfer learning, wherein a smaller NN is trained employing distilled data by initiating a large model. The large, intricate model is known as the mentor model, whereas the more compacted model is described as the student model. The student model gains advantages by assimilating knowledge from the teacher network.

- *Designing Condensed Layers*

In DNN, when weight values approach zero, computational resources are inefficiently utilized. A key strategy to address this issue involves creating a condensed layer in NN, excellently minimizing resource utilization such as memory and computation power. Christian and colleagues suggest addressing this by incorporating sparseness and substituting the entirely linked layers in GoogLeNet. In Residual-Net, an alternative approach is taken by replacing entirely linked layers through global regular merging to decrease resource demands. Substituting a large convolutional layer with several smaller and more compact layers can efficiently decrease the parameter count and subsequently lower computational requirements.

- *Network Pruning*

The fundamental concept behind network pruning involves the removal of less significant parameters, recognizing that not all parameters play a crucial role in extremely accurate DNN. As a result, associations between lower weights are eliminated, transforming a heavy network into a sparser one.

- *Parameter Quantization*

Achieving high performance in neural networks does not always require highly precise parameters, particularly when those parameters are unnecessary. Research has demonstrated that a relatively smaller quantity of parameters suffice for reconstructing a complete network.

2.3.3. Interpretation Acceleration

The primary concept behind accelerating models in interpretation is to decrease the runtime of interpretation on edge devices and achieve instantaneous replies for precise applications based on NN, all deprived of modifying the architecture of the trained model. There are two main categories of acceleration: hardware and software acceleration. The software acceleration approach is centered on enhancing resource management, pipeline structure, and compiler optimization.

- *Hardware Acceleration*

Methods for hardware acceleration concentrate on parallelizing inference tasks across accessible hardware, including CPU, GPU, and DSP. In recent times, the potency of mobile devices has seen a notable rise. A growing number of mobile platforms now feature GPUs. Given that mobile CPUs are less apt for deep neural network computations, leveraging embedded GPUs becomes a viable strategy to distribute computing tasks and expedite the inference process.

- *Software Acceleration*

Software acceleration primarily centers on enhancing resource allocation, refining pipeline design, and optimizing compilers. Methods for software acceleration aim to optimize the utilization of limited resources to achieve faster performance, but this can sometimes result in a reduction in accuracy in specific scenarios.

2.4. Edge Offloading

It is a distributed computation paradigm that furnishes computation services for edge caching, training, and interpretation. It allows tasks to be processed in cloud servers when edge hardware lacks capability. Four offloading strategies exist, including device-to-cloud (D2C), device-to-edge server (D2E), device-to-device (D2D), and hybrid offloading, each with its adaptiveness and resource utilization.

- *D2C Offloading Strategy*

In the D2C offloading strategy, devices transfer input data, such as audio or images, to a cloud server. Powerful computers perform high-accuracy inference using a large neural model, and the outputs are sent backward via the identical network. However, it has some primary drawbacks. Mobile devices need to communicate large volumes of information to the cloud, creating a bottleneck in the overall process [35]. The execution is reliant on internet connectivity. The transmitted information from mobile devices might inhibit users' personal information, such as personal photos, making it susceptible to attacks by mischievous hacks while the interpretation on the cloud server [36]. Various considerations, including energy efficiency, latency, and privacy, can guide the design of model partitioning and layer scheduling in this context.

- *D2E Offloading Strategy*

In contrast to D2C offloading, which involves transferring inferencing to a central server in the cloud, D2E offloading shifts inferencing to an Edge server. An Edge server, in this context, denotes robust servers that are physically close to mobile devices and possess greater processing power than typical edge devices.

- *D2D Offloading Strategy*

In the strategy of Device-to-Device (D2D) offloading, devices like smartwatches are connected to smartphones or home gateways and have the capability to delegate model interpretation tasks to more influential connected devices. Binary decision-based offloading and partial offloading exist in this context. Binary decision offloading involves deciding upon the execution of the task locally or offloading it. On the other hand, partial offloading entails breaking down the interpretation task into various subtasks and offloading a handful of them to connected devices.

- *Hybrid Offloading*

The hybrid computing framework efficiently leverages cloud services, edge computing, and mobile devices in a comprehensive approach. Distributed Deep Neural Networks (DDNNs) derived from this holistic computing architecture represent a hybrid offloading technique that strategically allocates portions of a DNN across a distributed computing hierarchy [36]. The collective training of these segments occurs in the cloud and aims to reduce communication and resource usage on edge devices. During the interpretation phase, individual edge devices perform local computations, and the resultant outputs are combined to generate the final results.

3. Edge Intelligence: Advantages and Applications

In recent times, there has been a noticeable trend towards enhanced intelligence in various aspects of life, ranging from smartwatches to automobiles, agriculture to industrial processes, and even urban environments, and additionally benefiting from the general benefits of edge intelligence, like reduced latency and bandwidth consumption. Edge intelligence can help enterprises make quicker data-driven decisions responding to the requirements of their clients. Reduced storage requirements in edge intelligence led to improved operational cost savings for enterprises.

3.1. Advantages

- *Enriching AI with Richer Data and Application Scenarios:*

Recent advancements in deep learning have been driven by four factors: algorithms, hardware, data, and application scenarios. Data plays a pivotal role in enhancing AI performance. As IoT grows, vast quantities of information will be spawned at the edge, challenging cloud-based processing due to bandwidth constraints. Edge intelligence addresses this challenge by enabling low-latency data processing closer to the data source, potentially boosting AI performance. Edge intelligence and AI counterparts one another technically and also with regard to application and adoption [37].

- *Key Infrastructure for AI Democratization:*

AI has made significant strides in digital products and services, from online shopping to self-driving cars. Major IT companies envision democratizing AI, making it accessible to everyone and everywhere. Edge intelligence is well-suited to this goal and offers diverse application scenarios. Thus, Edge intelligence serves as a crucial enabler for ubiquitous AI [38].

- *Popularizing Edge Intelligence with AI Applications:*

Edge intelligence is already bringing about significant transformations across various industries, such as manufacturing, energy, healthcare, agriculture, logistics, and transportation. [39–41]. Real-time video analytics, built on computer vision, emerges as a killer application for Edge intelligence due to its high computational demands, bandwidth requirements, privacy concerns, and low-latency needs [42]. Multiple benefits of Edge intelligence have created a path for expanded progression in the near future [43].

3.2. Applications of Edge Intelligence for Electric Vehicles

Gartner anticipates a significant surge in the adoption of edge intelligence use cases in the coming years. The projection is that by 2024, over fifty percent of the potential enterprises will have implemented a minimum of six edge intelligence use cases. This marks a remarkable expansion in comparison with the scenario in 2019, in which merely one percent of larger enterprises had very few edge intelligence deployments. Presently, some key applications of edge computing include:

3.2.1. Smart Vehicles

An intelligent vehicle is defined as a vehicle with computing capabilities, storage, and communication facilities that enables learning from its environment and making conclusions consequently. Sensors and multi-interface cards are used for equipping vehicles inside and outside. The increasing prevalence of smart vehicles endowed with onboard wireless devices and sensors like radar and lidar has led to a focus on efficient management and transportation applications. The goal is to improve traffic flow by reducing travel time and preventing jamming.

Smart vehicles possess a range of novel features, including information exchange and location info. These functionalities assist specialized applications, such as security communication and warnings. Vehicles inside a Vehicular Edge Computing (VEC) system typically have onboard wireless devices, particularly Onboard Units (OBUs). During disaster alarm schemes, sensors play a crucial role in verifying if airbags were deployed during an accident.

3.2.2. Smart Vehicle Services

Smart vehicles offer a diverse range of services. Some key services, such as assisted driving, autonomous vehicles, platooning, and parking solutions, are discussed below.

- *Assistant Driving*

In contemporary times, vehicles such as cars, buses, and trains are designed with the capability to convey valuable information, including details about accidents, road closures, and traffic congestion. This is achieved through the integration of sensors, actuators, and processors, enhancing safety and navigation for these vehicles. The data on traffic patterns, made available by these intelligent features, can prove advantageous for all types of organizations [44]. Intelligent vehicles are categorized into five layers by the National Highway Traffic Safety Administration [45].

- *Autonomous Vehicles*

As smart vehicles progress towards autonomous driving, establishing robust connectivity amongst smart vehicles becomes imperative. Vehicular networks, on the rise due to this evolution, play a pivotal role in shaping intelligent transportation systems and smart cities. These systems are anticipated to support a spectrum of advanced applications, ranging from road safety and enhanced traffic efficiency to automated driving and seamless admittance to Internet facilities [46,47].

The global acceptance of automated vehicles has sparked a transformation in the automobile sector. Nevertheless, challenges such as invulnerability, fidelity, and secrecy persist in realizing completely automated vehicle editions. Notably, the susceptibility of automated vehicles to security threats is a concern; a single attack on the software of an Autonomous Vehicle (AV) could lead to multiple mishaps. Additionally, interconnected systems on the Internet face risks of unauthorized access, presenting unknown threats. Vehicle design addresses safety-critical issues by enabling the vehicle to anticipate and respond to potential dangers while continuously monitoring road conditions throughout the journey. The assumption in the design is that the driver provides the destination or navigation. However, it may not be in regulation throughout the excursion, emphasizing the role of automated vehicular systems in ensuring safe operations [48]. While automated vehicular systems differ from connected vehicular technology, they share some similarities.

- *Platoon*

A platoon refers to a cluster of smart vehicles equipped with driving assistant schemes in which one vehicle follows another. The formation of a platoon involves several vehicles driven by technology, interconnected through shared communication. This collaborative driving concept, known as platooning, has become feasible because of the advancement of technologies. These technologies, fortified with sensors and actuators, enable modern vehicles to engage in cooperative platooning.

Cooperative platooning offers significant advantages, particularly in improving the fuel efficacy of heavy vehicles. By anticipating speed changes, the vehicles within a platoon can maintain a steady speed, leading to enhanced fuel efficiency. Since carbon dioxide emissions are directly linked to fuel consumption, cooperative platooning has the indirect effect of reducing environmental pollution. Additionally, this form of platooning contributes to the improvement of road safety. In emergencies, messages are transmitted to all vehicles in a platoon, triggering appropriate actions by the automated system [49].

- *Smart Parking*

In metropolitan regions, the number of vehicles parked in parking lots is substantial, distributed across various locations such as street parking and outer parking. Unlike moving vehicles, parked vehicles remain stationary for extended periods. Although they do not transport info from location to location, parked vehicles equipped with wireless communication devices and rechargeable batteries as part of Smart Street Vehicles (SSVs) serve as communication infrastructures with unique characteristics. This allows parked

SSVs to transmit information among themselves and also link to neighboring moving SSVs, functioning as static backbones to enhance communication amongst vehicles. The number of parked vehicles in a parking slot and their duration of stay are critical factors influencing their role as communication infrastructures [50]. Collaboration among parked SSVs, particularly in parking lots, enables the execution of heavy computation tasks under favorable communication conditions. Individual vehicles, constrained by limited resources, may struggle to meet substantial computation demands. Parked SSVs address this challenge by providing powerful and underutilized computation resources, efficiently accomplishing allocated tasks in less time. This environment can be likened to tiny data centers capable of handling intricate tasks that require significant computational capability.

3.2.3. Smart Vehicle Applications

The emergence of Vehicular Edge Computing (VEC) and the utilization of smart vehicles as infrastructures have paved the way for a multitude of associated vehicular applications. These applications span various domains, including driving safety, Augmented Reality (AR), infotainment services, and video streaming. Particularly in scenarios wherever higher computational processing is essential, VEC networks plays a crucial role in accelerating computing processes, in this manner curtailing delays. For instance, in the event of an accident, quick computations are needed to formulate solutions such as rescheduling traffic lights and efficiently dissipating a large traffic backlog. Meeting such demands places an exceptional requirement on computational resources [51]. In this context, applications are categorized into two groups: safety and non-safety. VEC proves to be supportive of both types of applications, as discussed below:

- *Safety Applications*

It emphasizes enhancing security by minimizing the likelihood of accidents. These applications monitor the driving environment and alert drivers to potentially harmful situations to prevent accidents. One such application involves the use of a Global Camera Sensor mounted at a traffic monitoring signal, capable of detecting movement in its region by recognizing number plates within the detection field. This sensor records the location and vehicle number, sending this information to the local edge server. A smart Local Camera Sensor (LCS) positioned at the front of the vehicle observes the driver's activities. The LCS issues warning messages to the driver for such activities, aiding in accident prevention. Repetitive warnings at appropriate times help drivers avoid hazardous situations and ensure their safety. The LCS is equipped to generate these warning messages and, later broadcasting a specific number, informs the edge server about any interrupting activities involving the vehicle. This report involves activity evidence and vehicle identification [52–56].

Context-aware systems are also employed, utilizing information related to the user to adapt operations based on environmental conditions. Context-aware applications adjust their operations according to the user's context, sensing information specific to the environment. These applications involve Context Acquisition, Processing, and Acting [57]. Leveraging contextual knowledge allows the generation of concise, context-aware information, reducing the radio resource requirements for transmission. Users can extract coveted content from the context using suitable decoders and big-data analytics techniques such as Natural Language Processing (NLP) [58,59].

- *Non-safety Applications*

Applications of Vehicular Edge Computing (VEC) extend beyond safety services to include the development of non-safety applications, such as multimedia services like video streaming, Augmented Reality (AR), and infotainment. The surge in streaming applications has notably contributed to a significant portion of network traffic, particularly in IoT communication, where video streaming plays a pivotal role [60]. This is particularly evident in smartphone applications like video crowdsourcing [61]. The Internet of Vehicles (IoV) supports various applications, including intelligent transportation systems and mobile multimedia. In IoV, users connect their mobile devices to the internet to access multimedia content from remote servers. However, maintaining Quality of Service (QoS) becomes challenging, given factors like jitter, buffering, throughput, and transmission delays in video streaming applications, exacerbated by the high mobility of vehicles in IoV. A proposed solution in [62] introduces distributed reliable real-time streaming in vehicular cloud-fog networks. A utility function is utilized to improve QoS and fairness in resource reservation among mobile devices, considering content provision for streaming and the number of tokens for content reservation from service providers, edge, and cloud. Mobile devices in the network query their probable location, the amount of data for streaming, and required tokens for content provisioning, facilitating effective reservation of streaming content from computing service providers and enhancing streaming utility reliability.

Addressing the parking lot monitoring issue, [63] proposes an edge computing-based scheme where each vehicle uploads street contents collected by the camera for video analytics. This enables ParkMaster to estimate precise locations and track parked vehicles using information from the vehicle's camera, GPS, and inertial sensors.

Augmented Reality (AR) is an evolving multimedia application that seamlessly integrates real scenes into virtual scenes, overlaying virtual content onto the real environment to enhance traditional image information. AR can improve traffic awareness for vehicles or pedestrians near drivers, with the head-up display (HUD) reducing distractions and enhancing driving safety. An exploration of the HUD-based navigation system with AR-based content is detailed in [64], illustrating its potential for safety and convenience services [65]. A novel application, walk navigation, utilizes a camera and GPS for a car navigation system with AR technology, providing real-time navigation without compromising safety. The device's camera output is analyzed by an edge computing application to overlay viewed objects with AR content. Given the intricate storage and processing demands of AR, VEC is considered the optimal solution to meet the specific requirements of AR applications in a vehicular network, including mobility, location awareness, and low latency.

Although edge intelligence presents more intriguing opportunities compared to cloud computing, organizations encounter numerous challenges during its implementation. Opportunities and challenges associated with edge intelligence are discussed in the next section.

4. Challenges and Opportunities for Edge Intelligence

Edge intelligence is yet in its early stages, and currently, there is yet to be an established framework to strengthen it. Such frameworks must encounter specific requisites, including the ability to develop applications for instantaneous processing on edge nodes. While existing cloud computing frameworks can handle data exhaustive purposes, enabling instantaneous data treatment at the network edge remains an area of ongoing research [66–69]. Moreover, we must gain a deep understanding of installing application capabilities on edge nodes, including strategies for workload placement, policies for connecting to edge nodes, and management of distinct node types when deploying applications at the edge.

4.1. Challenges

To create such a framework, five key research challenges spanning the hardware, middleware, and software layers are identified.

4.1.1. Enabling Generic Computing on Edge Nodes

In principle, the concept of Edge intelligence involves utilizing various nodes linking the edge device and the cloud. For illustration, base stations are equipped with specialized Digital Signal Processors (DSPs) designed to manage specific tasks. Nevertheless, in practical terms, base stations might not be ideal for handling critical assignments due to the fact that DSPs are not devised for versatile computation tasks. Furthermore, the situation remains uncertain since these nodes may not be able to execute additional computations along with the primary functions.

Research is underway to enhance the computing capabilities of edge nodes to assist generic tasks. For instance, it is possible to upgrade a wireless home router to handle added tasks [70]. Intel's Smart Cell Platform17 utilizes virtualization to accommodate supplementary tasks. An alternative solution involves exchanging specific DSPs with equivalent generic CPUs, although it demands substantial investment.

Cutting-edge AI techniques like neural networks have shown great potential in solving various challenges using remarkable precision. Nevertheless, this often arises at the expense of higher computation and memory demands. As a result, NN typically executes these algorithms on high-powered GPUs, which consume significant power. In contrast, embedded processors and DSPs propose a more power-efficient remedy and are capable of fixed-point processes [71]. To make neural networks functional for deployment on mobile devices, we need less complex CNN models that can execute on embedded processors without sacrificing precision. Additionally, it is essential to enhance both the efficacy of the inherent processes executed by neural networks and their overall structure to make them more suitable for resource-competent procedures.

4.1.2. Exploring Edge Node Discovery

The exploration of resources and services within a distributed computing environment is an established field. It is accomplished in both tightly and loosely connected setups through various techniques integrated into monitoring tools [72,73] and service brokerages [74,75]. These methods, like benchmarking, create the foundation for yielding decisions about allocating tasks to highly suitable resources to enhance performance.

Nevertheless, the challenge arises when we aim to control the capability of the network's edge. In a decentralized cloud configuration, discovering appropriate nodes necessitates mechanisms that go beyond manual intervention due to the sheer number of available devices at this level. Additionally, such mechanisms should accommodate diverse devices from diverse generations and adapt to prevailing workloads like newly added exhaustive machine learning tasks. Benchmarking methods must rapidly communicate the attainability and capabilities of resources. It is also desirable for them to handle node failures consistently and independent recovery.

In this context, the conventional methods used in the cloud for discovering edge nodes, such as resource management and task scheduling, face limitations:

- *Resource Management*

  It entails guaranteeing an ample supply of resources within the edge network, as exemplified in smart parking setups in which sensor data is seamlessly and dependably transmitted to edge devices [76]. Essential components of resource management include dynamic load balancing [77] and the creation of platforms for resource allocation [78]. Nevertheless, addressing on-demand resource requirements, fluctuating workloads, and data streams originating from diverse devices across extensive geographical areas may require making trade-offs between computing power and communication speed [79].

- *Resource Management and Task Scheduling*

  Edge devices exhibit numerous models, diverse hardware architectures, various operating systems, and inconsistent creation environments. Established edge intelligence platforms struggle to effectively incorporate and administer such diverse edge devices, particularly when it comes to supporting AI workloads. Managing and orchestrating AI workloads, which have distinct characteristics compared to web loads, is a pressing challenge. Consequently, Edge intelligence platforms must introduce novel resource perceptions tailored to the challenges of AI workloads, including GPU support, capability extension, and task dependence handling.

- *Customized AI Algorithms*

  While model compression preserves to foster AI execution on the edge, usually leads to a deficit of model precision. Static model compression methods fail to amend the dynamical hardware configurations and loads of edge nodes. Thus, there is a growing need for dynamic compression methods tailored to the complex conditions of edge nodes. Additionally, current model-splitting techniques utilize the hierarchical constitution of deep learning models. Future research should focus on developing partitioning methods tailored to the specific characteristics of AI applications.

Furthermore, data availability presents another formidable challenge for edge devices attempting to process raw data for edge training. The usability of data is crucial, and raw data captured from edge devices often cannot be directly used for model training and inference due to potential bias. While federated learning offers a partial solution, the synchronization of training procedures across devices and communication remains a challenging aspect. In conclusion, discovering and effectively utilizing edge nodes in a decentralized cloud setup poses unique challenges that necessitate innovative approaches beyond traditional cloud-based methods.

4.1.3. Dividing and Delegating Discovery

The advancement of distributed computation environments leads to the creation of various methods for dividing tasks to be implemented in numerous geographical localities [80]. One instance is the partitioning of workflows to be executed in different places [81]. Task splitting is typically conveyed obviously using a semantic or administration tool. However, using edge nodes for offloading computations presents the challenge of efficiently dividing computational assignments inevitably, deprived of essentially compelling appropriate definitions of the competencies or locations of edge nodes.

With the increasing global demand for mobile applications, mobile devices face growing constraints such as limited resources and reduced battery life. Mobile fog architectures have been discussed in the context of mobile cloud computing and code offloading mechanisms. Existing investigations have predominantly depended on simulation techniques for examining task offloading. However, this methodology has limitations because it cannot accurately depict the authentic characteristics of AI workloads in industrial settings. AI algorithms utilized in diverse industrial sectors exhibit distinct model structures and processing steps.

Consequently, when devising a task offloading algorithm, it is crucial to customize the offloading strategy to align with the processing steps of the AI application and its model structure. Current research is transitioning towards a synergy between cloud and Edge intelligence. In the future, the emphasis may pivot towards cooperative computing among edge nodes, where multiple edge nodes can collect information from various perspectives, contributing to heightened analysis and decision-making capabilities.

4.1.4. Unwavering Quality of Service and User Experience

The value delivered by edge nodes can be measured using Quality of Service (QoS), while Quality of Experience (QoE) assesses the quality experienced by users. In the realm of Edge intelligence, a critical principle to embrace is the avoidance of overburdening nodes with computationally demanding tasks [82,83]. The difficulties lie in ensuring that these nodes maintain high throughput and reliability while accommodating surplus workloads from data centers or other edge devices. Even though an edge node is fully utilized, users of edge devices and data centers rightfully expect a baseline stage of service. For instance, overloading a base station can negatively impact the service given to connected edge devices. It is imperative to have a comprehensive understanding of peak usage hours for edge nodes so that tasks can be effectively divided and organized in a versatile manner. While an administration framework could be beneficial, it also advances concerns associated with supervising, scheduling, and rescheduling at all levels.

Collaborative training is also an important task to be considered. Edge intelligence employs two AI training methods; distributed training and federated learning. In distributed machine learning, data analysis tasks are performed on nodes that create data, with models and data exchanged among different nodes [84,85]. Google has introduced federated learning as a privacy-preserving technique, which has found applications in sensitive areas like healthcare and finance. Federated learning and distributed training are distinct. Generally, distributed training emphasizes utilizing data at the network's edge, whereas federated learning places a greater emphasis on safeguarding data privacy. When dealing with diverse edge devices that vary in computing power and communication protocols, adapting models and ensuring serviceability poses challenges. The same methods may yield different learning outcomes when applied to different device clusters. Establishing robust, flexible, and secure synchronization between edge devices, servers, and cloud resources at both hardware and software levels is of utmost importance. There are significant research opportunities in developing a standardized API/IO interface for edge learning across various ubiquitous edge devices.

4.1.5. Utilizing Edge Nodes Safely and Publicly

Hardware assets held by data centers, supercomputing facilities, and private entities using virtualization have the potential to be repurposed to provide computing services as a utility. This approach involves assessing the associated risks for both providers and users, ultimately enabling pay-as-you-go computing. Consequently, a competitive market has emerged, offering a multitude of options to cater to computing consumers while adhering to Service Level Agreements (SLAs) [86].

Nevertheless, when contemplating the use of alternative devices like switches, routers, and base stations as publicly accessible edge nodes, several challenges must be confronted. Firstly, there's a necessity to clearly delineate and communicate the associated risks for both public and private organizations owning these devices and those planning their deployment. Secondly, it is imperative to ensure that the primary function of the device, such as a router managing internet traffic, remains unaltered when repurposed as an Edge intelligence node. Thirdly, realizing multi-tenancy on edge nodes demands technology that prioritizes security; for instance, containers, which are potentially lightweight technology for edge nodes, must exhibit more robust security features [87]. Fourthly, a baseline level of service must be assured for users of the edge node. Lastly, diverse factors like workloads, computation, data location and transfer, maintenance costs, and energy expenses need consideration when formulating appropriate pricing models for facilitating access to edge nodes.

The significance of data privacy and security cannot be overstated. Artificial Intelligence (AI) serves as an effective tool in identifying malicious attacks and preventing privacy breaches. However, edge devices face constraints in computing resources, presenting a substantial challenge in designing lightweight and efficient AI algorithms suitable for Edge intelligence (EC).

4.2. Opportunities

Despite the difficulties that arise in the implementation of Edge intelligence, several promising opportunities exist. We have identified five such opportunities.

4.2.1. Establishing Standards, Benchmarks, and a Marketplace

The practical realization and public accessibility of Edge intelligence hinge on the clear articulation of duties, associations, and consequences among all involved parties. Various efforts have been made to define cloud standards, including those by organizations such as the National Institutes of Standards and Technology (NIST) in 2021, the IEEE Standards Association, the International Standards Organization (ISO), the Cloud Standards Customer Council (CSCC), and the International Telecommunication Union (ITU). However, these standards must now be revisited to account for added stakeholders, such as public and private organizations that acknowledge edge nodes, to address the communal, legitimate, and moral aspects of edge node utilization. This is undeniably a complex task that demands devotion and investment from both public and private organizations as well as academic institutions.

The implementation of standards relies on the ability to benchmark the performance of edge nodes beside established metrics. Benchmarking efforts for the cloud have been undertaken by organizations like the Standard Performance Evaluation Corporation (SPEC) and several academic scientists. In an environment as unpredictable as the cloud, benchmarking presents substantial difficulties. Benchmarking edge nodes will pose even greater challenges but will open up additional opportunities for research.

Utilizing edge nodes becomes an enticing prospect when duties, associations, and consequences are well-defined. Much like a cloud marketplace, the creation of an Edge intelligence marketplace offering a heterogeneity of edge nodes on a pay-as-you-go basis is reasonable. Research is needed to establish Service Level Agreements (SLAs) for edge nodes and develop pricing models to facilitate the creation of such a marketplace.

4.2.2. Frameworks and Languages

There are numerous possibilities for running applications within the cloud paradigm. Besides widely used programming languages, there's a diverse range of offerings available for deploying cloud-based applications. In scenarios where resources beyond the cloud are utilized, such as running a bioinformatics workload on the public cloud with data sourced from a private database, a typical approach involves the use of workflows. Research has extensively explored software frameworks and toolkits for creating extensive workflows in a distributed environment [88]. However, as edge nodes capable of supporting general-purpose computing become more prevalent, the development of new frameworks and toolkits becomes necessary.

The potential applications of edge analytics are expected to vary significantly from established workflows, which have mainly been explored in scientific fields such as bioinformatics [89] or astronomy [90]. As edge analytics becomes relevant in user-driven scenarios, the current frameworks may not be well-suited for representing edge analytics workflows. The programming model created to leverage the capabilities of edge nodes should be capable of handling task and data-level parallelism while executing workloads across various hierarchical levels of hardware.

Additionally, the programming language that supports this model should take into account the diverse hardware landscape and resource capacities present in the workflow. In cases where edge nodes are highly specific to a particular vendor, the frameworks supporting the workflow must be adaptable. This level of complexity goes beyond that of existing models designed to make cloud computing accessible.

4.2.3. Utilizing Lightweight Libraries and Algorithms

In contrast to large servers, edge nodes face limitations in supporting resource-intensive software due to hardware constraints. For instance, consider a small cell base station equipped with Intel's T3K Concurrent Dual-Mode system-on-chip (SoC). This device typically features a 4-core ARM-based CPU and limited memory, making it inadequate for executing complex data processing tools like Apache Spark. Apache Spark demands a minimum of 8 CPU cores and 8 gigabytes of memory for optimal performance. In the context of edge analytics, there is a need for lightweight algorithms capable of performing reasonable machine learning or data processing tasks [91].

One example of a lightweight library is Apache Quarks, which can be utilized on compact edge devices such as smartphones to enable real-time data analytics. However, Quarks primarily supports basic data processing functions, such as filtering and windowed aggregations, which may not suffice for advanced analytical tasks like context-aware recommendations. There is a demand for machine learning libraries that consume less memory and disk space, benefiting data analytical tools designed for edge nodes.

TensorFlow is another framework to consider, supporting deep learning algorithms and heterogeneous distributed systems, although its potential for edge analytics remains to be explored.

AI algorithms play a pivotal role in extracting valuable insights from big data. Nevertheless, the information extracted by existing algorithms is somewhat limited. In the case of supervised learning, manual data labeling can introduce unknown errors. Furthermore, the future data acquisition systems for smart medical applications will predominantly rely on wearable devices. The rapid analysis and response to collected data on these wearables present a significant challenge in terms of energy supply. Balancing the accuracy and lightweight nature of AI models is an area that warrants further investigation.

4.2.4. Micro Operating Systems and Virtualization

Research into micro-operating systems or microkernels presents a potential avenue for addressing challenges associated with deploying applications on diverse edge nodes. These nodes, unlike traditional servers, typically have limited resources. Therefore, it is essential to optimize the general-purpose computing environment at the edge by conserving resources. Advantages such as rapid deployment, shorter boot-up times, and resource isolation are highly desirable [92]. Initial studies suggest that mobile containers, which distribute device hardware functions among multiple virtual devices, can offer performance comparable to native hardware [93]. Container technologies, such as docker, are advancing and enabling swift application deployment on various platforms. However, further research is needed to establish containers as a suitable method for deploying applications on edge nodes.

Virtualization plays a vital role in the evolution of IT technologies, enabling the simultaneous operation of multiple operating systems or numerous applications on a single server [94]. Its key function is to diminish the reliance on physical servers, leading to substantial reductions in power consumption and cooling costs. The growing prevalence of IoT, mobile devices, and sensors has amplified the requirement for remote data centers [95]. Consequently, there exists an opportunity to relocate applications and intelligence from the cloud to the edge network. This transition can usher in a new type of virtualization at the edge, wherein a physical server can provide adaptable and dedicated storage and cache resources.

4.2.5. Energy Efficiency

The rapid proliferation of edge devices in urban areas has worsened the global energy crisis and the issue of global warming. One potential method to mitigate this problem involves harnessing renewable energy sources to power these edge devices. Given that these devices are dispersed throughout the city, adopting distributed renewable energy generators can significantly reduce the reliance on conventional energy sources. However, this approach is not without its challenges. It must address issues like minimizing the use of conventional energy while ensuring the uninterrupted operation of edge devices and establishing a complementary power system for various edge devices [96,97]. In the context of an Energy Internet system, the energy router, which serves as a control center, requires a certain level of computational capacity [98]. Hence, a plausible avenue for future research is to explore the integration of energy routers with edge intelligence.

5. AI Solutions for Optimizing Computation of Edge Intelligence

As discussed, major problems in computing for edge Intelligence are computing offload, resource allocation, privacy, and security. Enhancement in conventional approaches or hybridization can help improve and optimize computing, resource allocation, privacy, and security for Edge Intelligence. A detailed review of the techniques proposed by the researchers to address these objectives is presented in Table 3.

Table 3. Summary of Solutions for Optimizing Computing for Edge Intelligence.

Problem Addressed	Objective	Technique/Algorithm	Details of Technique/Algorithm
Computing offloading optimization	Reduction in energy consumption and latency.	Deep Reinforcement Learning (DRL) based offloading scheme [99]	Lack of prior familiarity with transmission delay and energy consumption models reduces the complexity of the state space by employing Deep Reinforcement Learning (DRL) to augment the understanding speed. Additionally, consider the Energy Consumption (EC) scenario involving energy harvesting.
		Deep Reinforcement Learning (DRL) based computing offloading algorithm [100]	Utilizes a Markov decision process for the portrayal of computational offloading, employing Deep Reinforcement Learning (DRL) to acquire insights into network dynamics.
		A hybrid approach based on Q-function breakdown and double DQN [101]	Utilized a double deep Qnetwork for the attainment of optimal computing offloading in the absence of prerequisite, employed a novel function approximator founded deep neural network (DNN) model which is designed to address high-dimensional state spaces.
		Reinforcement learning utilizing neural network architectures [102]	A continuous-time Markov decision process with an infinite horizon and average rewards is employed to model the optimization issue. Additionally, a novel value function approximator is introduced to address the challenges posed by high-dimensional state spaces.
	Optimization of the hardware structure for edge devices	Binary Weight Convolutional Neural Network based Algorithm [103]	Static random-access memory (SRAM) is designed for binary weight convolutional neural networks (CNNs) with the aim of minimizing memory data output, facilitating parallel implementation of CNN operations.
		Approach based on DNN and FPGA [104]	Expeditor for weed species categorization utilizes a binarized deep neural network, which is employed on field programmable gate arrays (FPGA).
	Reduction in energy consumption	Distributed Deep Learning based offloading technique [105]	Built a model by adding the cost of varying local implementation assignments in the cost function.
	Reduction in latency	DL based Smart-Edge-CoCaCo [106]	An approach based on joint optimization of wireless communication, combined filter caching and computation offloading, is developed to reduce the latency.
		A heuristic offloading technique [107]	Using electronic communication networks to estimate the distance between origin and destination, along with heuristic searching, to identify the most effective scheme for reducing the communication lag of deep learning tasks.
		Cooperative Q-learning [108]	Noticeable improvement in the searching pace of the conventional Q-learning approach.
		TD learning involves a method that incorporates post-decision states and utilizes a semi-gradient descent approach [109]	Utilized approximate dynamical planning as a strategy to tackle the difficulties established by the curse of dimensionality.
		Online Reinforcement Learning [110]	Unique arrangements of state transitions are designed to address the difficulties raised by the curse of dimensionality. Moreover, it takes into account the energy-harvesting aspect of Edge intelligence.

Table 3. *Cont.*

Problem Addressed	Objective	Technique/Algorithm	Details of Technique/Algorithm
Security of Edge Intelligence		Hypergraph clustering [111]	Improves the identification rate by modeling the association among edge nodes and DDoS through hypergraph clustering.
		Extreme Learning Machine [112]	Demonstrate quicker convergence rates and enhanced generalization capabilities of the Extreme Learning Machine classifier compared to the majority of traditional algorithms.
		Distributed Deep Learning [113]	Eases the load of training the model while enhancing its accuracy.
		An algorithm based on restricted Boltzmann machines [114]	Enhances the ability to identify unfamiliar attacks through the incorporation of active learning features.
		Deep PDS-Learning [115]	Accelerate the training process by incorporating supplementary details, such as the energy consumption of edge devices.
Resource allocation optimization		Actor-critic RL [116]	Introduced an additional Deep Neural Network (DNN) for expressing a parameterized stochastic policy, aiming to enhance both performance and convergence speed. Additionally, incorporated a natural policy gradient approach to mitigate the risk of local convergence.
		DRL-based resource allocation scheme [117]	Enhanced the Quality of Service (QoS) through the integration of supplementary SDN
		Multi-task DRL [118]	Modifies the final layer of a Deep Neural Network (DNN) responsible for estimating the Q-function to accommodate action spaces with increased dimensions.
Privacy protection		Generative adversarial networks (GAN) [119]	An algorithm for objective perturbation and another for output perturbation, both ensuring adherence to the principles of differential privacy.
		Edge Sanitizer: A deep inference framework [120]	Proposes maximum utilization of data while ensuring privacy protection.
		Deep Q-learning [121]	Generate trust values through uncertain reasoning and prevent local convergence by regulating the learning rate.
Other ways to reduce energy consumption	Control device operational condition	DRL-based joint mode selection and resource management approach [122]	Minimizes energy consumption in the medium and long term by managing the communication mode of the operator apparatus and regulating the active state of processors.
	Merging into energy Internet	Model-based DRL [123]	Solves the energy supply issue of the multi-access edge server.
		Reinforcement Learning [124]	A fog computing device operating on energy generated from a renewable source.
		Minimax-Q learning [125]	Gradually learns the optimal strategy by raising the spectral efficiency throughput.
		Online learning [126]	Minimized bandwidth utilization by selecting the server with the highest reliability.
		Multiple Artificial Intelligence based algorithms [127]	Developed a mechanism for selecting AI algorithms intelligently to choose the most suitable algorithm for a given task.

6. Architectural Layers in the Roadmap for Edge Intelligence

The architectural layers in the Roadmap for edge intelligence, distinguishing between two main directions, viz. AI for the edge and AI on the edge as shown in Figure 4. Using a bottom-up strategy, our focus in Edge intelligence research is on dividing efforts into topology, content, and service segments, all of which stand to gain advantages from AI technologies. Conversely, a top-down method dissects AI research on the edge into model adaptation, framework design, and processor acceleration. Prior to exploring AI for the edge and AI on the edge as distinct entities, it is essential to establish a shared objective, termed Quality of experience (QoE), which consistently takes precedence. The detailed discussion of QoE, AI for the edge, and AI on the edge is discussed further in this section.

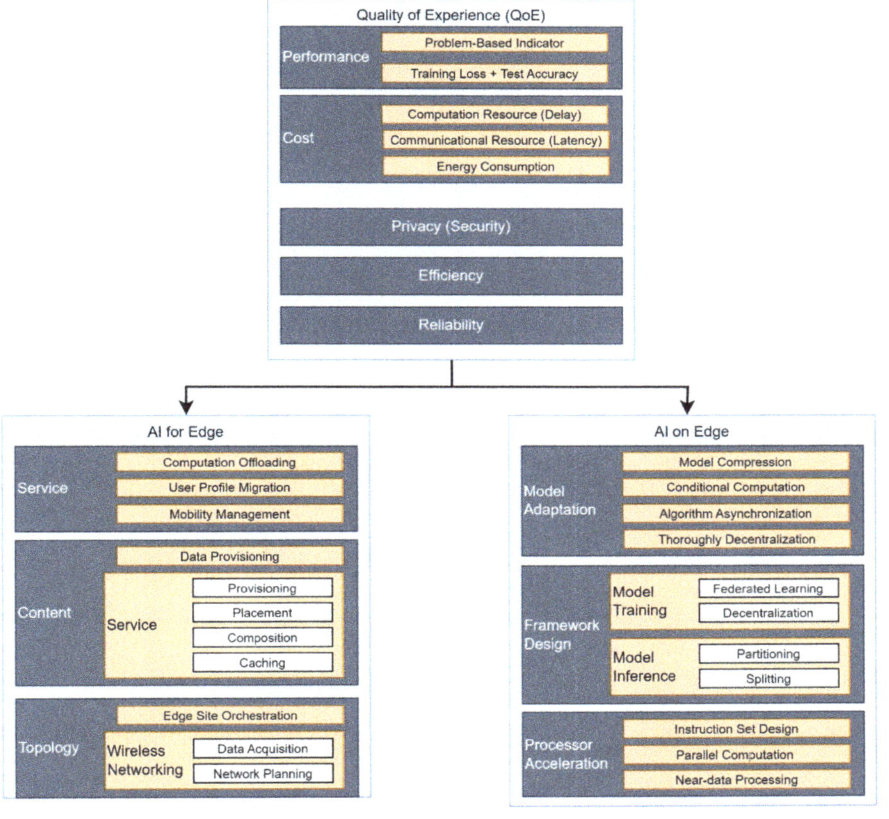

Figure 4. The architectural layers in the Roadmap.

6.1. Quality of Experience

We believe that QoE should be tailored to specific applications and should be established by contemplating multiple criteria: performance, cost, privacy (security), efficiency, and reliability.

- *Performance*

Performance criteria differ between AI for the edge and AI on the edge. In the case of the former, performance metrics are tailored to specific problems. For instance, it might encompass metrics like the successful offloading ratio in computation offloading challenges or the efficient optimization of revenue and hiring costs for base stations in service placement issues. On the other hand, for the latter, performance primarily centers around training loss and inference accuracy, both critical for AI models. Despite the transition from cloud clusters to a system integrating devices, edge, and cloud, these criteria continue to be significant.

- *Cost*

Cost considerations generally encompass computation cost, communication cost, and energy consumption. Computation cost indicates the need for computing resources, including factors like CPU cycle frequency and allocated CPU time. Communication cost deals with the resource requirements for communication, considering aspects like power, frequency band, and access time, with a focus on minimizing delays arising from the allocation of computation and communication resources. Energy consumption is particularly crucial, especially for mobile devices with restricted battery capacity. The importance of cost reduction cannot be overstated, as Edge intelligence holds the potential for substantial decreases in delay and energy consumption, simultaneously addressing critical challenges in realizing 5G capabilities.

- *Privacy and Security*

With growing apprehensions about data leaks, safeguarding privacy has gained significant attention. Consequently, Federated Learning has emerged as a solution involving the aggregation of local machine-learning models from distributed devices while actively preventing data leakage [128]. Security is intricately linked with privacy preservation and holds implications for the resilience of middleware and edge systems.

- *Efficiency*

Whether in the realm of AI for the edge or AI on the edge, achieving high efficiency is paramount to achieving outstanding performance with minimal overhead. The pursuit of efficiency is crucial for the improvement of existing algorithms and models, especially in the context of AI on the edge. Numerous strategies, including model compression, conditional computation, and asynchronous algorithms, have been suggested to enhance the efficiency of training and inference processes for deep AI models.

- *Reliability*

System reliability plays a pivotal role in ensuring the continuous operation of Edge intelligence over specified durations, a critical element for user experience. In the domain of edge intelligence, system reliability holds particular importance for AI on the edge. This is especially true when considering that model training and inference frequently take place in a distributed and synchronized manner, and local users may encounter obstacles related to wireless network congestion when attempting to complete model uploads and downloads.

6.2. Edge Intelligence/Intelligent Edge Computing

The Roadmap, as illustrated in Figure 4, pertains to AI for Edge intelligence, which we refer to as intelligent edge computing (IEC). AI offers potent tools for addressing intricate challenges in learning, planning, and decision-making. We adopt a bottom-up approach to categorize the primary concerns in Edge intelligence into three layers: topology, content, and service.

- *Topology*

In terms of topology, our attention is directed towards orchestrating edge sites (OES) and wireless networking (WN). Within our framework, an edge site is defined as a micro data center hosting deployed applications and connected to a small-cell base station (SBS). OES focuses on the deployment and configuration of wireless telecom equipment and servers. Notably, recent years have seen a surge in interest surrounding the management and automation of unmanned aerial vehicles (UAVs). These UAVs, equipped with a small server and access point, can be viewed as mobile edge servers with exceptional maneuverability. Consequently, numerous studies explore scheduling and trajectory planning challenges with the aim of minimizing UAV energy consumption.

For instance, Chen et al. [129] investigated power consumption by caching popular content based on predictions, introducing a conceptor-based echo state network (ESN) algorithm to learn user mobility patterns. Leveraging this efficient machine learning technique, their algorithm significantly outperforms benchmarks in terms of transmission power and user satisfaction. On the other hand, WN encompasses data acquisition and network planning. The former focuses on swiftly acquiring data from widely distributed sources at edge devices, while the latter concentrates on network scheduling, operation, and management. Fast data acquisition involves elements such as multiple access, radio resource allocation, and signal encoding/decoding. Network planning explores efficient management through protocols and middleware.

Significantly, recent years have witnessed a rising trend in intelligent networking, utilizing AI technologies to construct intelligent wireless communication mechanisms. As an example, Zhu et al. [130] proposed learning-driven communication, exploiting the synergy between communication and learning in edge systems.

- *Content*

The focus lies on several vital aspects, including data provisioning, service provisioning, service placement, service composition, and service caching. In the realms of data and service provisioning, resources can be drawn from remote cloud data centers and edge servers. Recent initiatives have concentrated on developing lightweight Quality of Service (QoS) aware service-based frameworks. Alternatively, shared resources may originate from mobile devices with suitable incentive mechanisms in place.

Service placement complements service provisioning and delves into the location and method of deploying complex services on potential edge sites. In recent times, numerous studies have approached service placement from the perspective of application service providers (ASPs). For instance, Chen et al. [131] endeavored to deploy services within a limited budget on fundamental communication and computation infrastructure. Subsequently, they applied the multi-armed bandit (MAB) theory, a branch of reinforcement learning, to optimize service placement decisions.

Service composition involves the selection of candidate services for composition, taking into account energy consumption and Quality of Experience (QoE) for mobile end users. This domain presents opportunities for leveraging AI technologies to generate more effective service selection strategies. Service caching, akin to service provisioning, revolves around designing a caching pool to store frequently accessed data and services. It can also be explored in a cooperative manner, offering research prospects for applying multi-agent learning to enhance QoE in large-scale Edge intelligence systems.

- *Services*

Regarding services, our focus is on computation offloading, user profile migration, and mobility management. Computation offloading addresses the load balancing of various computational and communication resources, involving edge server selection and frequency spectrum allocation. Recent research has concentrated on dynamically managing radio and computational resources for multi-user, multi-server Edge intelligence systems, utilizing Lyapunov optimization techniques. Computation offloading decisions are also being optimized through Deep Q-Network (DQN), modeling the problem as a Markov Decision Process (MDP) to maximize long-term utility performance, encompassing the aforementioned Quality of Experience (QoE) indicators.

User profile migration involves adjusting the location of user profiles, encompassing configuration files, private data, and logs, as mobile users are constantly on the move. User profile migration is often intertwined with mobility management. For instance, the JCORM algorithm, proposed in [132], optimizes computation offloading and migration through cooperative networks, presenting research opportunities for the application of more advanced AI technologies to enhance optimality.

Mobility management, viewed through the lens of statistics and probability theory, is another area of interest. There is a notable inclination toward realizing mobility management with the assistance of AI.

6.3. AI on the Edge (AIE)

The right side of the Roadmap focuses on AI on edge, referred to as AIE, involving the study of implementing AI model training and inference on the network edge. This area is categorized into four research endeavors: framework design, model training, inference and adaptation, conditional computation, and processor acceleration. Given that model adaptation builds upon existing training and inference frameworks, the initial focus will be on introducing framework design.

- *Framework Design*

Framework design aims to establish improved training and inference architecture for the edge without altering existing AI models. Researchers strive to create new frameworks for both model training and model inference.

- *Model Training*

Presently, the predominant frameworks for model training are largely distributed, with the exception of those based on knowledge distillation. Distributed training frameworks can be categorized based on data splitting and model splitting methods [133]. Data splitting involves master-device, helper-device, and device-device splitting, differing in how training samples are sourced and how the global model is aggregated. Model splitting involves separating neural network layers onto various devices, relying on complex pipelines. Knowledge distillation-based frameworks may or may not be decentralized, utilizing transfer learning technologies to enhance the accuracy of shallower student networks [119]. This process entails training a basic network on a standard dataset and transferring the learned features to student networks, which are then trained on their respective datasets by multiple mobile end devices. There is significant potential for exploration in knowledge distillation-based frameworks for model training at the edge.

The leading approach in model training is Federated Learning, designed to preserve privacy while training Deep Neural Networks (DNNs) in a distributed manner [134]. It trains local models on multiple clients, optimizing a global model by averaging trained gradients. Due to limited resources in edge nodes, training a comprehensive model there is impractical, making distributed training more feasible. However, coordination between edge nodes becomes essential. The challenge lies in optimizing the global gradient from distributed local models. Regardless of the learning algorithms used, stochastic gradient descent (SGD) plays a crucial role in model training. Edge nodes utilize SGD to update local gradients based on their datasets, sending updates to a central node for global model enhancement. Balancing model performance and communication overhead is critical. Selective transmission of local gradients showing significant improvements can ensure global model performance while reducing communication overheads, preventing network congestion caused by simultaneous transmissions from all edge nodes.

- *Model Inference*

While splitting a model during training poses challenges, it is a preferred method during model inference. Model splitting, or partitioning, serves as a framework for model inference, and various techniques, including model compression, input filtering, early exit, etc., are adaptations from existing frameworks. A well-described example of model inference on the edge can be found in reference [129], where a Deep Neural Network (DNN) is divided into two parts, each processed collaboratively. The computationally intense segment operates on the edge server, while the other part functions on the mobile device. The challenge lies in determining the optimal layer to split and when to exit the intricate DNN while maintaining inference accuracy.

- *Model Adaptation*

Model adaptation is a process that fine-tunes existing training and inference frameworks, particularly in the context of Federated Learning, to better suit Edge intelligence. While Federated Learning can potentially operate on the edge, its traditional version demands significant communication efficiency, as complete local models are transmitted back to the central server. Consequently, many researchers are concentrating on developing more efficient model updates and aggregation policies. Their endeavors aim to minimize costs, enhance robustness, and ensure system performance. Approaches to achieving model adaptation include techniques such as model compression, conditional computation, algorithm synchronization, and comprehensive decentralization.

Model compression exploits the inherent sparsity structure of gradients and weights. Potential strategies encompass quantization, dimensional reduction, pruning, precision downgrading, components sharing, and cutoff, among others. Implementation methods involve techniques like Singular Value Decomposition (SVD), Huffman coding, Principal Component Analysis (PCA), and similar approaches.

- *Conditional Computation*

Conditional computation serves as an alternative approach to reduce computations by selectively disabling less crucial calculations in Deep Neural Networks (DNNs). Feasible methods include component shutdown, input filtering, early exit, and results caching. This concept can be likened to block-wise dropout [135,136]. Additionally, random gossip communication can help reduce unnecessary calculations and model updates. Asynchronization in algorithms aims to aggregate local models asynchronously, mitigating the inefficiency and lengthy synchronous steps of model updates in Federated Learning. Thorough decentralization involves eliminating the central aggregator to prevent potential data leaks and address issues stemming from the central server's malfunction. Strategies for achieving complete decentralization encompass blockchain technologies, game-theoretical approaches, and similar methods.

- *Process Acceleration*

Processor acceleration aims to optimize the structure of DNNs, specifically targeting the frequently used computation-heavy multiply-and-accumulate operations for improvement. Strategies for enhancing DNN computations on hardware involve various methods, such as Creating specialized instruction sets for DNN training and inference, developing highly parallel computing paradigms, and implementing near-data processing to bring computation closer to memory. Highly parallelized computing paradigms can be categorized into temporal and spatial architectures. Temporal architectures like CPUs and GPUs can be accelerated by reducing the number of multiplications and increasing throughput. Spatial architectures, on the other hand, can be accelerated by enhancing data reuse with data flows.

7. Conclusions

Autonomous vehicles represent a significant milestone in the latest wave of technological advancements, serving as a crucial indicator of technological evolution. AVs can significantly contribute to decreasing traffic accidents and enhance road safety by eliminating hazardous driving behaviors like fatigued driving. This article extensively examines the essential technologies indispensable for realizing autonomous vehicles, highlighting the interconnections between these technologies and the development of AVs. Edge intelligence, a fusion of AI and edge computing, plays a pivotal role in empowering vehicles to make intelligent decisions swiftly. The critical components of edge intelligence include edge caching, edge training, edge inference, and edge offloading are well evaluated and analysed in this research. Efficient deployment of cache is essential to make optimal use of both base station and device resources, and it is crucial to optimize the cache replacement strategy to enhance overall performance. In the context of edge training, the focus lies on collaborative training, accelerating and optimizing communication frequency and costs, all while maintaining a high standard of security and privacy. While AutoML techniques, specifically architectural search, play a role in edge interpretation, human expertise remains vital for designing optimal models. Techniques such as knowledge distillation and pruning are employed to achieve model compression. Leveraging a distributed computation paradigm through edge offloading can significantly improve both model training and interpretation performance.

The efficient deployment of AI to the network's edge hinges on improving the efficacy of AI algorithms with limited computing and energy resources, necessitating the design of lightweight AI models. Beyond the significance of individual technologies, this study also delves into the challenges that must be overcome and the areas that require reinforcement. Investigating the incorporation of important technologies, issues, opportunities, and roadmap in this study will be a valuable resource for the community engaged in research on edge intelligence in EV.

Author Contributions: All the authors have contributed equally towards the conceptualization, methodology, analysis, investigation, and resources. The original draft and changes were also performed by all the authors. All authors have read and agreed to the published version of the manuscript.

Funding: This research received no external funding.

Data Availability Statement: No new data were created or analyzed in this study. Data sharing is not applicable to this article.

Conflicts of Interest: The authors declare no conflict of interest. Sachin B. Chougule and Sheetal N. Ghorpade are employees of Rubiscape Private Limited, Pune, India. The paper reflects the views of the scientists and not the company.

References

1. Elvas, L.B.; Ferreira, J.C. Intelligent Transportation Systems for Electric Vehicles. *Energies* **2021**, *14*, 5550. [CrossRef]
2. Ahmad, K.; Khujamatov, H.; Lazarev, A.; Usmanova, N.; Alduailij, M.; Alduailij, M. Internet of Things-Aided Intelligent Transport Systems in Smart Cities: Challenges, Opportunities, and Future. *Wirel. Commun. Mob. Comput.* **2023**, *2023*, 7989079. [CrossRef]
3. Ghorpade, S.N.; Zennaro, M.; Chaudhari, B.S. GWO Model for Optimal Localization of IoT-Enabled Sensor Nodes in Smart Parking Systems. *IEEE Trans. Intell. Transp. Syst.* **2021**, *22*, 1217–1224. [CrossRef]
4. Khayyam, H.; Javadi, B.; Jalili, M.; Jazar, R.N. Artificial intelligence and internet of things for autonomous vehicles. In *Nonlinear Approaches in Engineering Applications*; Springer: Berlin/Heidelberg, Germany, 2020; pp. 39–68.
5. Pradhan, N.M.; Chaudhari, B.S.; Zennaro, M. 6TiSCH Low Latency Autonomous Scheduling for Industrial Internet of Things. *IEEE Access* **2022**, *10*, 71566–71575. [CrossRef]
6. Zhang, K.; Zhu, Y.; Leng, S.; He, Y.; Maharjan, S.; Zhang, Y. Deep learning empowered task offloading for mobile edge computing in urban informatics. *IEEE Internet Things J.* **2019**, *6*, 7635–7647. [CrossRef]
7. Chaudhari, B.; Borkar, S. Design Considerations and Network Architectures for Low-Power Wide-Area Networks. In *LPWAN Technologies for IoT and M2M Applications*; Elsevier: Amsterdam, The Netherlands, 2020; pp. 15–35, ISBN 9780128188804. [CrossRef]
8. Dai, Y.; Xu, D.; Maharjan, S.; Qiao, G.; Zhang, Y. Artificial intelligence empowered edge computing and caching for internet of vehicles. *IEEE Wirel. Commun.* **2019**, *26*, 12–18. [CrossRef]
9. Mendez, J.; Bierzynski, K.; Cuéllar, M.; Morales, D.P. Edge Intelligence: Concepts, architectures, applications and future directions. *ACM Trans. Embed. Comput. Syst. (TECS)* **2022**, *21*, 1–41. [CrossRef]
10. Ghorpade, S.N.; Zennaro, M.; Chaudhari, B.S.; Saeed, R.A.; Alhumyani, H.; Abdel-Khalek, S. Enhanced Differential Crossover and Quantum Particle Swarm Optimization for IoT Applications. *IEEE Access* **2021**, *9*, 93831–93846. [CrossRef]
11. Li, Y. Deep reinforcement learning. *arXiv* **2018**, arXiv:1810.06339.
12. Shakya, A.K.; Pillai, G.; Chakrabarty, S. Reinforcement learning algorithms: A brief survey. *Expert Syst. Appl.* **2023**, *231*, 120495. [CrossRef]
13. Stefenon, S.F.; Yow, K.-C.; Nied, A.; Meyer, L.H. Classification of distribution power grid structures using inception v3 deep neural network. *Electr. Eng.* **2022**, *104*, 4557–4569. [CrossRef]
14. Ghorpade, S.N.; Zennaro, M.; Chaudhari, B.S. IoT-based hybrid optimized fuzzy threshold ELM model for localization of elderly persons. *Expert Syst. Appl.* **2021**, *184*, 115500. [CrossRef]
15. Ghorpade, S.N.; Zennaro, M.; Chaudhari, B.S. Binary grey wolf optimization-based topology control for WSNs. *IET Wirel. Sens. Syst.* **2019**, *9*, 333–339. [CrossRef]
16. Drolia, U.; Guo, K.; Tan, J.; Gandhi, R.; Narasimhan, P. Cachier: Edge-caching for recognition applications. In Proceedings of the 2017 IEEE 37th International Conference on Distributed Computing Systems (ICDCS), Atlanta, GA, USA, 5–8 June 2017; pp. 276–286.
17. Drolia, U.; Guo, K.; Narasimhan, P. Precog: Prefetching for image recognition applications at the edge. In Proceedings of the Second ACM/IEEE Symposium on Edge Computin, San Jose, CA, USA, 12–14 October 2017; p. 17.

18. Guo, P.; Hu, B.; Li, R.; Hu, W. Foggy Cache: Cross-device approximate computation reuse. In Proceedings of the 24th Annual International Conference on Mobile Computing and Networking, MobiCom 2018, New Delhi, India, 29 October–2 November 2018; pp. 19–34.
19. Xu, M.; Zhu, M.; Liu, Y.; Lin, F.X.; Liu, X. Deep Cache: Principled cache for mobile deep vision. *arXiv* **2017**, arXiv:1712.01670.
20. Li, T.; Xiao, Z.; Georges, H.M.; Luo, Z.; Wang, D. Performance analysis of co-and cross-tier device-to-device communication underlaying macro-small cell wireless networks. *KSII Trans. Internet Inf. Syst.* **2016**, *10*, 1481–1500.
21. Xiao, Z.; Li, T.; Ding, W.; Wang, D.; Zhang, J. Dynamic PCI allocation on avoiding handover confusion via cell status prediction in LTE heterogeneous small cell networks. *Wirel. Commun. Mob. Comput.* **2016**, *16*, 1972–1986. [CrossRef]
22. Xiao, Z.; Liu, H.; Havyarimana, V.; Li, T.; Wang, D. Analytical study on multi-tier 5g heterogeneous small cell networks: Coverage performance and energy efficiency. *Sensors* **2016**, *16*, 1854. [CrossRef]
23. Xiao, Z.; Li, T.; Cheng, W.; Wang, D. Apollonius circles based outbound handover in macro-small wireless cellular networks. In Proceedings of the 2016 IEEE Global Communications Conference (GLOBECOM), Washington, DC, USA, 4–8 December 2016; pp. 1–6.
24. Ji, M.; Caire, G.; Molisch, A.F. Wireless device-to-device caching networks: Basic principles and system performance. *IEEE J. Sel. Areas Commun.* **2016**, *34*, 176–189. [CrossRef]
25. Chen, W.; Li, T.; Xiao, Z.; Wang, D. On mitigating interference under device-to-device communication in macro-small cell networks. In Proceedings of the 2016 International Conference on Computer, Information and Telecommunication Systems (CITS), Kunming, China, 6–8 July 2016; pp. 1–5.
26. Ioannou, A.; Weber, S. A survey of caching policies and forwarding mechanisms in information-centric networking. *IEEE Commun. Surv. Tutor.* **2016**, *18*, 2847–2886. [CrossRef]
27. McMahan, B.; Ramage, D. Federated learning: Collaborative machine learning without centralized training data. *Google Res. Blog* **2017**, *3*, 355–359. [CrossRef]
28. Valery, O.; Liu, P.; Wu, J.-J. CPU/GPU collaboration techniques for transfer learning on mobile devices. In Proceedings of the 2017 IEEE 23rd International Conference on Parallel and Distributed Systems (ICPADS), Shenzhen, China, 15–17 December 2017; pp. 477–484.
29. Valery, O.; Liu, P.; Wu, J.-J. Low Precision Deep Learning Training on Mobile Heterogeneous Platform. In Proceedings of the 2018 26th Euromicro International Conference on Parallel, Distributed and Network-Based Processing (PDP), Cambridge, UK, 21–23 March 2018; pp. 109–117. [CrossRef]
30. Xing, T.; Sandha, S.S.; Balaji, B.; Chakraborty, S.; Srivastava, M. Enabling edge devices that learn from each other: Cross modal training for activity recognition. In Proceedings of the 1st International Workshop on Edge Systems, Analytics and Networking, Munich, Germany, 10 June 2018; ACM: New York, NY, USA, 2018; pp. 37–42.
31. Bonawitz, K.; Ivanov, V.; Kreuter, B.; Marcedone, A.; McMahan, H.B.; Patel, S.; Ramage, D.; Segal, A.; Seth, K. Practical secure aggregation for privacy-preserving machine learning. In Proceedings of the 2017 ACM SIGSAC Conference on Computer and Communications Security, Dallas, TX, USA, 30 October–3 November 2017; ACM: New York, NY, USA, 2017; pp. 1175–1191.
32. Liu, H.; Simonyan, K.; Yang, Y. Darts: Differentiable architecture search. *arXiv* **2018**, arXiv:1806.09055.
33. Howard, A.G.; Zhu, M.; Chen, B.; Kalenichenko, D.; Wang, W.; Weyand, T.; Andreetto, M.; Adam, H. Mobilenets: Efficient convolutional neural networks for mobile vision applications. *arXiv* **2017**, arXiv:1704.04861.
34. Choudhary, T.; Mishra, V.; Goswami, A.; Sarangapani, J. A comprehensive survey on model compression and acceleration. *Artif. Intell. Rev.* **2020**, *53*, 5113–5155. [CrossRef]
35. Ghorpade, S.N.; Zennaro, M.; Chaudhari, B.S. Towards Green Computing: Intelligent Bio-Inspired Agent for IoT-enabled Wireless Sensor Networks. *Int. J. Sens. Netw.* **2021**, *35*, 121. [CrossRef]
36. Raval, N.; Srivastava, A.; Razeen, A.; Lebeck, K.; Machanavajjhala, A.; Cox, L.P. What you mark is what apps see. In Proceedings of the 14th Annual International Conference on Mobile Systems, Applications, and Services, Singapore, 26–30 June 2016; ACM: New York, NY, USA, 2016; pp. 249–261.
37. Wendelken, S.; MacGillivray, C. Worldwide and U.S. IoT Cellular Connections Forecast, 2021–2025. Available online: https://www.idc.com/getdoc.jsp?containerId=US47296121 (accessed on 17 February 2022).
38. Wong, T.-T. Performance evaluation of classification algorithms by k-fold and leave-one-out cross validation. *Pattern Recognit.* **2015**, *48*, 2839–2846. [CrossRef]
39. Mittal, S. A survey of FPGA-based accelerators for convolutional neural networks. *Neural Comput. Appl.* **2018**, *32*, 1109–1139. [CrossRef]
40. Manokaran, J.; Vairavel, G. An Empirical Comparison of Machine Learning Algorithms for Attack Detection in Internet of Things Edge. *ECS Trans.* **2022**, *107*, 2403.
41. Watson, D.S. On the Philosophy of Unsupervised Learning. *Philos. Technol.* **2023**, *36*, 28. [CrossRef]
42. Thomos, N.; Maugey, T.; Toni, L. Machine Learning for Multimedia Communications. *Sensors* **2022**, *22*, 819. [CrossRef]
43. Chaudhari, B.S.; Zennaro, M. Introduction to low-power wide-area networks. In *LPWAN Technologies for IoT and M2M Applications*; Elsevier: Amsterdam, The Netherlands, 2020; pp. 1–13. [CrossRef]
44. Atzori, L.; Iera, A.; Morabito, G. The internet of things: A survey. *Comput. Netw.* **2010**, *54*, 2787–2805. [CrossRef]
45. Cheng, X.; Chen, C.; Zhang, W.; Yang, Y. 5G-enabled cooperative intelligent vehicular (5GenCIV) Framework: When Benz Meets Marconi. *IEEE Intell. Syst.* **2017**, *32*, 53–59. [CrossRef]

46. Liang, L.; Peng, H.; Li, G.Y.; Shen, X. Vehicular communications: A network layer perspective. *IEEE Trans. Veh. Technol.* **2017**, *66*, 10647–10659. [CrossRef]
47. Ye, H.; Li, G.Y.; Juang, B.-H.F. Deep reinforcement learning based resource allocation for V2V communications. *IEEE Trans. Veh. Technol.* **2019**, *68*, 3163–3173. [CrossRef]
48. A Look at the Future of 5G. Available online: https://spectrum.ieee.org/computing/software/a-look-at-the-future-of-5g/ (accessed on 28 November 2023).
49. Gaudet, B. Review of cooperative truck platooning systems. *Natl. Res. Counc. Can.* **2014**, *10*, 1–79.
50. Hou, X.; Li, Y.; Wu, D.; Chen, S. Vehicularfog computing: A viewpoint of vehicles as the infrastructures. *IEEE Trans. Veh. Technol.* **2016**, *65*, 3860–3873. [CrossRef]
51. Eltoweissy, M.; Olariu, S.; Younis, M. Towards autonomous vehicular clouds. In Proceedings of the Springer Conference on International Conference on Ad Hoc Networks, Edmonton, AB, Canada, 20–22 August 2010; pp. 1–16.
52. Hong, K.; Lillethun, D.; Ramachandran, U.; Ottenw, B.; Koldehofe, B. Mobile fog: A programming model for largescale applications on the internet of things. In Proceedings of the ACM SIGCOMM Workshop on Mobile Cloud Computing, Hong Kong, China, 16 August 2013; pp. 15–20.
53. Kaur, S.; Dhillon, K.K.; Manvi, M.; Singh, R. An automatic system for detecting the vehicle registration plate from video in foggy and rainy environments using restoration technique. *Int. J. Comput. Appl.* **2014**, *97*, 14–19. [CrossRef]
54. Roy, S.; Bose, R.; Sarddar, D. A fog-based DSS model for driving rule violation monitoring framework on the internet of things. *Int. J. Adv. Sci. Technol.* **2015**, *82*, 23–32. [CrossRef]
55. Vashitz, G.; Shinar, D.; Blum, Y. In-vehicle information systems to improve traffic safety in road tunnels. *Transp. Res. Part F Traffic Psychol. Behav.* **2008**, *11*, 61–74. [CrossRef]
56. Miah, S.J.; Ahamed, R. A cloud-based DSS model for driver safety and monitoring on Australian roads. *Int. J. Emerg. Sci.* **2011**, *1*, 634.
57. Vahdat-Nejad, H.; Ramazani, A.; Mohammadi, T.; Mansoor, W. A survey on context-aware vehicular network applications. *Veh. Commun.* **2016**, *3*, 43–57. [CrossRef]
58. Baldauf, M.; Dustdar, S.; Rosenberg, F. A survey on context aware systems. *Int. J. Ad Hoc Ubiquitous Comput.* **2007**, *2*, 263–277. [CrossRef]
59. Bogale, T.E.; Wang, X.; Le, L.B. Machine intelligence techniques for next-generation context-aware wireless networks. *Comput. Sci. Inf. Theory* **2018**. [CrossRef]
60. He, Q.; Liu, J.; Wang, C.; Li, B. Coping with heterogeneous video contributors and viewers in crowdsourced live streaming: A cloud-based approach. *IEEE Trans. Multimed.* **2016**, *18*, 916–928. [CrossRef]
61. Zhuo, G.; Jia, Q.; Guo, L.; Li, M.; Li, P. Privacy-Preserving Verifiable Set Operation in Big Data for Cloud-Assisted Mobile Crowdsourcing. *IEEE Internet Things J.* **2017**, *4*, 572–582. [CrossRef]
62. Huang, C.; Xu, K. Reliable real time streaming in vehicular cloud-fog computing networks. In Proceedings of the IEEE Conference on Communications in China, Chengdu, China, 27–29 July 2016; pp. 1–6.
63. Grassi, G.; Bahl, P.; Jamieson, K.; Pau, G. Park Master: An in vehicle, edge-based video analytics service for detecting open parking spaces in urban environments. In Proceedings of the ACM/IEEE Symposium on Edge Computing, San Jose, CA, USA, 18–21 April 2017; p. 16.
64. Cho, S.Y. Development of an IGVM integrated navigation system for vehicular lane-level guidance services. *J. Position. Navig. Timing* **2016**, *5*, 119–129. [CrossRef]
65. Park, H.S.; Park, M.W.; Won, K.H.; Kim, K.-H.; Jung, S.K. In-Vehicle AR-HUD system to provide driving-Safety information. *ETRI J.* **2013**, *35*, 1038–1047. [CrossRef]
66. Ghorpade, S.; Zennaro, M.; Chaudhari, B. Survey of Localization for Internet of Things Nodes: Approaches, Challenges and Open Issues. *Future Internet* **2021**, *13*, 210. [CrossRef]
67. Ghorpade, S.N.; Zennaro, M.; Chaudhari, B.S.; Saeed, R.A.; Alhumyani, H.; Abdel-Khalek, S. A Novel Enhanced Quantum PSO for Optimal Network Configuration in Heterogeneous Industrial IoT. *IEEE Access* **2021**, *9*, 134022–134036. [CrossRef]
68. Li, W.; Zhao, Y.; Lu, S.; Chen, D. Mechanisms and challenges on Mobility-augmented Service Provisioning for Mobile Cloud Computing. *IEEE Commun. Mag.* **2015**, *53*, 89–97. [CrossRef]
69. Hromic, H.; Le Phuoc, D.; Serrano, M.; Antonic, A.; Zarko, I.P.; Hayes, C.; Decker, S. Real Time Analysis of Sensor Data for the Internet of Things by Means of Clustering and Event Processing. In Proceedings of the IEEE International Conference on Communications, London, UK, 8–12 June 2015; pp. 685–691.
70. Meurisch, C.; Seeliger, A.; Schmidt, B.; Schweizer, I.; Kaup, F.; Muhlh, M. Upgrading Wireless Home Routers for Enabling Large-scale Deployment of Cloudlets. In *Mobile Computing, Applications, and Services*; Springer: Cham, Switzerland, 2015; pp. 12–29.
71. Shafique, M.; Theocharides, T.; Bouganis, C.S.; Hanif, M.A.; Khalid, F.; Hafz, R.; Rehman, S. An overview of next-generation architectures for machine learning: Roadmap, opportunities and challenges in the IoT era. In Proceedings of the 2018 Design, Automation Test in Europe Conference Exhibition (DATE), Dresden, Germany, 19–23 March 2018; pp. 827–832.
72. Povedano-Molina, J.; Lopez-Vega, J.M.; Lopez-Soler, J.M.; Corradi, A.; Foschini, L. DARGOS: A Highly Adaptable and Scalable Monitoring Architecture for Multi-Tenant Clouds. *Future Gener. Comput. Syst.* **2013**, *29*, 2041–2056. [CrossRef]

73. Perez-Espinoza, J.A.; Sosa-Sosa, V.J.; Gonzalez, J.L.; Tello-Leal, E. A Distributed Architecture for Monitoring Private Clouds. In Proceedings of the 2015 26th International Workshop on Database and Expert Systems Applications (DEXA), Valencia, Spain, 1–4 September 2015; pp. 186–190. [CrossRef]
74. Grozev, N.; Buyya, R. Inter-Cloud Architectures and Application Brokering: Taxonomy and Survey. *Softw. Pract. Exp.* **2014**, *44*, 369–390. [CrossRef]
75. Garg, S.K.; Versteeg, S. A Framework for Ranking of Cloud Computing Services. *Future Gener. Comput. Syst.* **2013**, *29*, 1012–1023. [CrossRef]
76. Singh, P.; Kaur, A.; Gill, S.S. Machine learning for cloud, fog, edge and serverless computing environments: Comparisons, performance evaluation benchmark and future directions. *Int. J. Grid Util. Comput.* **2022**, *13*, 447–457. [CrossRef]
77. Stacker, L.; Fei, J.; Heidenreich, P.; Bonarens, F.; Rambach, J.; Stricker, D.; Stiller, C. Deployment of deep neural networks for object detection on edge ai devices with runtime optimization. In Proceedings of the IEEE/CVF International Conference on Computer Vision, Montreal, BC, Canada, 11–17 October 2021; pp. 1015–1022.
78. Iftikhar, S.; Ahmad, M.M.M.; Tuli, S.; Chowdhury, D.; Xu, M.; Gill, S.S.; Uhlig, S. Hunterplus: Ai based energy-efficient task scheduling for cloud–fog computing environments. *Internet Things* **2023**, *21*, 100667. [CrossRef]
79. Mousavi, S.; Mood, S.E.; Souri, A.; Javidi, M.M. Directed Search: A New Operator in Nsga-Ii for Task Scheduling in IoT Based on Cloud-Fog Computing. *IEEE Trans. Cloud Comput.* **2022**, *11*, 2144–2157. [CrossRef]
80. Ghafariana, T.; Javadi, B. Cloud-aware Data Intensive Workflow Scheduling on Volunteer Computing Systems. *Future Gener. Comput. Syst.* **2015**, *51*, 87–97. [CrossRef]
81. Tang, W.; Jenkins, J.; Meyer, F.; Ross, R.; Kettimuthu, R.; Winkler, L.; Yang, X.; Lehman, T.; Desai, N. Data-Aware Resource Scheduling for Multicloud Workflows: A Fine-Grained Simulation Approach. In Proceedings of the IEEE International Conference on Cloud Computing Technology and Science, Singapore, 15–18 December 2014; pp. 887–892.
82. Beck, M.T.; Maier, M. Mobile Edge Computing: Challenges for Future Virtual Network Embedding Algorithms. In Proceedings of the International Conference on Advanced Engineering Computing and Applications in Sciences, Rome, Italy, 24–28 August 2014; pp. 65–70.
83. Simoens, P.; Van Herzeele, L.; Vandeputte, F.; Vermoesen, L. Challenges for Orchestration and Instance Selection of Composite Services in Distributed Edge Clouds. In Proceedings of the IFIP/IEEE International Symposium on Integrated Network Management, Ottawa, ON, Canada, 11–15 May 2015; pp. 1196–1201.
84. Gupta, S.; Chaudhari, B.S.; Chakrabarty, B. Vulnerable Network Analysis Using War Driving and Security Intelligence. In Proceedings of the 2016 International Conference on Inventive Computation Technologies (ICICT), Coimbatore, India, 26–27 August 2016; IEEE: Coimbatore, India, 2016; pp. 1–5. [CrossRef]
85. Valerio, L.; Passarella, A.; Conti, M. A communication efficient distributed learning framework for smart environments. *Pervasive Mob. Comput.* **2017**, *41*, 46–68. [CrossRef]
86. Baset, S.A. Cloud SLAs: Present and Future. *ACM SIGOPS Oper. Syst. Rev.* **2012**, *46*, 57–66. [CrossRef]
87. Bui, T. Analysis of Docker Security. *arXiv* **2015**, arXiv:1501.02967. Available online: http://arxiv.org/abs/1501.02967 (accessed on 3 June 2023).
88. Deelman, E.; Vahi, K.; Juve, G.; Rynge, M.; Callaghan, S.; Maechling, P.J.; Mayani, R.; Chen, W.; da Silva, R.F.; Livny, M.; et al. Pegasus: A Workflow Management System for Science Automation. *Future Gener. Comput. Syst.* **2015**, *46*, 17–35. [CrossRef]
89. Serrano-Solano, B.; Fouilloux, A.; Eguinoa, I.; Kalaš, M.; Grüning, B.; Coppens, F. Galaxy: A decade of realizing CWFR concepts. *Data Intell.* **2022**, *4*, 358–371. [CrossRef]
90. Ruiz, J.; Garrido, J.; Santander-Vela, J.; Snchez-Expsito, S.; Montenegro, L.V. Astro Taverna-Building Workflows with Virtual Observatory Services. *Astron. Comput.* **2014**, *78*, 3–11. [CrossRef]
91. Kartakis, S.; McCann, J.A. Real-time Edge Analytics for Cyber Physical Systems Using Compression Rates. In Proceedings of the International Conference on Autonomic Computing, Philadelphia, PA, USA, 18–20 June 2014; pp. 153–159.
92. Xu, L.; Wang, Z.; Chen, W. The Study and Evaluation of ARM based Mobile Virtualization. *Int. J. Distrib. Sens. Netw.* **2015**, *11*, 310308. [CrossRef]
93. Andrus, J.; Dall, C.; Hof, A.V.; Laadan, O.; Nieh, J. Cells: A Virtual Mobile Smartphone Architecture. In Proceedings of the ACM Symposium on Operating Systems Principles, Cascais, Portugal, 23–26 October 2011; pp. 173–187.
94. Ghorpade, S.N.; Zennaro, M.; Chaudhari, B.S. Localization approaches for internet of things. In *Optimal Localization of Internet of Things Nodes*; Springer: Cham, Switzerland, 2022; pp. 17–50.
95. Morabito, R.; Beijar, N. Enabling data processing at the network edge through lightweight virtualization technologies. In Proceedings of the 2016 IEEE International Conference on Sensing, Communication and Networking (SECON Workshops), London, UK, 27 June 2016; pp. 1–6.
96. Barker, A.; Varghese, B.; Ward, J.S.; Sommerville, I. Academic Cloud Computing Research: Five Pitfalls and Five Opportunities. In Proceedings of the USENIX Conference on Hot Topics in Cloud Computing, Philadelphia, PA, USA, 17–18 June 2014.
97. Liu, Y.; Yang, C.; Jiang, L.; Xie, S.; Zhang, Y. Intelligent edge computing for IoT-based energy management in smart cities. *IEEE Netw.* **2019**, *33*, 111–117. [CrossRef]
98. Liang, H.; Hua, H.; Qin, Y.; Ye, M.; Zhang, S.; Cao, J. Stochastic optimal energy storage management for energy routers via compressive sensing. *IEEE Trans. Ind. Inform.* **2022**, *18*, 2192–2202. [CrossRef]

99. Min, M.; Xiao, L.; Chen, Y.; Cheng, P.; Wu, D.; Zhuang, W. Learning-based computation offloading for IoT devices with energy harvesting. *IEEE Trans. Veh. Technol.* **2019**, *68*, 1930–1941. [CrossRef]
100. Cheng, X.; Feng, L.; Quan, W.; Zhou, C.; He, H.; Shi, W.; Shen, X. Space/aerial-assisted computing offloading for IoT applications: A learning-based approach. *IEEE J. Sel. Area. Comm.* **2019**, *37*, 1117–1129. [CrossRef]
101. Chen, X.; Zhang, H.; Wu, C.; Mao, S.; Ji, Y.; Bennis, M. Optimized computation offloading performance in virtual edge computing systems via deep reinforcement learning. *IEEE Internet Things J.* **2019**, *6*, 4005–4018. [CrossRef]
102. Lei, L.; Xu, H.; Xiong, X.; Zheng, K.; Xiang, W.; Wang, X. Multiuser resource control with deep reinforcement learning in IoT edge computing. *IEEE Internet Things J.* **2019**, *6*, 10119–10133. [CrossRef]
103. Biswas, A.; Chandrakasan, A.P. CONV-SRAM: An energy-efficient SRAM with in-memory dot-product computation for low-power convolutional neural networks. *IEEE J. Solid-State Circ.* **2019**, *54*, 217–230. [CrossRef]
104. Lammie, C.; Olsen, A.; Carrick, T.; Azghadi, M.R. Low-power and high-speed deep FPGA inference engines for weed classification at the edge. *IEEE Access* **2019**, *7*, 51171–51184. [CrossRef]
105. Huang, L.; Feng, X.; Feng, A.; Huang, Y.; Qian, L. Distributed deep learning-based offloading for mobile edge computing networks. *Mobile Netw. Appl.* **2018**, *66*, 6353–6367. [CrossRef]
106. Hao, Y.; Mian, Y.; Hu, L.; Hossain, M.S.; Muhammad, G.; Amin, S.U. Smart-edge-coCaCo: AI-enabled smart edge with joint computation, caching, and communication in heterogeneous IoT. *IEEE Netw.* **2019**, *33*, 58–64. [CrossRef]
107. Xu, X.; Li, D.; Dai, Z.; Li, S.; Chen, X. A heuristic offloading method for deep learning edge services in 5G networks. *IEEE Access* **2019**, *7*, 67734–67744. [CrossRef]
108. Kiran, N.; Pan, C.; Wang, S.; Yin, C. Joint resource allocation and computation offloading in mobile edge computing for SDN based wireless networks. *J. Commun. Netw.* **2020**, *22*, 1–11. [CrossRef]
109. Lei, L.; Xu, H.; Xiong, X.; Zheng, K.; Xiang, W. Joint computation offloading and multiuser scheduling using approximate dynamic programming in NB-IoT edge computing system. *IEEE Internet Things J.* **2019**, *6*, 5345–5362. [CrossRef]
110. Xu, J.; Chen, L.; Ren, S. Online learning for offloading and autoscaling in energy harvesting mobile edge computing. *IEEE Trans. Cogn. Commun. Netw.* **2017**, *3*, 361–373. [CrossRef]
111. An, X.; Su, J.; Lu, X.; Lin, F. Hypergraph clustering model-based association analysis of DDOS attacks in fog computing intrusion detection system. *J. Wirel. Comm. Netw.* **2018**, *1*, 249–258. [CrossRef]
112. Kozik, R.; Ficco, M.; Choraś, M.; Palmieri, F. A scalable distributed machine learning approach for attack detection in edge computing environments. *J. Parallel Distrib. Comput.* **2018**, *119*, 18–26. [CrossRef]
113. Abeshu, A.; Chilamkurti, N. Deep learning: The frontier for distributed attack detection in fog-to-things computing. *IEEE Commun. Mag.* **2018**, *56*, 169–175. [CrossRef]
114. Chen, Y.; Zhang, Y.; Maharjan, S.; Alam, M.; Wu, T. Deep learning for secure mobile edge computing in cyber-physical transportation systems. *IEEE Netw.* **2019**, *33*, 36–41. [CrossRef]
115. He, X.; Jin, R.; Dai, H. Deep PDS-learning for privacy-aware offloading in MEC-enabled IoT. *IEEE Internet Things J.* **2019**, *6*, 4547–4555. [CrossRef]
116. Wei, Y.; Yu, F.; Song, M.; Han, Z. Joint optimization of caching, computing, and radio resources for fog-enabled IoT using natural actor-critic deep reinforcement learning. *IEEE Internet Things J.* **2019**, *6*, 2061–2073. [CrossRef]
117. Wang, J.; Zhao, L.; Liu, J.; Kato, N. Smart resource allocation for mobile edge computing: A deep reinforcement learning approach. *IEEE Trans. Emerg. Top. Comput.* **2019**, *9*, 1529–1541. [CrossRef]
118. Chen, J.; Chen, S.; Wang, Q.; Cao, B.; Feng, G.; Hu, J. iRAF: A deep reinforcement learning approach for collaborative mobile edge computing IoT networks. *IEEE Internet Things J.* **2019**, *6*, 7011–7024. [CrossRef]
119. Du, M.; Wang, K.; Xia, Z.; Zhang, Y. Differential privacy preserving of training model in wireless big data with edge computing. *IEEE Trans. Big Data* **2020**, *6*, 283–295. [CrossRef]
120. Xu, C.; Ren, J.; She, L.; Zhang, Y.; Qin, Z.; Ren, K. EdgeSanitizer: Locally differentially private deep inference at the edge for mobile data analytics. *IEEE Internet Things J.* **2019**, *6*, 5140–5151. [CrossRef]
121. He, Y.; Yu, F.; He, Y.; Maharjan, S.; Zhang, Y. Secure social networks in 5G systems with mobile edge computing, caching, and device-to-device communications. *IEEE Wirel. Commun.* **2019**, *25*, 103–109. [CrossRef]
122. Sun, Y.; Peng, M.; Mao, S. Deep reinforcement learning-based mode selection and resource management for green fog radio access networks. *IEEE Internet Things J.* **2019**, *6*, 1960–1971. [CrossRef]
123. Munir, M.S.; Abedin, S.F.; Tran, N.H.; Hong, C.S. When edge computing meets microgrid: A deep reinforcement learning approach. *IEEE Internet Things J.* **2019**, *6*, 7360–7374. [CrossRef]
124. Conti, S.; Faraci, G.; Nicolosi, R.; Rizzo, S.A.; Schembra, G. Battery management in a green fog-computing node: A reinforcement-learning approach. *IEEE Access* **2017**, *5*, 21126–21138. [CrossRef]
125. Wang, B.; Wu, Y.; Liu, K.R.; Clancy, T.C. An anti-jamming stochastic game for cognitive radio networks. *IEEE J. Sel. Areas Commun.* **2011**, *29*, 877–889. [CrossRef]
126. Li, B.; Chen, T.; Giannakis, G.B. Secure mobile edge computing in IoT via collaborative online learning. *IEEE Trans. Signal Process.* **2019**, *67*, 5922–5935. [CrossRef]
127. Wang, Y.; Meng, W.; Li, W.; Liu, Z.; Liu, Y.; Xue, H. Adaptive machine learning-based alarm reduction via edge computing for distributed intrusion detection systems. *Concurr. Comput. Pract. Exp.* **2019**, *31*, e5101. [CrossRef]

128. McMahan, H.B.; Moore, E.; Ramage, D.; Hampson, S.; Arcas, B.A.Y. Communication-efficient learning of deep networks from decentralized data. In Proceedings of the 19th International Conference on Artificial Intelligence and Statistics, AISTATS 2016, Cadiz, Spain, 9–11 May 2016; pp. 1273–1282.
129. Li, E.; Zhou, Z.; Chen, X. Edge intelligence: On-demand deep learning model co-inference with device-edge synergy. In Proceedings of the Workshop Mobile Edge Communications (MECOMM@SIGCOMM), Budapest, Hungary, 20 August 2018; pp. 31–36.
130. Wang, J.; Zhang, J.; Bao, W.; Zhu, X.; Cao, B.; Yu, P.S. Not just privacy: Improving performance of private deep learning in mobile cloud. In Proceedings of the 24th ACM SIGKDD International Conference on Knowledge Discovery & Data Mining, London, UK, 19–23 August 2018; pp. 2407–2416. [CrossRef]
131. Chen, M.; Challita, U.; Saad, W.; Yin, C.; Debbah, M. Artificial neural networks-based machine learning for wireless networks: A tutorial. *IEEE Commun. Surv. Tutor.* **2019**, *21*, 3039–3071. [CrossRef]
132. Zhang, C.; Zhao, H.; Deng, S. A density-based offloading strategy for IoT devices in edge computing systems. *IEEE Access* **2018**, *6*, 73520–73530. [CrossRef]
133. Park, J.; Samarakoon, S.; Bennis, M.; Debbah, M. Wireless network intelligence at the edge. *arXiv* **2018**, arXiv:1812.02858. [CrossRef]
134. Arulkumaran, K.; Deisenroth, M.P.; Brundage, M.; Bharath, A.A. Deep reinforcement learning: A brief survey. *IEEE Signal Process. Mag.* **2017**, *34*, 26–38. [CrossRef]
135. Srivastava, N.; Hinton, G.; Krizhevsky, A.; Sutskever, I.; Salakhutdinov, R. Dropout: A simple way to prevent neural networks from overfitting. *J. Mach. Learn. Res.* **2014**, *15*, 1929–1958.
136. Lee, J.; Eshraghian, J.K.; Cho, K.; Eshraghian, K. Adaptive precision CNN accelerator using radix-X parallel connected memristor crossbars. *arXiv* **2019**, arXiv:1906.09395.

MDPI AG
Grosspeteranlage 5
4052 Basel
Switzerland
Tel.: +41 61 683 77 34

World Electric Vehicle Journal Editorial Office
E-mail: wevj@mdpi.com
www.mdpi.com/journal/wevj

www.ingramcontent.com/pod-product-compliance
Lightning Source LLC
Chambersburg PA
CBHW061501210926
51850CB00008B/1785
9783725814695